BUILD A PERSONAL EARTH STATION FOR WORLDWIDE SATELLITE TV RECEPTION

2ND EDITION

BUILD A PERSONAL EARTH STATION FOR WORLDWIDE SATELLITE TV RECEPTION

2ND EDITION

ROBERT J. TRAISTER

TAB BOOKS Inc.
Blue Ridge Summit, PA 17214

FIRST EDITION

THIRD PRINTING

Printed in the United States of America

Library of Congress Cataloging in Publication Data

Traister, Robert J.
Build a personal earth station for worldwide satellite TV reception.

Includes index.
1. Earth stations (Satellite telecommunication)—Amateurs' manuals. 2. Direct broadcast satellite television—Amateurs' manuals. I. Title.
TK9962.T73 1985 621.388'8 85-14721
ISBN 0-8306-0909-1
ISBN 0-8306-1909-7 (pbk.)

This book is dedicated to the memory of the late William Cyrus Gilmore, Jr., my best friend. His contributions to the entertainment field as a radio/television manager and actor, along with his production of the U.S. Air Force "Serenade in Blue" program, touched many. His sense of humor and unparalleled story-telling abilities reached millions through the Ed Sullivan Show and as the Brooklyn cabbie in several television commercials. He will be remembered by many as the funniest comedian who ever lived and by those who knew him well as their finest friend.

Contents

Introduction ix

1 The Television Age 1

2 Satellite Broadcasting 9

Transmitting Stations—Receiving Stations—Radio Waves—Line of Sight Broadcasts—Skywave Communications—Space Communications—Summary

3 The Personal Earth Station 29

Satellite Receiver—A Slightly Different Design—Basic TVRO Earth Station—Enjoyment of a TVRO Earth Station—Questions and Answers—Summary

4 Obtaining Surplus Components 63

5 Principles of TVRO Antennas 71

Waveguides—Horn Antennas—Paraboloid Reflectors—Spherical Antennas—Parabolic/Spherical Comparison—Other Types of Antennas—Summary

6 Commercially Available Antennas 101

Channel Master Antennas—Paraclipse Antennas—Assembling Your Antenna—Antenna Actuator Installation—Summary

7 **Commercially Available Receiving Equipment** 141

Gillaspie Model GCI-8300 Satellite Receiver—Gillaspie Model 2001R Satellite Receiver—Gillaspie Model 2001C Antenna Controller—GCI 9600 Receiver—Channel Master—ICM Video—AVCOM—Uniden—UST 1000 Satellite Receiver—M/A-COM Receivers—Low Noise Amplifiers—Am I Smart Enough to Install My Own Earth Station?—Summary

8 **Satellite Location Techniques** 205

9 **A Commercial Cable Television System** 225

10 **Specific Site Selection** 237

11 **Future Applications** 249

Direct Broadcast Satellite Television—New Satellite Industries—Upgrading to DBS—Satellite Signal Scrambling

12 **The Earth Station Assembly** 271

Site Determination—Equipment Selection—The Equipment Arrives!—The Checkout—Problems—Calling the Manufacturer—Interference—Summary

Appendix A List of Satellite Video Programming Sources 281

Appendix B List of Geostationary Satellites by Orbital Positions 287

Appendix C List of Satellites Launched in 1980 293

Appendix D Charts and Tables 307

Appendix E Maps 317

Appendix F Manufacturers and Distributors 321

Glossary 347

Index 367

Introduction

The first edition of this book was written in late 1981, when the idea of a personal satellite TVRO (television-receive-only) Earth station was still quite exotic. Personal Earth stations were a rarity then, but they were beginning to catch on. Prices for complete Earth stations ranged in price from a low of about $3,000 to a high of $12,000 or more. A few adventurous souls purchased Earth stations, while others sat back and watched, waiting for the prices to drop.

Today, the TVRO Earth station is a common sight in many neighborhoods, especially in areas where standard Earth-based television reception is poor or nonexistent. Businesses that cater exclusively to the TVRO crowd have sprung up from coast to coast, and many are doing quite well. In 1981, most businesses dealing in these products treated them as a sideline to supplement the sale of television receivers and other conventional entertainment products.

I was well aware, when writing the first edition, that the market was in for some drastic changes in the coming years. I even went so far as to predict that within the decade, most television broadcast services would broadcast via satellite. I still hold with this prediction, although none of the major networks are currently available by this means.

This book was intended to be an introduction to satellite television reception, and this is the main purpose of the second edition as well. Fortunately, the greater part of the information imparted in the pages of the first edition was very basic and still applies today. Other portions of the book dealt more specifically with manufacturers' products, many of which have been updated and made more sophisticated. This has rendered those portions of the first edition relatively obsolete.

The first edition devoted a sizable number of pages to describing the products of a company that was rated as the leader in satellite television back in 1981. This company was written up in nearly every electronics magazine and its president appeared on several television talk shows to discuss satellite television. To my chagrin, this company was bankrupt by the time the first edition came off the presses. This speaks to the uncertainty involved in satellite television at that time and the direction the market would eventually take. Fortunately, many other companies that received mention are now entrenched in a highly lucrative field and are offering new and better products regularly.

The price of TVRO Earth stations has indeed dropped significantly, but not to the level many industry experts predicted. In 1981, there was a general feeling that Earth stations would drop to the $1,000 price range, making them only a little more expensive than a good color television receiver. This event may indeed come about in the next five years or so, but I wouldn't bet on it. Earth stations still fall into the "specialty" category and can be compared with giant-screen color television receivers and sophisticated video recorder/camera outfits. As long as a product remains in this "specialty" category, technological advances, more so than utility, will probably dictate which product will be most popular. In other words, the fancier an Earth station seems to be, the more popular it will be. In order to recoup their investments for this fanciness, the manufacturers will necessarily charge more for their products. I'm not speaking of fanciness in a frivolous light, but in the area of advanced technology and added convenience. This would address motorized positioning units for the antennas, highly attractive digital readout receivers, and so on. However, Earth stations will eventually fall out of the "specialty" category and into the realm of the "can't do without it" category. This may occur when some major networks finally start broadcasting via satellite. It is then that we may see the price of Earth stations drop to the $1,000 level.

Since 1981, prices have dropped considerably. It is now possible to purchase a generic Earth station for as little as $1,250. Most, however, will fall somewhere between $2,500 and $3,500. The actual price will be dependent upon the user's geographical location and the signal strength in that area. Those persons who live in the central portion of the United States and Canada can often get by with a smaller antenna and less sensitive low noise amplifier than those persons living on either the East or West Coast. The antenna and low noise amplifier represent a substantial portion of the overall cost of a TVRO Earth Station.

The second edition of this book has been updated in such a way that the references to manufacturers and products that are now hopelessly out of date are no longer included and have been replaced with descriptions of the most current market offerings. However, the "meat and potatoes" of the first edition, which is still current and probably will be for quite some time, remains intact.

If you're just getting started in the satellite television pursuit, I think you will find this book to be an excellent introduction to satellite television. It is written on a level that is fairly nontechnical and offers many practical suggestions and shortcuts that can save both time and money. For those readers who already own an Earth station, I believe the general product review will be most helpful.

To all readers, let me say that the personal Earth station is here to stay. Within this decade, it will be classified as a product that will impact nearly every household and will carry over into business as well. Certainly, there are those who have fought its coming, just as there were those who scoffed at the idea of television several decades ago. In the end, the absence of a personal Earth station for satellite television reception in a homeowner's backyard may be as big a rarity as the presence of one was considered to be at the beginning of this decade.

Chapter 1

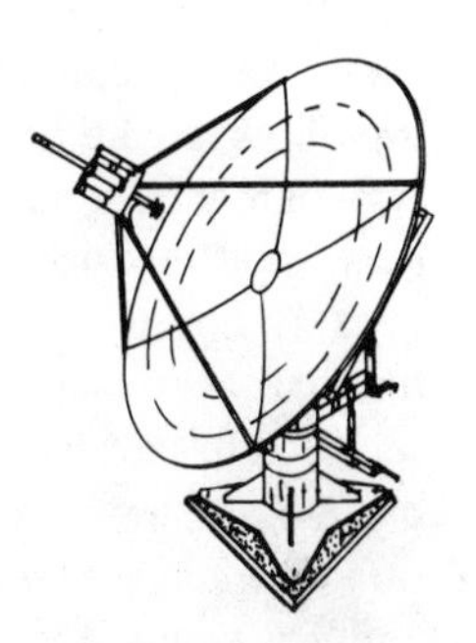

The Television Age

MOST OF US ON PLANET EARTH HAVE SEVERAL THINGS in common. One of these is the television age. We are all a part of it and here in the United States, we depend upon our personal television receivers for more than we consciously realize. Think of a world without television. Admittedly, it's pretty hard to conceive of the idea. This is how dependent we have become on this (now) not so modern electronic invention.

Certainly, there are those who will say that the American people would be a lot better off without the TV receiver. This is difficult to imagine, because through this electronic marvel, the entire world is at our fingertips. The television receiver brings information into every home. Information breeds knowledge if properly mated with the human mind. Sure, television viewing can be abused, but this fault lies with the viewer not with the devise itself.

I can still remember the first television set I ever viewed. This occurred through a department store window, against which a youthful face was flatly pressed in order to get as close as possible to that miniature picture tube. And what wonders that tube displayed! Cartoon figures with huge heads and little bodies scurrying here and there in comic disarray! How many readers can still remember the cartoon drama (lasting all of three minutes) entitled "Barnyard Bunk," and starring the little farmer? The Mickey Mouse Club? Or even Winky Dink (starring Jack Barry)? Sure, these were all "kid shows," but in the middle fifties, I was just that, a pre-school kid.

Now, the education radicals will ask what good was derived from these senseless programs? I say, "bunk" (or should I say Barnyard Bunk?) These shows stimulated youthful imaginations which lead to other endeavors (some avid viewers even became writers).

Reflecting again on the past, the first television receiver my family purchased was about a year after my first contact with the device through the department store window. What a pleasure it was to return home and find white-coveralled men climbing to the top of our peaked roof, precariously erecting a mast with stacked two-element, yagi antennas at the top. The immediate rush of tiny legs into the living room brought a grand sight to ever-widening eyes which threatened to burst from their sockets and fly straight for that large picture tube (21 inches). There on a hastily uncluttered table sat the biggest "TV" the world had ever known. Made by Dumont, if my memory serves me correctly, it boasted full reception of channels two through thirteen, had a massive black case, and must have weighed two hundred pounds!

The installation took a bit longer than anticipated. The workmen stayed until past 7:00 P.M. Labor was included in the price of the receiver, mast, and antenna, with the total cost running to nearly one hundred dollars. "What extravagance," my mother must have thought. That night, my father did the honors by cautiously flipping the control to the on position, amid the cheers of a gathering audience which consisted of half the neighborhood (children and adults alike). "Make it play! Make it play!" was an oft-heard cry.

And play it did. I vividly remember watching a documentary on what is now Saudi Arabia. Thinking back, the television announcer was wrong, because I distinctly remember pyramids along with the camels and the people, so the country was obviously Egypt. But this didn't matter then. Television was so new and wonderful that it didn't make any difference what was on, as long as it was something.

This last statement is not as ridiculous as it might sound. In those days, we didn't have cable television service. While our location was near Washington D.C. and Baltimore (close enough for television reception), TV stations didn't broadcast into the wee hours of the morning. Some of them didn't start to transmit until late morning or even early afternoon. Most were off the air by 10:00 in the evening. With these conditions, we were lucky to get an hour or so of viewing in each day.

But this was the peak of the television's growth from being an oddity to serving as a major appliance in the home. More channels started to become active as new stations went on the air. Others increased transmit power and were suddenly available to us. Often, the receiver was activated just to see what new station had sprung up from nowhere. Americans wanted television, and they sure got it!

The amateurish notions of many locally oriented programs began to take on a more professional air. We children saw the latest toys which were on sale, "Just in time for Christmas giving." We became aware that there were people on the face of the earth who did not look exactly like we did. We even discovered that there were all sorts of funny looking tubes which ran through our bodies and performed a myriad of services for us, without which life would be impossible.

But again, there were the cartoons. Those glorious animations which showed elves, dragons, dwarfs, fairy princesses, and occasionally, real people. Buffalo Bob and Howdy Doody were daily treats. We cheered on the heroes and hissed the villains who never seemed to do any real harm. We learned right from wrong and that you were always punished in some form when you chose the wrong path. Let's face it, most of us received a large part of our ideas about life itself from television. . . in partnership with parental guidance. So, to say television is bad is also to say knowledge is bad. Knowledge is always good. What we choose to do with it is another thing altogether.

It is, I feel, totally appropriate to mention that the old Dumont which I devoured whenever possible as a child for the cartoon wonders it held was the same set that, ten years later, I watched the coverage of the late John F. Kennedy assassination. I even saw his accused assassin, Lee Harvey Oswald, mortally wounded while being covered in a live broadcast. The same rapt attention was there as when viewing the cartoons years earlier, and I and millions like me were still learning. However, we were not overjoyed by what we saw. Rather, we were repelled by a true fact of life.

Television was becoming more and more true to life. Some of the entertainment or amusement aspects were beginning to fade away. Their place was taken by news, information, and definite realities. Ideal lifestyles were still presented, but this was evenly balanced by harsh events which were happening on a daily basis. Some will blame television for these disgusting realities, but I don't agree.

You are probably wondering at this point what these maudlin

reflections and opinionated statements have to do with a book on satellite television Earth stations. Admittedly, it may be a rather unusual way to start a book; but it is vastly important to realize just why people are beginning to turn to increased coverage of the world and, indeed, outer space through the use of satellite transmissions. It all boils down to the importance we all place on information. Television is the most convenient method available to us today for the gathering and dissemination of information. It has irrevocably changed the world and certainly for the better. It continues to instigate changes.

Take our American political system, for example, and our methods of electing a President. Our individual votes do not directly elect a person to this high office (or remove him from it). Rather, we elect representatives to the Electoral College. These electorates vote the overall wishes of the people. When a candidate has enough electorate votes he has won the election. As a matter of fact, it is possible for the majority of the voters in the United States to vote for one man for President and still have his opponent (who received less popular votes) win. Albeit, the majority would have to be a very small one and would have to come from those smaller states with a proportionately smaller number of electoral votes. Some people have been wondering why we have an Electoral College instead of electing the President by the popular vote.

The answer is simple.... Television (or rather, a lack of this medium). When the Electoral College system was first established, the United States was spread far and wide. The same distances separate us today (in statute miles), but television has brought us all closer together. Now, we know on the East Coast what the voters are doing in the West. Commentators may even tell us why they're voting the way they are. We know instantly and accurately. Often, the news commentators predict the winner long before the loser concedes. Sometimes they're wrong, but the vote tabulation rarely is.

But back when the Electoral College system was established,we didn't have television coverage (or radio either). There was simply no practical way to keep tabs on what the entire nation of voters was doing from a central location. The only thing left to do was to throw the burden on each individual state to let the popular votes, which could more easily be kept track of on a state basis, determine who the electorates would vote for and then let state population determine the number of electorate votes each state was permitted. The system was certainly sensible for the times.

Today, however, many persons (some in office, some not) are saying that the Electoral College system is outmoded and obsolete due to the tremendously efficient communications systems that are now available to us here in the United States. Television has played a great part in this area, because almost anyone can learn, first hand, by sight and by sound, just how good our ability to communicate with each other, near and far, really is.

Americans are more opinionated today, again, mainly due to television. We see more events and hear every side of different arguments. We are better able to form opinions and see others whom we admire share those opinions (and those we don't admire usually disagree). No, television shouldn't become an opiate which runs our lives, but rather an information source which helps us enjoy life and stimulates our activities.

Let's discuss another aspect of television, one which introduces some negative aspects. Television supplies information; therefore, it is an active system or media. We receive information; therefore, television viewers are often said to be passive. I guess there is a bit of truth in this statement, although there is a good argument that the passive reception of information often leads to the active application of it. But assuming the former statement is true, even this is changing. Already, a few cable companies with local programming have installed special interfaces on their home hookups which allow the television audience to respond directly to programs by pushing a button. Local elections, referendums, and commonly-shared ideas and problems are discussed on the air. The audience is then asked to respond with a yes or no by depressing the appropriate button on their home consoles. Now, information is taking a two-way street. The information transmitted by the cable company or television station on the cable is passively received by the home viewers. When they respond, they are sending information back to the studio which assumes the passive role in this cycle of operation. This allows for an accurate measurement of the feelings of a large portion of a local market. This is already being used to advantage by local governments who can now receive information from many more people than the few who will usually show up at weekly or monthly meetings. This helps government to do a better job of serving the people. It also allows the private citizen to take a bigger part in government.

At this point, we have come full circle, from cartoon shows to audience-participation television. Let's begin again by going back to a cartoon show. You may remember this program from your youth if you are in your early thirties; but if not, you will be slightly amazed at the parallel which can be drawn between a child's program and audience-participation TV.

The show referred to was aired in the middle 1950s and was mentioned a bit earlier in this chapter. I believe it may have been partially or wholly developed by Jack Barry, who was the human star. If not, my apologies to the genius who dreamed this idea up. The real star of the show was a real star, the five-pointed variety, and he went by the name of Winky Dink. The cartoon character was a five-pointed star with arms and legs, and he got himself involved in all sorts of cartoon adventures. How can this possibly compare in any way, shape or form with audience-participation TV? Read on!

This would have been just an average cartoon show if not for the basic concept of the entire program. You had to actively participate or the show was completely incomprehensible. This participation also required the expenditure of fifty cents for a Winky Dink Magic Screen, Magic Crayons, and Magic Eraser. The Magic Screen was simply a piece of clear, flexible plastic which would adhere to the television screen without glue. I guess the static created by the picture tube performed an adhesive function. The crayons were standard crayola types, although they were bigger than usual. The Magic Cloth was an ordinary cotton rag.

Now that you're completely confused, let's move on to the system. Shortly before the program started, the Magic Screen was applied to the surface of the picture tube. Jack Barry always told us to rub it flat with the Magic Cloth so that it didn't have any wrinkles. Then, with crayons in hand, we watched for the entrance of the "star Star" Winky Dink. Jack would interview him for a minute or two; then the adventure would start. A special guest was announced and Jack Barry would draw a line on a clear glass panel which lay directly in front of him and was invisible to the TV audience. It seemed as though he were writing in midair. We at home would then trace this line on our magic screens. With the camera locked in place, the first line would disappear and Jack would draw another. A few minutes and many lines later, we kids had traced a perfect kangaroo or other such animal on the Magic Screen. Those home viewers without this attachment saw nothing. Then a mouth would appear at the proper place on the home-drawn character's face (this was produced at the studio through video effects). The animal would begin to talk and the mouth would move with the speech pattern. Again, the audience without a Magic

Screen only saw a moving mouth suspended in space with no body around it.

With Jack Barry's help, we drew bridges for Winky Dink to cross raging rivers on and boats for him to sail the ocean. We even drew little rocks which seemed to be hurled across a room as the actual TV scene was quickly moved from right to left or vice versa. We got our first driving experience by sitting behind an automobile's instrument panel which we drew ourselves (ala Jack Barry) and watched the television roadway and scenery go by at a rapid rate. By golly, we participated! Without loyal viewers (and artists) such as myself, ol' Winky Dink would have been a self-destroying supernova in no time at all. We felt as if we were right there with him through all his trials and tribulations, and we were the heroes that saved the day. That is true audience participation and direct involvement.

I don't know why this program finally left the air. It couldn't have been because of lack of an audience. Perhaps the FCC began to take a dim view of programs which could appeal only to those who could afford the fifty cents for the Magic Screen. Of course, long before my Magic Screen arrived, I was cheating the system by using a piece of plastic wrap taped to the TV screen and my coloring book crayons.

Satellite television reception will open up yet another world of entertainment, information, and educational programming which was never before possible. This is certainly the "television age," but we are entering a new phase which will significantly change the lives of many millions of people.

Through the use of a personal Earth station, you will be able to select programs that were never before available and to search out new programs that are offered on a regular basis. You may even become a satellite "hound" who takes pleasure in homing in on exotic foreign orbiters (Russian spy satellites, for example) which you have seen listed in your elevation/azimuth charts.

You, quite possibly for the first time in your life, will be receiving signals directly from outer space, from a tiny sphere or oblong object which lies about 23,000 miles in altitude over the earth's equator. The only major problem you will have will be in deciding which one of the hundreds of programs you want to watch.

It's an exciting world and even more exciting universe, and there are things which are literally out of this world which you will be depending on and making use of daily. And all of this can be had from the comfort of your easy chair with your personal satellite Earth station snuggled on a small patch of land in your backyard.

Like those early years of black and white television receivers for the home, the owner of a personal Earth station is somewhat of an interesting oddity, one which all the neighbors will want to be friendly with in order to obtain a personal demonstration of this Captain Midnight-like device.

Yes, we are in a new age of television, and those of us who are at the forefront will be able to take advantage of the increased services satellite television has to offer. As wonderful as satellite TV is now, it can only get better as the state of the art develops. Just like that old Dumont black and white that grew up with the author, your satellite Earth station will grow up in a new age with you.

Chapter 2

Satellite Broadcasting

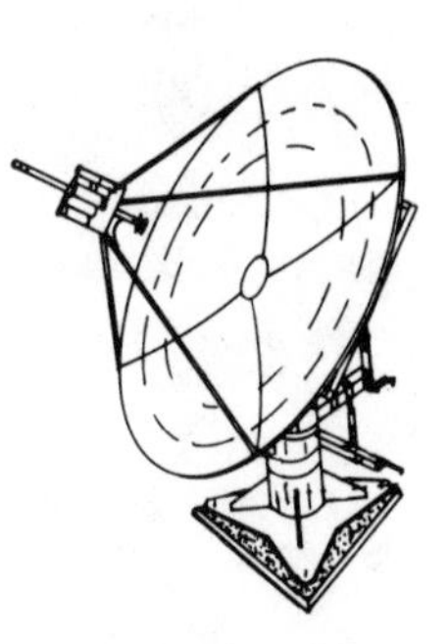

WE ARE ALL FAMILIAR WITH THE MANY STANDARD BROADcasting services. We can also include such facilities as amateur radio stations, CB stations, and a multitude of other services which use the air waves to transmit and receive information. Any radio system is composed of two basic parts, the transmitting station and the receiving station. A brief discussion of these will better enable us to push onward to satellites and the equipment associated with their uses.

TRANSMITTING STATIONS

Transmitting stations have one purpose: the gathering of information which is *discernible* to the human auditory and visual systems and then modifying this information to enable it to be broadcast over generally long distances. Let's take your local radio station, for example. In Fig. 2-1, the announcer speaks into a microphone. The microphone is a transducer. It takes energy from one system and transfers it to another. The human voice is composed of audible energy. The frequency, level, and other aspects of the human voice convey information. This is received at the input to the microphone. This device transfers the information to an electrical system. The vibrations of the human voice cause a diaphragm within the microphone element to vibrate sympathetically. This movement sets up an electrical current flow that is the equivalent of the human voice.

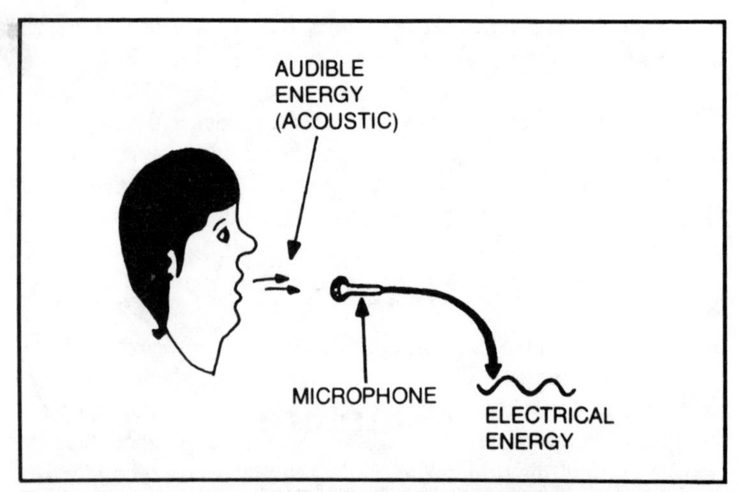

Fig. 2-1. As the announcer speaks into the microphone his voice, which is audible energy, is transferred into electrical energy by vibrations in the microphone element.

The output from the microphone is of very low level or amplitude. The next stage in the broadcast chain is to step up this energy, or amplify it. This process is accomplished by audio amplifiers, which accept a small voltage at their inputs and produce a much larger voltage at their outputs. It is important to note here that while the output voltage may be higher, it changes in such a manner as to directly correspond with the changing levels at the input. For example, if the audio input to an amplifier containing a bit of information which was originally applied to the microphone by a human voice caused the voltage to swing from 1 to 10 volts, then 10 volts is the maximum applied input. Let's assume that the output from the amplifier has a maximum level of 100 volts. Based upon the same amplifier input, the output voltage would swing from 10 volts to 100 volts. The output is much higher, but the relationship of the highest value to the lowest is still 10 to 1. As long as this relationship is maintained, the detected audio information from the output will be the same as that at the input, only much louder. You should note that this is a very simplistic explanation of the functioning of an audio amplifier and that there are many other criteria involved in audio reproduction and amplification.

After the voice information has been converted to an electrical system and amplified, it is then passed on to the radio transmitter. If we were to connect a speaker to the output of the amplifier, we would again hear the human voice. The speaker is a transducer which acts like a microphone in reverse. It accepts information from an electrical system and converts it to sound waves. But, we are not ready to demodulate the information contained in the system. We must first use the transmitter to alter the electrical energy from the output of the audio amplifier and place it into a system which will allow this information to be carried long distances. The transmitter effectively takes the low frequency audio information and steps it up to a high frequency. This is done because high frequency waves travel further than low frequency waves. The human voice normally produces frequencies between 300 and 3000 hertz (cycles per second). The broadcast distance of the unaided human voice is very limited (as far as we can shout). If we were to take this same information and step it up to a high frequency, then the broadcast range would be greatly increased. A good way of demonstrating this principle of high frequencies traveling farther than low frequencies is to use a high school band as an example. The piccolo, which produces the highest frequency output of any band instrument, is easily heard, while the tuba, a low frequency instrument, is not. The level or loudness of the piccolo's output is probably far less than that of the powerful tuba, but since the high frequency travels further, the piccolo stands out. You would have to get very close to the tuba to be able to hear it as well.

The radio transmitter also contains an audio amplifier. In AM transmitters, this amplifier is often called the modulator and actually consists of several audio amplifiers in series. At a 1000-watt station, the final audio amplifier output of the modulator will be around 500 watts. The announcer's voice, which contained less than a watt of power as he was speaking, now packs a walloping 500 watts of power.

The second section of the transmitter produces a high frequency carrier wave. Its frequency is on the order of 1,000,000 hertz (1 megahertz). The carrier serves as a reference. When the amplified audio information reaches the modulator, it is applied directly to the high carrier frequency. Let's assume (to make things simpler) that the announcer is humming a musical note. The frequency of this note is 1,000 hertz (1,000 cycles per second). When this is applied to the carrier, two sidebands are generated. One will be 1,000 cycles per second higher than the carrier frequency of 1,000,000 cycles. The other will be 1,000 cycles less than the carrier frequency. But the entire transmitted wave is no longer in the audio spectrum but in the high frequency range at around 1,000,000 cycles. The upper sideband is broadcast at a frequency of 1,000,000 cycles plus 1,000 cycles. The lower sideband is generated at 1,000,000 cycles minus 1,000 cycles. The upper sideband frequency is then 1,001,000 hertz and the lower sideband is 999,000 hertz.

Now that we have converted from audio frequency to radio frequency, the transmission is broadcast from the radio station's antenna. This is how all basic communication systems operate.

RECEIVING STATIONS

Like transmitting stations, receiving stations have one major purpose: to detect the information from the transmitted wave and convert it to information which can be directly detected by a human being.

Referring to where the previous discussion left off, we now have a radio wave being broadcast from the station which contains audio information (the 1,000 hertz tone). The radio wave spreads out from the transmitter site until it strikes the receiving antenna. The receiving system is shown in in Fig. 2-2. Here, the antenna is designed to respond to the frequency of the transmitted wave. It transfers a small portion of this wave to the receiver circuitry. The carrier, having served its purpose, is eliminated and the audio information is retrieved. The purpose of the carrier is to carry the audio information; hence, its name. It also serves as a reference. Without it, the audio frequencies cannot be deciphered and would simply look like a garble of radio frequencies. But remember, the audio information is spaced a specific distance (in frequency) from the carrier. If the input was 1,000 hertz, then the audio information is 1,000 hertz either side of the frequency of the carrier.

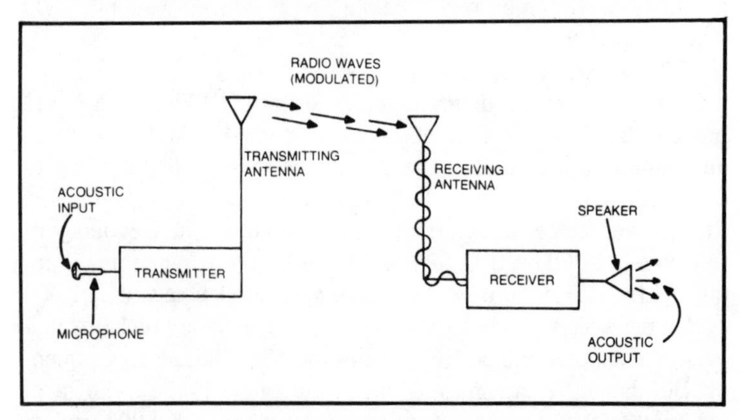

Fig. 2-2. In a receiving system the antenna responds to the frequency of the transmitted wave.

The receiver circuits utilize the carrier as a reference. Simply put, the carrier frequency is equal to 0 hertz. The sideband frequencies which are 1,000 hertz either side of zero are then passed on as a 1,000 hertz audio tone. We are still sending information in an electrical manner and it becomes necessary to convert this to sound wave output that can be directly detected by the human ear. A speaker is attached to the electrical system and its output is the same as what was produced at the microphone by the announcer.

This complicated process occurs at the speed of light and is ongoing throughout the entire cycle of transmitting and receiving. An earlier discussion in this chapter stated that if the voltage swing at the output of an audio amplifier is equivalent to the swing at the input as far as percentage is concerned, then the retrieved information will be identical at both points, only at different levels. The same can be said of transmitted information on a radio wave. As long as the frequency swing between the carrier and the sideband is identical to the audio frequency swing, then the information detected at the receiver will be identical to that which was originally put into the transmitter. The frequency of transmitter operation makes no difference whatsoever. Since the audio frequencies of the average human voice cover a range of approximately 300 to 3,000 hertz, the sidebands produced at the output of an AM transmitter will swing from 300 to 3,000 hertz either side of the carrier frequency. This is why you can hear your announcer exactly as he sounds when talking in the studio.

RADIO WAVES

While the frequency of the transmission of radio waves makes no difference to the demodulating process (retrieving the audio information), it is very significant when speaking of how these waves will travel across the surface of the Earth and, indeed, through the Earth's atmosphere and out into space. Most of your AM radio station's transmitted power goes to the ground wave. This means that the transmitted signal travels along the surface of the earth. Mountains, buildings, towers, and other obstacles interfere with the wave and sap power from it. This, along with free space attenuation (the lessening of the amplitude of the signal as it travels farther and farther from the transmit site), means that very, very weak signals are applied to the receiving antenna. Low-powered AM stations of 1,000 watts or less have ranges that are typically less than 20 miles, although this will depend upon flatness of terrain and other criteria. If you are located outside of the broadcast area, the transmitted signals are too weak for good reception or perhaps for any reception at all.

Frequency is the great determining factor in whether or not the output of a transmitter will produce a ground wave. The antenna also will play a part in this. Ground wave communications are limited. This is why there are so many radio stations in the United States.

LINE OF SIGHT BROADCASTS

To partially overcome the limited range of AM broadcast stations whose transmitted power is mostly tied up in ground wave transmissions, there is the FM broadcast band, which lies between 88 and 108 megahertz. This band is at about 100 times the frequency of most AM broadcast stations. Due to the high frequency of operation, FM broadcast band stations normally transmit by line of sight. This means that the signal will be received when the receiving station lies within a direct line of the transmitter. This is shown in Fig. 2-3. Mountains and other obstructions will still create signal losses, but in many areas, reception is more dependable. Line of sight broadcasts are typical of all radio transmissions but especially of those above 50 megahertz.

A major problem comes into the picture when discussing line of sight broadcasts and there's really very little than can be done to overcome it. Figure 2-4 shows two stations on the surface of the

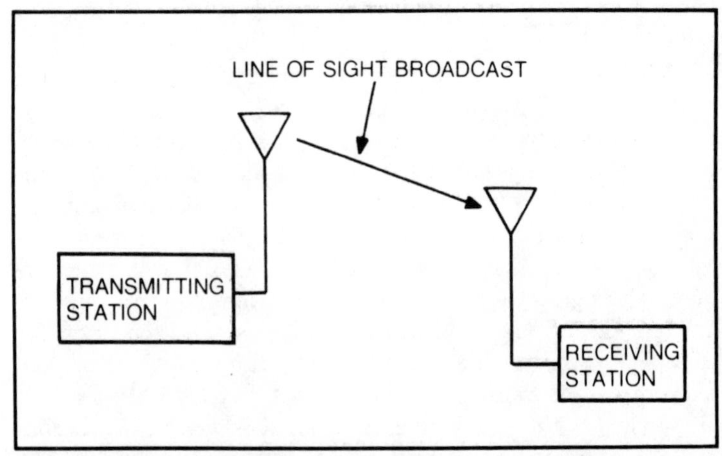

Fig. 2-3. In line of sight broadcasting systems the receiving station must lie within a direct line of the transmitting station.

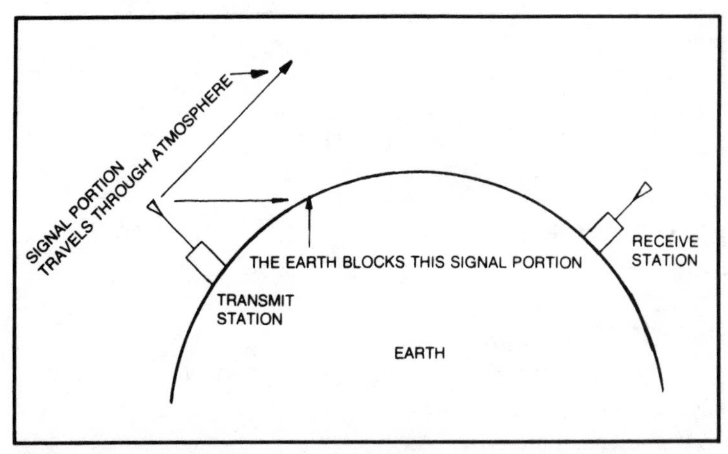

Fig. 2-4. Since the Earth is not flat, the curvature obstructs transmitted signals and prevents them from reaching the intended receiver.

Earth, one for transmitting and one for receiving. The Earth is not flat (something we all know) and its curvature is a direct block to line of sight communications when the two stations are separated by enough distance so that their antennas fall behind this curvature. The Earth itself blocks the transmitted signals from the receiver.

Figure 2-5 shows one method of overcoming the curvature problem. A passive reflector has been installed between the transmit and receive stations. This device is within line of sight of the transmit and receive antennas. Now, when the transmitter produces its output, it strikes the passive reflector which reflects the signal

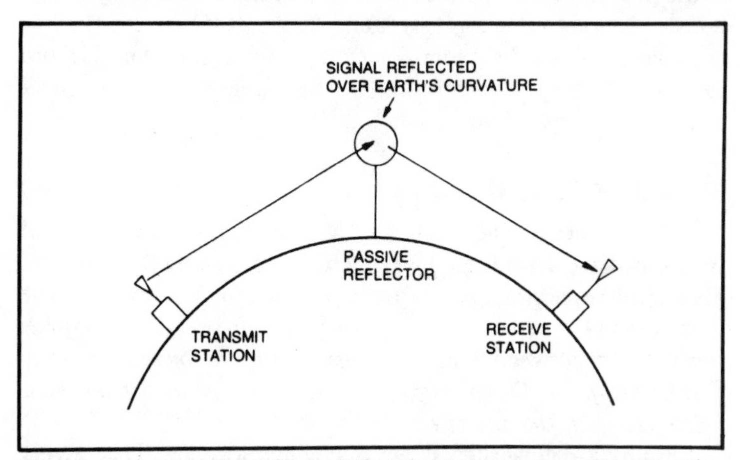

Fig. 2-5. Here, a passive reflector is installed between transmit and receive stations.

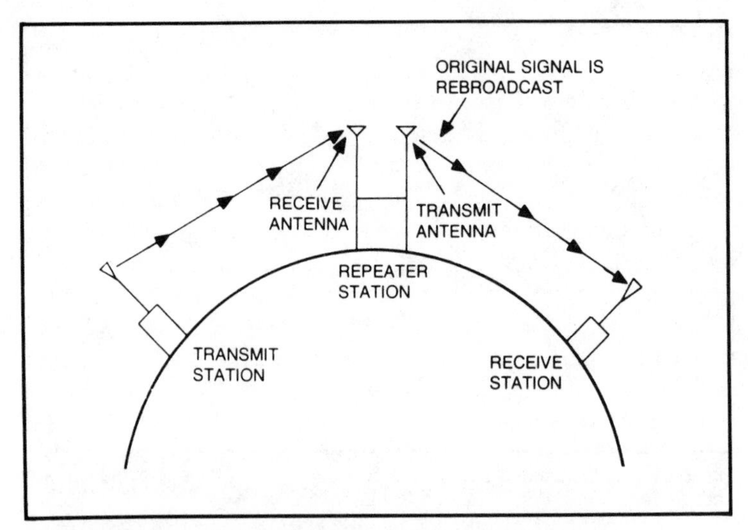

Fig. 2-6. Another solution might be to install an active repeater.

in the direction of the receiving antenna. Such installations are quite expensive but are often used for microwave communications purposes. The alignment of the reflector with the other two antennas is extremely critical.

Another method uses an active repeater between the two stations. This is shown in Fig. 2-6. When the transmitted wave strikes the repeater, a receiving antenna detects the signal and feeds its output to the input of another transmitter. Its antenna broadcasts to the final receive station located behind the curvature. What this boils down to is another receiver and transmitter between the two sites which are used for communicating information. This is more expensive that a passive reflector, but the signal strength and quality at the final receive location is far better.

SKYWAVE COMMUNICATIONS

The frequencies between the AM broadcast band and the FM broadcast band are utilized by many services. Among them are amateur radio operators, some business band communications, the citizens band network, and a myriad of foreign broadcast stations. Frequencies between 3 and 30 megahertz travel by ground wave and by skywave. Often, a greater portion of broadcast power is wrapped up in the skywave.

Figure 2-7 illustrates skywave communications. This is very similar to the previous example of line of sight broadcasting, where a passive reflector was used. Skywave communications also utilize the passive reflector concept, but here, the reflector is a natural one-the Earth's atmosphere. Referring to Fig. 2-7, signals leave the transmitting antenna and travel skyward. The wave strikes the ionosphere (an ionized layer in the Earth's upper atmosphere) and is reflected back to the Earth. This is often called *skip* because the radio waves skip off the ionosphere. The reflected signal is aimed back toward the earth and may touch down at a receiving site thousands of miles away from the transmitting site. The actual distance covered will depend upon the type of antenna used at the transmitter, the frequency of transmission, and the condition of the ionosphere.

Multiple hop transmission may also occur as shown in Fig.2-8. Here, the reflected wave from the ionosphere strikes the Earth, which also serves as a reflector at these frequencies. The signal leaves the Earth, traveling skyward again to be reflected once more from the ionosphere. This type of skip can take the form of a multihop transmission, which will often travel completely around the Earth. I recall several occasions where I have talked from Virginia with a station in California, using the amateur radio band, by multihop transmission. Due to the transmitting conditions and the state of the ionosphere, the signals did not travel across the United States. Rather, they traveled around the world, being received in California after several skip cycles. The amateur operator in California was transmitting by the same path. The audio quality of the

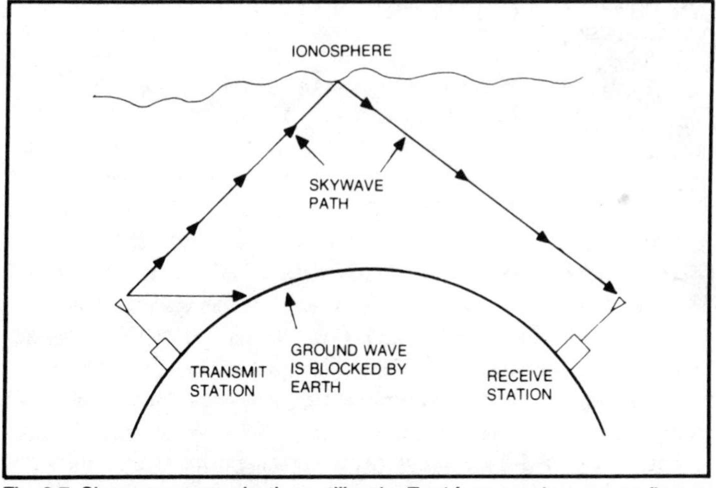

Fig. 2-7. Skywave communications utilize the Earth's atmosphere as a reflector.

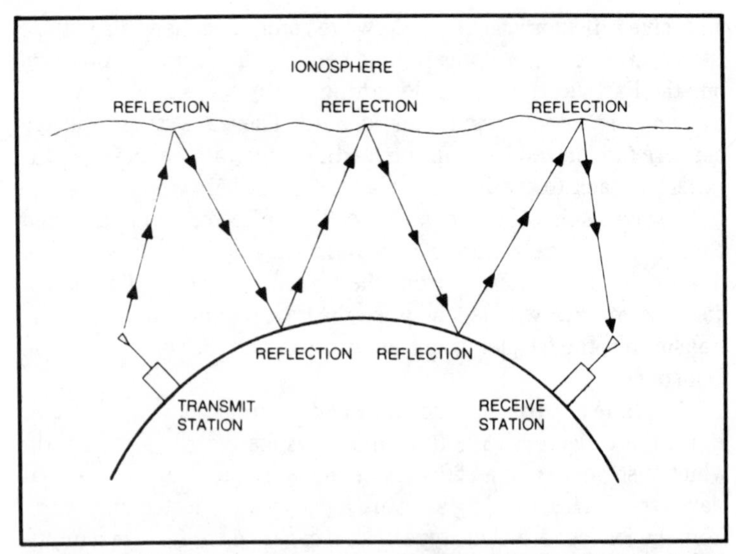

Fig. 2-8. In a multi-hop transmission, the reflected wave strikes the Earth, travels upward to the ionosphere, and then back again.

received signal, along with the reception of foreign amateur operators, were telltale signs that we were communicating by long path DX, a term used to describe these conditions. This is not a tremendously rare occurrence but is exceptional enough to be noted here.

SPACE COMMUNICATIONS

As radio frequencies increase, line of sight becomes the rule rather than the exception. Skip effect begins to disappear. This is due to the makeup of the ionosphere. During skip transmission, only those signals which lie below a specific frequency will be reflected back to earth. Higher frequency waves will pass through the ionosphere and out into space. They are lost forever unless reflected back by some means. Meteors can do this, as can the Moon. Another favorite practice of amateur radio operators is called *Moon Bounce*. Here, an ultra-high frequency is transmitted from the Earth, strikes the Moon, and is reflected back to the Earth again. In this case, the Moon is used as a passive reflector.

Orbiting satellites may also serve as passive reflectors, and there is less space loss than when using the Moon simply because the transmitted and reflected distances are far shorter. Figure 2-9 shows a passive satellite system whereby signals from an earthside transmitter travel through the ionosphere and out into space. Here, they strike the surface of a stationary satellite and are reflected back to Earth again. This is still a line of sight system, in that the satellite must be within direct line of sight with the transmitter as well as the receiver. Due to the altitude of the orbiter, many points on the Earth are within direct line of sight. Without the satellite, the signal which is transmitted from the Earth would travel out into space and never return again.

When satellites are used as passive reflectors, many of the same problems associated with Earthbound reflectors are incurred and often multiplied. There is a large signal loss between the transmitter and the satellite and an equal loss between the satellite and the receiving station. It requires tremendously high transmitter power in order to assure a strong enough signal back on earth after the reflection process has taken effect.

In an earlier discussion, an earth-based passive reflector was used to enable line of sight communications to take place around the curvature of the Earth. To overcome the problems of signal strength, the passive reflector was replaced with a radio repeater. This device received the transmissions from the sending station and retransmitted them to the receiving station. The same can be done in space communications.

The reflecting satellite just discussed is also called a passive communications satellite. This compares with the Earthbound passive reflector. The space communications equivalent of an Earth-based repeater is called an active communications satellite. The

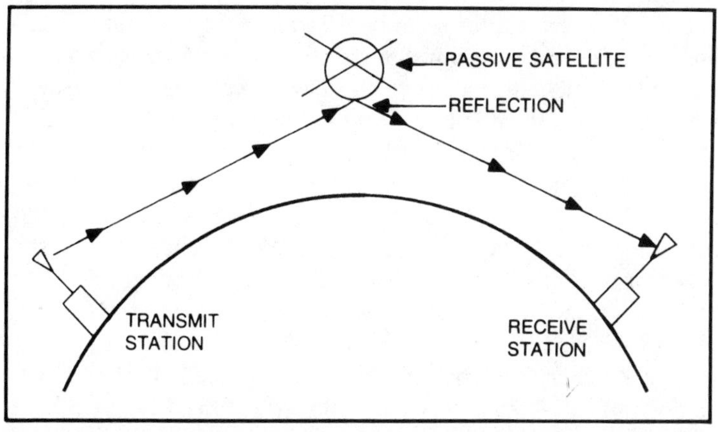

Fig. 2-9. In a passive satellite system, signals strike the satellite and are reflected back to Earth again.

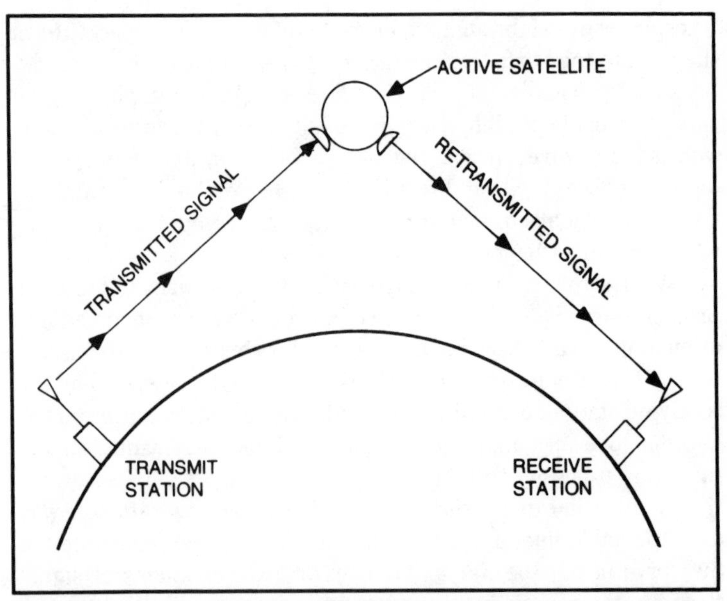

Fig. 2-10. An active communications satellite is equipped with a transmitter, which serves to amplify the transmission before sending it back to Earth.

principle of operation is shown in Fig. 2-10. By studying this, you can see that this type of offered operates in exactly the same manner as an Earth-based repeater.

A high-powered signal is transmitted from the sending station on the Earth. This line of sight microwave broadcast travels through the atmosphere and to the receiver of the active satellite. The output of this receiver is fed directly to a transmitter within the satellite and a new signal which contains the same information as the old one is transmitted back to the Earth. Even when very high amounts of power were used by the Earth transmitter in a passive satellite system, the signals that were received back on Earth were extremely weak. The same high amounts of power must still be used with an active communications satellite system, but since the signal is retransmitted out in space within the satellite proper, the received transmissions back on Earth are far stronger.

The active satellite system is the one we are interested in as far as television receive-only Earth stations are concerned. The original broadcasts are transmitted from very high-powered Earth stations whose antennas are aimed directly at the satellite. These stations transmit on a frequency of around 6 gigahertz. Out in space, the active satellite is equipped with transponders. A transponder is simply a transmitter and receiver which are connected to one another. The detected information from the receiver is fed to the input of the transmitter and beamed back to Earth. One satellite may have a few transponders or possibly even twenty or more. When receiving satellite TV at home, each transponder serves as a separate television channel which can be selected in much the same manner as is done today with your present set.

Satellites are usually physically small devices and space must be conserved wherever possible. For this reason, a satellite with many different channels (transponders) may have only two antennas, one for transmit and one for receive. Each channel shares these antennas by using them for small fractions of a second. To provide an easily understood example, let's assume that a satellite has two channels which must share the same antennas. Channel A may use the antennas for 100 milliseconds and then be switched off for an equal amount of time while channel B uses the system for 100 milliseconds. When channel B is switched off, channel A uses the antennas again. Channels A and B will be switched on and off many times during the blink of an eye, but you could never tell this by watching a received picture at your Earth station because of the speed with which the switching occurs. If you were watching channel A, you would never know when channel B was using the same antennas to transmit on another frequency which you were not receiving. Using standard television sets and earthside broadcast stations, the video picture you receive at home in one second is actually about thirty different pictures. Without getting too technical, the television picture can be likened to the easel of a very fast artist. Every picture that is painted is stationary. It has no movement whatsoever. But when the next picture is painted, the figures have changed slightly and the change is even greater when the next picture is painted. When thirty different pictures are painted per second (as is the case on your home television), things happen so fast that we do not sense the time intervals between picture tube sweeps. Therefore, we view the television picture as a continuous sequence of events which happen in periods of seconds rather than milliseconds. Human beings are incapable of sensing or responding to events which occur in small fractions of a second. Because of this, we can take a look at a television screen where many thousands of things are taking place in these small fractions of a second and automatically see what has occurred by averaging these events over time periods of seconds.

To clear this up a bit more, refer to Fig. 2-11 which shows a ball being thrown. This picture shows the position of the ball at many points along its trajectory. Any one of these points could represent a fraction of a second of the time it took for the ball to leave the pitcher's hand and strike its target. We humans do not perceive things in this manner, however. In viewing the pitch, we would average the events which took place during this throw over a time span of one or two seconds. We would see the total event rather than the tiny segments which are shown in the drawing. Computers and other electronic circuits can be said to think in millionths of a second. They can receive information, perhaps a thousand times faster than us. We could say that computers are very fast, but computers could say we humans are very slow. If a computer had self-awareness, it would see human beings as extremely slow-moving objects, because we think and act in a completely different realm of time. Think of it. If it takes you one minute to tie your shoes, it would take a computer (if it were capable of this type of motion) one-thousandth of a minute. On the other hand, if you were the computer, one-thousandth of a minute would seem to you to be about as long as a full minute is to normal human beings. Therefore, after you tied your computer shoes, you would seem to wait a full one thousand minutes for your human counterparts to tie his.

In any event, the on/off nature of satellite transponders is totally unnoticeable by human beings who cannot think in small fractions of a second. While our previous example used a satellite with only two channels, most have more than this and each shares the receive/transmit antennas for correspondingly smaller periods of time.

As was stated earlier, most television stations which use satellites for their broadcasts use very high-powered transmitters on the ground. They transmit at a frequency of approximately 6 gigahertz, but these signals do not come back to Earth at the same frequency. Referring to Fig. 2-12, the 6 gigahertz transmission

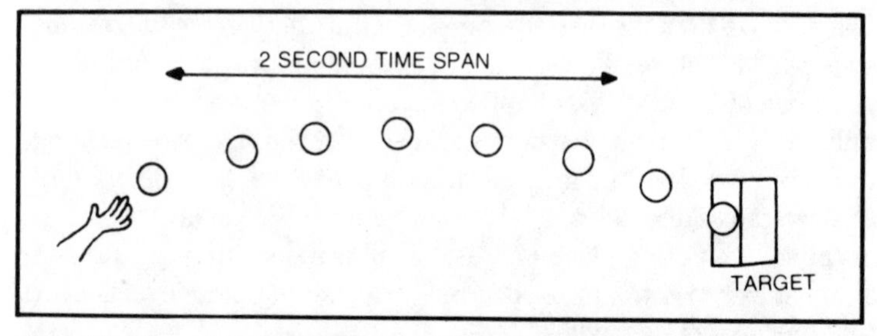

Fig. 2-11. Although in this drawing the ball is seen at all the positions it travels, humans perceive this in a single motion rather than as fractional segments.

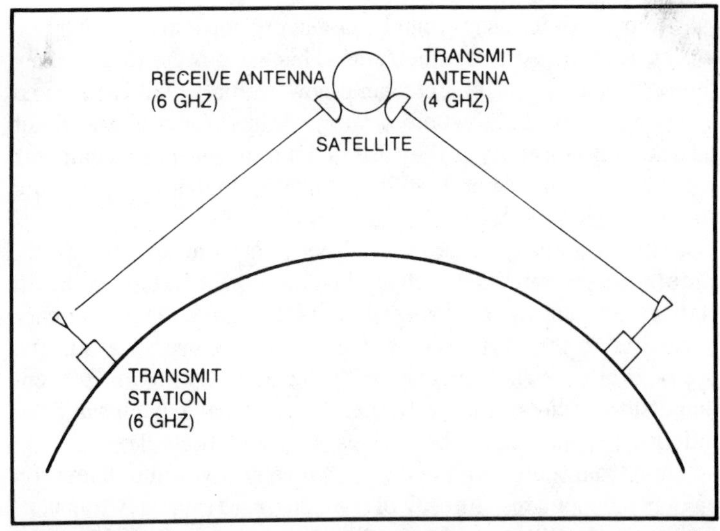

Fig. 2-12. Signals which leave the Earth at one frequency do not return at the same frequency.

leaves the Earth and travels to the satellite. The orbiter's receiver is designed to detect a six gigahertz signal. It pulls the audio and video information from the transmission and then feeds it directly to the satellite transmitter, which has an output at a frequency of about 4 gigahertz. This is the frequency which the earth-based television receive station is set up to detect. A further explanation of frequencies will be found in future chapters.

Since we know that the satellite signal is originally transmitted on the Earth, travels into space, is retransmitted by the satellite at a different frequency and is finally received back on Earth again, it can be said that the signal that is ultimately received by an Earth station is a function of:

- The signal transmitted to the satellite at 6 gigahertz.
- Signal processing in the satellite.
- The signal transmitted at 4 gigahertz from the satellite.
- Directivity of the satellite antennas gain.
- Path loss.
- Gain of the receiving antenna.
- Noise temperature of the antenna.
- Low noise amplifier noise temperature and gain.
- Cable loss to the receiver.
- Receiver noise figure.

All of these terms may not be familiar to you and later chapters will clear up many of the questions. This list is presented to show the many factors which determine how well a signal is received by your personal Earth station. Many of these functions need not be dealt with directly, as they are taken into account in computer reports which can be ordered inexpensively to describe receiving conditions in your area.

There are many different satellites in orbit around the Earth. Entertainment satellites such as the ones used by personal Earth stations enthusiasts are in special orbits that make them stationary in relation to the Earth. To us, these satellites are always in the same location. The Appendices provide a listing of present domestic satellites, along with their longitudinal locations and additional information on the various programs they carry.

It is from these satellites that a whole new world of television enjoyment emanates. Sure, all of the signals are originally transmitted here on Earth, but with a satellite deep in space, these transmissions can be received by so many more persons. The wonderful thing about receiving signals from satellites is found in the fact that we tend to pay very little attention to these multi-million dollar orbiters. We know they're up there and we know their positions in order to properly aim our antennas; but other than this, the Earth station equipment is the part of this complex system which gets the most attention. The satellites are reliable, always present, and simply do not require the attention that our personal Earth stations require. Billions upon billions of dollars of research have gone into making our satellite program as dependable and useful as it is, and while most of the reader's interest will be taken up in the building of his own personal Earth station and not concentrated on the satellite itself, an understanding of what has been done over the years to enable that satellite to be overhead is important.

A new era for all mankind opened in 1961 when Major Yuri Gagarin of the Soviet Union was rocketed into space. He was followed shortly thereafter by Commander Alan B. Shephard, Jr. of the United States. Since that time, men have been launched into Earth orbit so many times as to barely warrant a news story. Americans have been to the Moon and space communication equipment was more than adequate to send back beautiful color pictures for all the world to see. None of this would have been possible had it not been for the many years of experimentation and testing of satellites that preceded manned flights. The first satellite was Sputnik, which was launched into orbit by the Soviets in 1957. This tiny device did little more than beep a weak signal back to Earth. It weighed 184 lbs. and was quickly followed by Sputnik II, which weighed 6,610 lbs. The first Sputnik was programmed to record temperature and pressure. The second orbiter read solar and cosmic radiations, along with temperature and pressure. The Sputnik was followed by the American satellite Explorer I. This satellite weighed 31 lbs. and was used to measure temperature, cosmic rays, and micrometeorites. All of these satellites and the many more to follow had one thing in common; they sent radio signals and television signals back to Earth.

It was long known that rockets could be tracked by radio. This simply involved the placing of a radio transmitter with a very stable output in the nose cone of the rocket. Receiving stations on the Earth could detect motion of the rocket from the condition of the radio signal. There are two very different purposes for radio transmissions in rockets and satellites: tracking and telemetry. Tracking simply tells you where the offered is in relationship to the Earth or other objects. Telemetry is a laboratory measurement, only very rarely concerned with position finding. The telemetry transmitter receives its input from instruments carried in the satellite. The detected results at the Earth receiving station are converted back to the original information.

It can be said that satellites used for television reception are a form of telemetry. The sensed information from space is the incoming transmissions from the Earth sending station. The signal is processed within the satellite and then transmitted back to Earth again, where it is detected by the personal Earth receiving station. The information is pulled from the signal and converted to audio and video output which feeds the television set. This is the same manner in which we received color television pictures from the Moon during the last lunar landing, except the transmitted signals originated on the Moon instead of on the Earth.

As satellite experimentation continued, the equipment became much more dependable and highly efficient. In the early 1960s, the communications satellite Telstar was launched into orbit and it suddenly became possible for us to receive pictures directly from the other side of the world. Before this time, television programs were recorded at distant points on the globe and played back in the United States at a later time. Most of us today take for granted coverage of the Olympics, for instance, which we receive via satellite. It seems like no great thing that we can see the swimming competition here in the United States at the same time this event is occurring in Germany. Today, any point on the globe can serve as a temporary television studio because of the satellites orbiting the Earth. Such portable communications would have been totally unknown just twenty years ago when television broadcasts had an average range of less than 150 miles. Today, the same broadcasts can circle the earth.

I saw the first telecast via Telstar when I was a boy. The picture quality was not tremendously good, nor was the audio. Back in those days, you could still tell the difference between a standard broadcast and one that used satellites—today, you cannot. Space communications technology has made vast strides since the 1960s and a picture received from 23,000 miles out in space is just as clear as one received from your local television station across town. This, of course, assumes a properly installed and operating satellite television Earth station.

This is truly a wonderful aspect of a technological age. Satellites open up many possibilities for all types of people. Satellites have been designed and built by amateur radio operators working in conjunction with space communications technicians and then blasted into orbit as part of a satellite package by the U.S. government. Amateur operators can now communicate via satellites. Telephone conversations to the other side of the world no longer use expensive cables lying on the ocean floor. They use satellites. Anything which can be sent by hard wire can also be sent by means of satellites. And this is just the beginning. Think of the strides which have been made in space communications over the past twenty years. You will see that the last five years represent an even faster technological growth which is bound to accelerate in the coming years. We are bound to see more and more satellite services being opened.

One of the recently accomplished strides is the Space Shuttle program. Satellites have limited effective life spans of anywhere from three to eight years. After this time, their orbits begin to decay until they finally burn up in the Earth's atmosphere. But the Space Shuttle now gives us the capability of going out into space, picking the aging satellites up and servicing them inside the ship. After everything has been checked, the satellite is placed back into orbit again. This corresponds to your television repairman coming to your home to make repairs on your set. You don't throw away your television set when it breaks, so why do it with satellites which cost millions upon millions of dollars? The Space Shuttle opens up many thousands of possibilities for satellite users. This ship is quite capable of placing many satellites in orbit during a single trip. Its massive cargo hold is adequate for carrying satellites to and from the Earth. Instead of launching one or two satellites with huge rockets which are used only one time, the Space Shuttle is used again and again and will carry a larger payload. This is bound to decrease the costs associated with building communications satellites and placing them in orbit. The Shuttle should also be able to assure the large corporations which own these satellites that they won't be lost after five or six years. From the space communications standpoint, we are entering the most exciting era of all time. Those of us interested in setting up our own satellite television Earth station are on the edge of a new frontier.

SUMMARY

Satellite broadcasting has opened up limitless opportunities for the average person. We have been realizing the advantages of space technology for many years. Previously, satellites were used strictly for government, military, and scientific purposes. In the early 1960s Telstar brought satellite communications into every home. But even then, and for many years thereafter, the average person had little control over what he could receive. Now, with personal satellite television Earth stations being offered at affordable prices, a tremendous variety of programs can now be received from space. This field can only continue to grow. New programs and programming services are being offered on a regular basis. All the average individual has to do is properly equip his receiving site with the electronic devices which will enable him to tap this bountiful resource. Throughout this book, we will be exploring the many methods being used by average persons to effectively use what is being made available to all of us. As the state of the art advances, equipment prices should drop and programmers will become more and more competitive by offering information, entertainment, and a myriad of other services which will be aimed directly at a mass of individuals rather than to a few scientific, government, and military installations.

As additional satellites are placed in orbit, the personal Earth station will become more and more valuable. It has been predicted that within this decade, the satellite television service will become more and more of a replacement for conventional television broadcasts. The versatility of space communications is far superior to Earth-only transmissions since more people are able to take advantage of the many broadcasts. Development in areas such as this tend to become aligned directly with user needs and demands. As more individuals assemble their own Earth receiving stations, the industry will respond. Already, there are plans for more satellites and more programs. You, the consumer, are the one who will reap the benefits of these achievements.

Chapter 3

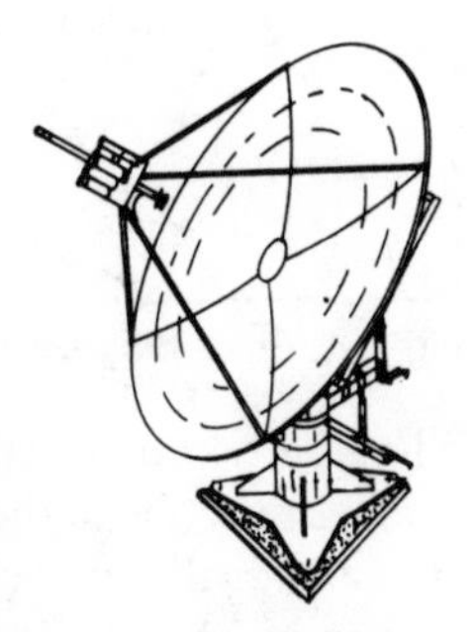

The Personal Earth Station

YOU ARE PROBABLY IN A BIT OF A DAZE AS TO EXACTLY what is involved in receiving entertainment channels which originate on Earth, travel over 22,000 miles into space, and then back again. Surely, you must be thinking, satellite reception must involve extremely complex electronic circuits and probably requires an engineer or two to be properly set up and adjusted. This is only half true. Yes, the reception of signals from stationary satellites does involve highly complex electronic circuits, but no, a personal Earth station does not require an engineering degree to be set up, aligned, and maintained correctly.

Figure 3-1 will be helpful in explaining a basic Earth station designed to receive satellite transmissions and convert them into audio and video signals which are applied to the television receiver by means of standard coaxial cable. While this is a simplistic diagram, it takes in the units which you will purchase or build and then assemble yourself by interconnecting them in the order shown.

Let's start first at the front of the receiving end in a personal Earth terminal. As shown in Fig. 3-1, this is the antenna system. The antenna picks up the transmitted signal from the satellite in exactly the same manner as your present television antenna picks up signals from the television station. This antenna will look different from your TV antenna because a completely different set of frequencies is used. Figure 3-2 shows a typical TVRO (television receive-only) antenna system. We refer to this as an antenna system because several elements are included. The actual sam-

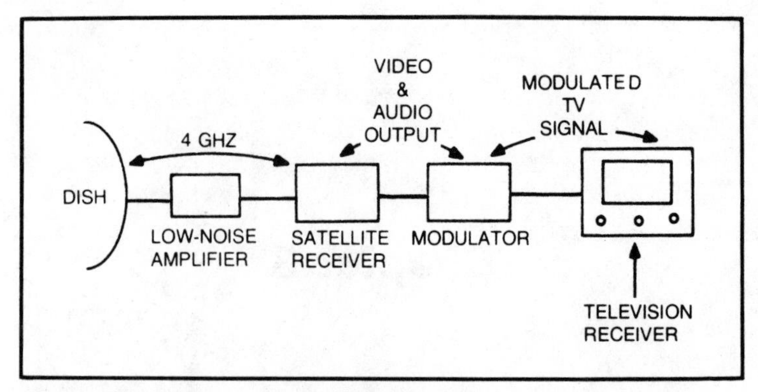

Fig. 3-1. Block diagram of a typical Earth station.

pling portion of the antenna is also called the feed horn and is the smallest portion of the entire structure. The feed horn is where the actual signal is detected. This, in turn, is relayed on to the electronic equipment which will convert the satellite signal into a signal which is usable by the television receiver. What's that massive dish-shaped structure for then? This is the passive part of the antenna system, and it is the portion which the satellite transmissions first come in contact with. The large dish is able to sample a much larger portion of the transmitted signal than is the tiny feed horn. The construction of the dish allows all of the satellite signals which come in contact with it to be focused at the feed horn input. This allows the TVRO system to pickup a great deal more of the signal than

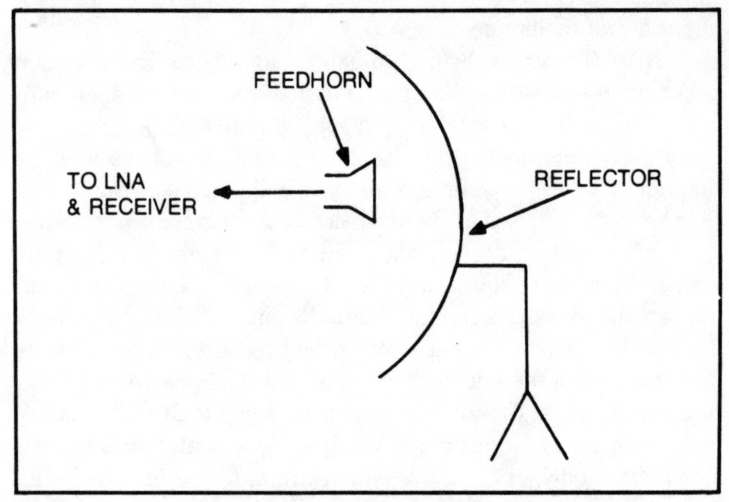

Fig. 3-2. A typical TVRO (Television Receive Only) antenna system.

would be the case if the feed horn were used without the dish.

The construction of the dish is rather critical. It must be aligned in such a manner as to reflect all incoming signals to a very fine point a short distance away from the dish center. It is at this exact point that the feed horn is mounted. The point where all incoming signals are focused by the dish is called the focal point, a term often used in astronomy and when dealing with lenses in general. For this reason, microwave dish antennas are often called microwave lenses. Figure 3-3 shows a pictorial drawing of the basic operation of the dish and feed horn portions of the antennas. A more thorough explanation of microwave antenna operations will appear in Chapter 5.

This drawing shows that the dish acts as a sort of funnel, in that every signal which strikes it is brought to a small point. In this manner, much more signal is able to be detected at the feed horn. We can make a further comparison by placing two empty soda bottles outdoors on a rainy day. One is fitted with a funnel, while the other is left as is. After a few minutes, it will be obvious that the one using the funnel fills much faster. The same amount of rain is being applied to both bottles, but the funnel allows for more collection. If you think of the unadorned soda bottle as a feed horn and the one with the funnel attached as the same feed horn with a dish, then thinking of the rain as satellite transmissions, it can be seen that the latter one is able to sample more of the downpour. Every drop of rain which hits the enlarged surface of the funnel is channeled to a focal point at its narrow end. This should provide a good idea of just how a dish antenna operates. The sole purpose of a TVRO antenna is to collect as much radiated energy as possible.

Carrying this last statement further, it is easy to understand

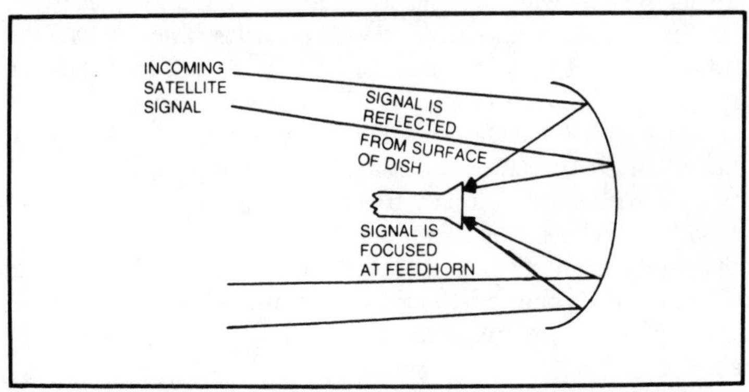

Fig. 3-3. This drawing illustrates the operation of the dish and feed horn.

that the bigger the dish, the more signal that is applied to the receptor or feed horn. As a general rule of thumb, antennas used for satellite TV reception should be at least ten feet in diameter, but larger diameters are to be preferred, since they will supply more signal strength to the feed horn and thus to the receiving equipment. Some manufacturers make dish antennas in the ten foot range which can later be converted with a special kit to a diameter of twelve to fourteen feet or more.

Once we have gotten as much signal as possible to the feed horn, another set of events are put into motion. We must now provide a means for the signal to travel on to the receiver. But before doing this, we must take another fact into account. While the properly designed dish antenna does an excellent job of concentrating satellite signals, we still do not have enough signal strength to provide adequate input to even the finest solid-state receiver. It is necessary to amplify the voltage which is induced at the feed horn by the incoming transmissions. Here, almost every TVRO Earth station uses an LNA (low noise amplifier). To put this in terms which are more familiar, the LNA is a preamplifier. By definition, this device is an amplifier circuit which responds to very low inputs and faithfully reproduces them at its output. The strength of the output signal is not great but it is adequate to drive a standard receiver. Preamplifiers are used in many PA systems in order to bring the level of the very low signal voltage from the microphone up to a point where it can be easily used by the main or power amplifier. All modern radios and television receivers incorporate preamplifiers in their designs. These act directly upon the signals which are supplied by the antenna lead-in cable.

While we are making comparisons here with standard broadcasting services which most of us are at least partially familiar with, specific problems develop at the frequencies used for satellite transmission. A commercial AM broadcast station transmits at a frequency of between 0.6 and 1.6 MHz (1 million cycles per second). Commercial FM stations transmit their signals between 88 and 108 MHz. Satellite transmissions for purposes of home television reception are transmitted within a frequency range of 3.7 to 4.2 thousand MHz. This latter term is shortened to gigahertz, meaning a thousand megahertz. The abbreviation for gigahertz is GHz. Frequencies in this range are referred to as microwave transmissions. The previous two frequency ranges lie within the shortwave spectrum.

At microwave frequencies, physical materials tend to behave

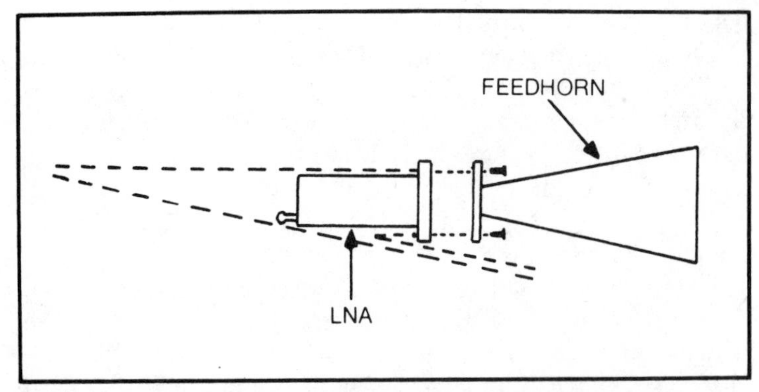

Fig. 3-4. Most feed horns are designed for direct LNA attachment.

differently than they do in the shortwave spectrum. A short length of copper wire which serves as a simple conductor of electricity in an AM radio receiver might look like a very large coil or inductor to incoming microwave signals. Most persons who have experimented with circuits are far more familiar with low frequency operation than those used with TVRO Earth stations. The same theoretical rules apply, but it's a completely different world when it comes to actual practice.

This brief explanation of microwave frequencies is necessary in order to properly explain the attachment of the LNA and, for that matter, the rest of the receiving equipment as well. Whereas most communications receivers incorporate their preamplifiers within the major part of the circuitry, this is not done at microwave frequencies. Coaxial cable, which is very efficient in the shortwave spectrum for relaying signals between the antenna and receiver, is very inefficient at microwave frequencies. A great deal of signal loss will be introduced into the system by channeling these frequencies through even the best types of coax. The LNA is designed to operate from very low signal levels which serve as its input, but the signal loss between the feed horn and this amplifier might be enough to render it completely useless, especially if a sizable length of cable is involved.

This problem is easily overcome by inserting LNA directly at the feed horn instead of at the receiver. When a direct attachment is made between the feed horn and the LNA, your signal loss is cut to a minimum. Most TVRO antennas are designed so that the LNA is attached directly. This is shown in Fig. 3-4. The LNA, then, becomes a physical part of the antenna structure.

The output of the LNA contains the same information as the

output of the feed horn. The signal strength, however, is greatly amplified and can better withstand the rigors of traveling over an inefficient cable to the receiver. This cable serves two purposes, the first being to carry the microwave signals to the receiver, and the second to supply power to the LNA circuitry. The power supply is located within the receiver unit. The two signals intermix quite well due to the fact that one is dc in nature, while the other is at the opposite end of the frequency spectrum.

Now, we have sampled the satellite transmissions, collected them in the dish, focused them to the feed horn, and amplified them with the LNA. The coaxial cable attached to the output of the LNA channels the signal to the next major stage of the TVRO Earth Station.

SATELLITE RECEIVER

The end of the cable opposite the LNA is attached directly to the input connector of the satellite receiver. This device is shown in Fig. 3-5 and corresponds to the standard radio or television receiver you presently have in your home. We cannot connect the LNA output directly to your color TV console, because it is not designed to operate at microwave frequencies. The satellite receiver

Fig. 3-5. A satellite receiver is similar in operation to that of a radio or television set.

is. While it contains no picture tube or speaker system, this device takes the incoming signal and converts it into video and audio channels.The receiver is a detector which is able to grasp the video and audio information from the microwave signal. This is called demodulation. Just like your home television receiver, the satellite receiver will respond to many different channels and is switched accordingly. One satellite may transmit ten or more different channels, so it is necessary to be able to choose the one you want. This is done in the same manner as with your home television—by turning a dial.

The detected audio and video signals could be applied directly to a television receiver, but it would be necessary to bypass the tuner and other demodulation circuits which it contains for the direct conversion of standard TV signals. This is not practical, as it would require circuit modifications to your television receiver. Instead, the video and audio outputs are connected to another portion of the TVRO Earth station.

The video and audio information which are output by the receiver are in detected form. It is no longer traveling on a radio wave. Your standard home television receiver is intended to accept a radio frequency at its input and then to demodulate it and pass it on to the video and audio circuitry. Since the satellite receiver has already performed the demodulation processes, it is necessary to put this information on a radio wave within the frequency range at which the home television receiver is designed to operate. This is accomplished using a modulator, which is sometimes called a *down converter.* This latter terminology is not exactly accurate, as the modulator actually steps up the frequency of the detected satellite transmission to standard television frequencies.

The modulator is actually a transmitting system. It produces a radio frequency output or carrier wave at standard vhf or uhf TV frequencies. The demodulated output from the satellite receiver serves as an input to the modulator. Now, the radio frequency output of this latter unit is modulated with the demodulated output of the satellite receiver. Figure 3-6 explains this a little more clearly. Most of us are familiar with simple modulators. These are often sold as wireless microphones and are devices which transmit a low-powered carrier frequency which falls within the range of commercial AM or FM broadcast receivers. The input of this device is usually a microphone, and when noises are detected, the carrier is modulated. This is basically how the TVRO modulator works, although instead of a microphone, the circuit gets its input from

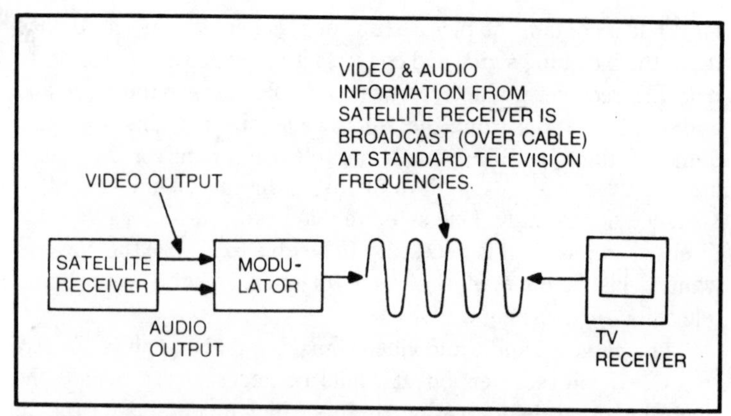

Fig. 3-6. A TVRO modulator is actually a transmitting system.

the audio and video outputs of the satellite receiver.

Some manufacturers incorporate the modulator portion of a TVRO system within the satellite receiver itself. In these instances, the output from the LNA is connected to the receiver in the usual manner, but the output of the receiver may be connected directly to the home television set. Combination receiver/modulators offer the advantage of less component and wiring complexity in regard to connecting the various major components of a TVRO system. There are less interconnecting cables as well and signal loss due to longer travel paths and connector mismatches may be decreased.

After leaving the modulator, the transmission that originally came from a satellite in space at microwave frequencies now looks (to the television receiver) like any ordinary television signal that is broadcast here on Earth. The receiver happily takes this input, which is attached by a standard length of RG/59 70-ohm coaxial cable. It is necessary to tune the home television receiver to the channel which corresponds to the frequency output of the modulator. Often, channels 3 and 4 are used as outputs in vhf modulators. Uhf modulators may be adjustable over the entire television uhf spectrum.

That's basically all there is to a TVRO Earth station. While appearing to be physically different from an Earth-only television transmission/reception operation, the actual working system is very much the same but operates at higher frequencies. In order to emphasize this point, let's look at a standard television station/television receiver system. Referring to Fig. 3-7, a video image and sound are detected in the studio of the television station and serve to modulate the video and audio carriers in the transmitter. This information is broadcast at vhf or uhf frequencies through an appropriate transmitting antenna. When the signal leaves the antenna, this corresponds to a signal leaving the transponder of a satellite. Now, the Earth-based transmission travels through the atmosphere until it is detected by a television antenna. The satellite signal does the same thing, except it must travel through open space and the Earth's atmosphere and in a more direct path to the area of reception.

When the Earth-based television signal crosses the home antenna, an electrical voltage is induced which travels down the transmission line cable to the back of the television receiver. Here, the receiver preamplifier increases the strength of the tiny voltage which was induced in the antenna. This is passed on to audio and video demodulator circuits, which extract the original information that was picked up by cameras and microphones at the television studio.

The satellite system compares with this, in that the signal from the orbiter is detected by the dish antenna. It is not passed directly to the television receiver but must first be increased in amplitude

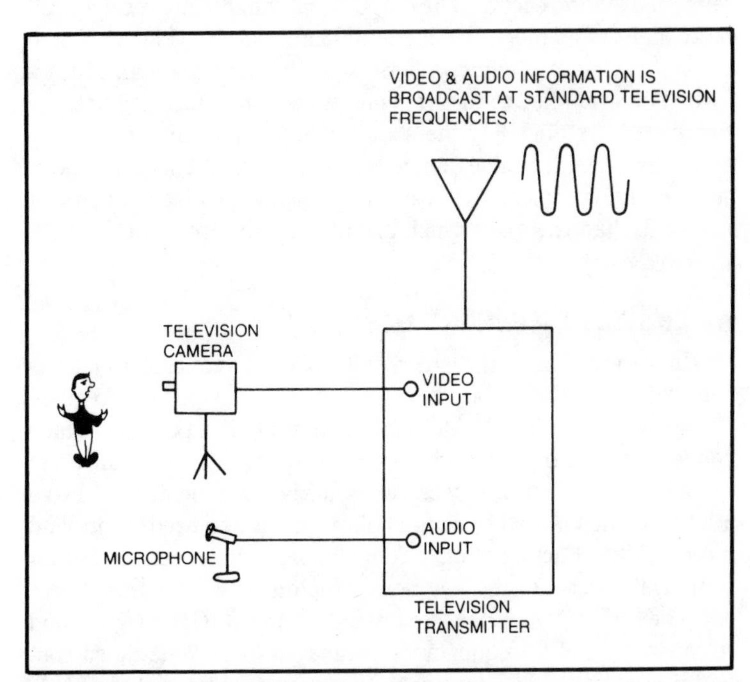

Fig. 3-7. In a standard TV station/receiver system, video and sound are detected in the studio and broadcast by means of a transmitter.

by the LNA. It's interesting to note that many fringe area standard Earth television installations for normal reception may use preamplifiers attached to the yagi antennas that are most often used for conventional television reception. You have probably seen these offered by many different manufacturers. Television preamplifiers are the vhf and uhf equivalents of the LNAs which operate at microwave frequencies.

After leaving the LNA, the satellite signal passes through a transmission line cable to the satellite receiver. This exactly parallels what happens in a conventional television-receive system. The satellite receiver also has a preamplifier which boosts the incoming signal to a level which may by used by the video and audio demodulation circuitry. The only difference between the satellite receiver and your home television receiver is that the former does not have a means of making the detected audio and video information directly usable by human beings. In other words, the satellite receiver does not have a speaker and a picture tube.

The reader may be surprised to learn that at this point in the satellite system, a repeat of a complete conventional Earth-based television station occurs. The output from the satellite receiver can be thought of as the video and audio information which is picked up on the camera and microphone in the television studio. This information is applied to the transmitter section of the modulator in the TVRO system. Only the standard antenna is eliminated here and replaced with a cable that goes directly to the back of the television set. As was mentioned earlier, the signal received at your set is exactly like the ones which are transmitted from the studio.

A SLIGHTLY DIFFERENT DESIGN

In recent years, the trend with TVRO Earth stations has changed due to the problems encountered in transporting frequencies at the 4 GHz level. Although high-grade heliax cable, which connects to the output of the low noise amplifier and transports the 4 GHz signal to the receiver is fairly efficient, it does leave a lot to be desired when long cable runs are required. More and more, TVRO Earth stations are including a down converter that connects directly to the low noise amplifier.

The down converter simply takes the 4 GHz signal and retransmits it at a frequency of around 70 MHz. The down converter is actually like a portion of the receiver located at the LNA. The output from the down converter is at a much lower frequency, and therefore, this signal can be passed via standard coaxial cable with much better efficiency. The 70 MHz signal contains the same video and audio information as the original 4 GHz signal. This means that receivers that operate with down converters are far less expensive than those which are designed to accept the 4 GHz signal. However, the added price of the down converter usually offsets any savings at the receiver end.

Systems that use down converters are ideal in situations where long cable runs are necessary from the personal Earth station site to the actual viewing location. Highest efficiency is achieved because the down converter is connected directly to the low noise amplifier, avoiding the cabling of any microwave signals. The transporting of signals via cable has been accomplished at the 70 MHz frequency, where far less loss takes place. This also avoids the extra expense involved when a receiver is mounted at the antenna site and is controlled by a remote box at the television receiver. Such systems normally involve a modulator, which can add a certain amount of signal distortion under certain circumstances.

BASIC TVRO EARTH STATION

While TVRO Earth stations are made up of some fairly sophisticated electronic devices, the actual installation of one as well as the operation is fairly simple and can be accomplished by almost anyone. True, many of the components seem unconventional, but this is only because most persons are accustomed to dealing with the reception of television signals that are transmitted at vhf or uhf frequencies and not in the microwave region. Some of the antennas and components used for vhf and uhf frequencies would also seem quite unconventional to persons accustomed to dealing only with dc or high frequency circuits. The high frequency spectrum falls between 3 and 30 MHz. Vhf frequencies are from 30 to 300 MHz, while the uhf spectrum begins at 300 MHz. As frequency increases the physical dimensions and shapes of components such as inductors, capacitors and antennas can change drastically. From a theoretical standpoint, an antenna, for example, works in exactly the same manner at 3 MHz as it does at 3 GHz. Why don't you see any dish antennas designed for operation at 3 MHz? This is due mainly to physical limitations. As a general rule of thumb, when frequencies increase, the physical size of antennas and some other components decrease. A quarter wavelength at 3 MHz is 82 feet. Therefore, a quarter wave antenna for this frequency would require an element 82 feet long. But at microwave frequencies, the size decreases drastically. At 3 GHz, a quarter wavelength antenna would be 0.984 inches in length. Therefore, a 10-foot dish antenna at 3 GHz represents a structure which is many, many wavelengths in diameter. But at 3 MHz, the same dish would be equal to less than .04 wavelengths. To get the equivalent gain at 3 MHz that we do at 3 GHz would require a dish antenna that would span nearly two miles. A wavelength is the physical distance in meters that a radio signal will travel during one cycle. At 1 MHz, a cycle is generated one million times per second. So the wavelength at 1 MHz is the physical distance the radio wave will travel in one-millionth of a second. Radio waves travel at the speed of light regardless of frequency. Now, at 1 GHz, single cycle is generated in one-billionth of a second. Therefore, it can be seen that since all radio waves travel at the speed of light, a single cycle will travel less distance in a billionth of a second than it will in a millionth of a second, just like your family car traveling at sixty miles per hour will not go as far in one minute as it will in two minutes.

Back to the previous discussion, a ten foot dish which offers X amount of gain 3 GHz would offer the same amount of gain divided by one thousand at 3 MHz. You would have to increase the size of the dish by a thousand times to equal the same amount of gain at the lower frequency. In other words, a dish antenna at 3 MHz which would equal the gain of a 10 foot antenna at 3 GHz would have a diameter of nearly 10,000 feet.

The purpose of this comparison is to show that physical limitations in size dictate the use of different types of antennas than the ones we are accustomed to seeing. From an antenna standpoint alone, we can easily build models with extremely high gains which fill a relatively small space. Unfortunately, this attribute is offset by the complexities involved in building efficient circuits to handle these microwave frequencies. As wavelength becomes smaller and smaller, physical circuit components such as resistors, capacitors, and even conductors begin to assume the qualities of other components. As will be mentioned several times in this book, a single short length of copper conductor can appear as a very sizable inductance at microwave frequencies. Most transistors are completely useless due to their physical construction and will not be efficient much above the vhf and uhf range. The construction of inductors is highly critical. Those of you who have had the opportunity to wind coils for high frequency circuits know that a slight variation in spacing the turns will have little effect. At microwave frequencies, circular coils are not used for inductors, but the equivalent of a minor spacing error would be disastrous in this upper spectrum.

At 3 MHz, a difference of a couple of inches in the length of an antenna which is supposed to be 82 feet long will not have a great effect on its operation, since three inches is an extremely small percentage of the total length. But at microwave frequencies where the same antenna would be measured in fractions of an inch, a minute error could render the antenna useless. At microwave frequencies, we are dealing with extremely small physical sizes in frequency-determining sections of receivers, transmitters, and LNAs. It is difficult to achieve micrometer-like accuracy in building many circuits. This is one reason why some types of microwave equipment cost so much and also why home building of microwave circuits is not nearly as applicable as it is in the lower frequency ranges.

So that you're not scared to death, it should be understood that rarely are dimensions of one quarter wavelength used in the microwave spectrum. Most receiving and transmitting components use frequency-determining elements with dimensions of one wavelength or more. Using our previous example of a quarter wavelength antenna, theory in this area will indicate that a 3/4 wave antenna or any radiator with an odd multiple of a quarter wavelength will act very much like the original quarter-wave design. For practical purposes, it would behoove design engineers to use an antenna element 21 quarter wavelengths long, as an example, than one of one-quarter wavelength. They would both behave in the same basic manner, but the larger one would be easier to design to tight tolerances because of its length and the lower percentage of error that would result. By this we mean that an error of one one-hundredth of an inch would be a lower percentage of the larger antenna than it would be if we used a tiny quarterwave design.

Fortunately, anyone interested in their own TVRO Earth station will not have to worry too much about the fine details of the principles we have lightly touched on here. This discussion was offered in order to show you that we are dealing with a whole new ball game at microwave frequencies. It is absolutely essential to bear in mind that, at these frequencies, losses can develop by some of the tiniest errors that are completely unnoticeable in other types of systems.

SINCE the final output of a TVRO Earth station will be at

standard television frequencies, it is desirable to get from the microwave range to the vhf spectrum in as short a space as possible. A TVRO system operates at microwave frequencies from the antenna to the LNA and finally, to the satellite receiver. From this point on, we are back in the realm of longer wavelengths.

We know that it is easier to incur greater signal losses within the TVRO system at microwave frequencies than at vhf frequencies. Therefore, special attention must be given to the connections between the antenna, the LNA, and the satellite receiver. After this point, the lower vhf frequency to the television receiver prevents few loss difficulties. Short interconnecting cable lengths must always be used between the LNA and the receiver. This is where a great deal of loss can enter a system. Some manufacturers of satellite equipment offer receivers which must be located at the television receiver in order to change frequency. This means that the transmission line between the LNA and the receiver must travel a longer distance than if the receiver were located right at the dish antenna. This may work well in areas where strong satellite signals are received and where a special low-loss transmission line is used between the LNA and the receiver. But in marginal areas, the loss incurred in the cable may result in poor reception. Figure 3-8 shows the installation under discussion.

Most manufacturers also offer the same receivers as discussed above but with a remote tuning option. Figure 3-9 shows a satellite receiver which must be mounted near the television receiver, while Fig. 3-10 shows the same receiver with a remote control option. Both contain nearly identical circuits, but the latter one allows the microwave portion of the receiver to be mounted at the antenna site while a separate control box is used to remotely tune to different channels. The latter system is more efficient, as it incurs

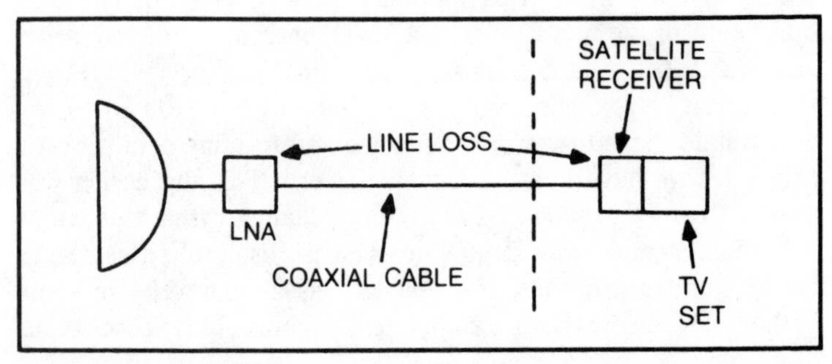

Fig. 3-8. Block diagram of a TVRO Earth station showing a receiver mounted near the television set.

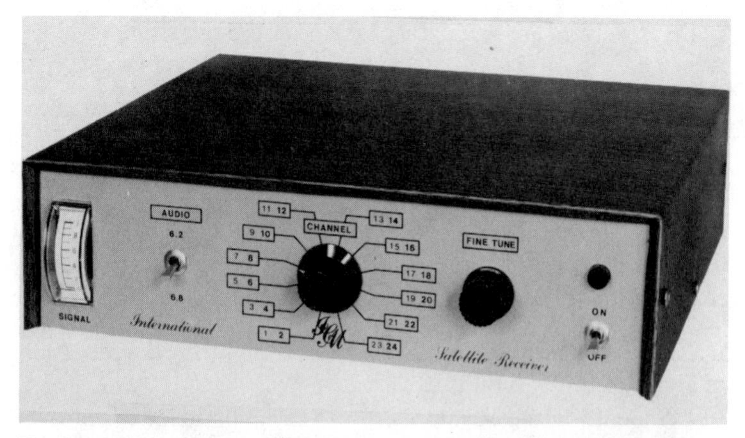

Fig. 3-9. In this type of installation, the satellite receiver must be located at the television receiver.

less losses in signal strength. Using this system, the receiver portion would be mounted in back of the dish antenna, possibly in a waterproof enclosure, although most are enclosed for protection from the elements. Some manufacturers of dish antennas offer a special platform at the mounting base for the location of these receivers.

Once the receiver is installed, its output may be connected directly to the modulator or if one is included in the main circuit, the output line is connected directly to the television set. The remote control unit sets atop the television and is connected by a multiconductor cable. When the frequency knob is turned, the remote receiver automatically switches frequency. Alternately, the output of the remote receiver may be connected by coaxial cable to the modulator, which also sets atop the television receiver. Remote con-

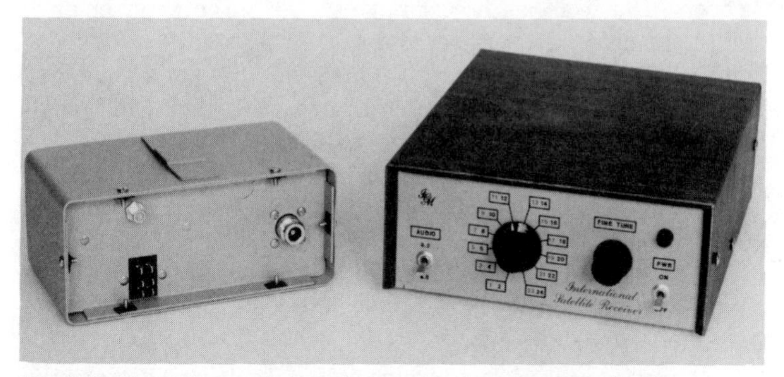

Fig. 3-10. Here, the receiver is shown with a remote control option.

trolled receivers are mandatory whenever the television receiver is located a great distance from the antenna site. In some cases, the antenna may be mounted in a backyard just a few feet away from the receiver. Here, you may get away with having the entire receiver directly at the television set without having to go to the expense of obtaining the remote control option. In other installations, it may be necessary to locate the antenna at a site some distance from the home which offers a better window on the satellite. This is where a remote control receiver must be used. Some antenna manufacturers even offer remote antenna adjustment controls which will rotate the dish in order to allow for the reception of signals from several different satellites.

While this discussion has stressed the importance of preventing losses in the microwave sections of the TVRO Earth station and indicated that there was not as much concern about losses in the vhf transmission portion, losses can occur at vhf frequencies as well. It is simpler to prevent losses at these lower frequencies, but it is quite important to pay attention to this area with as close an inspection as is mandated for the microwave section. The Earth station owner must make certain that all cables throughout the system are of high quality and in good repair. The manufacturers of the various pieces of equipment for Earth stations usually make recommendations as to what types of cable should be used with their products. Some coaxial cable is intended for indoor use only, while others may be used with care outdoors. By this we mean that any cable, regardless of what it was designed to do, will induce great losses into any system if moisture is allowed to get within its protective outer layer. If connectors are improperly attached to the line or if tears are left unpatched in the plastic outer insulator, the line will quickly become infiltrated with moisture and reception will be poor. Always take as direct a route as possible between the LNA and the receiver, the receiver and modulator, and the modulator and the television receiver with cable runs. It is not a good idea to coil the coaxial cable midway through the line. This adds length to the signal path and causes greater loss.

After years of use, even the best coaxial cable can offer degraded performance. You should plan to replace all cables after a specific period of time. Manufacturers' ratings and specifications can help in this area, but a received picture which steadily gets worse and worse over days, weeks and months of operation is a pretty good indication of cable problems.

The connectors (devices which attach the coaxial cable to the various system components) can also present loss problems. Make certain that you use only high quality hardware in this area and that all connector-to-cable attachments are made in accordance with the manufacturer's specifications. Some people have been known to wire connectors to cables in an expert manner and then to attach the connectors to the mating fitting of a piece of equipment poorly. Make sure all connections are tight. A loose connector will allow for moisture to invade the system, bringing along the inherent losses that we are trying to avoid.

One other area of loss which is dealt with in Chapter 6 and 10 is that produced by antenna movement. The signals from the satellite are very narrow in comparison with normal broadcast transmissions. We have already discussed obtaining a window on satellites and the importance of a good, properly aligned path between satellite and dish antenna. If your antenna is not mounted properly, even moderate winds can blow it slightly off track. This will degrade reception and possibly eliminate it entirely. The mounting structure must be firm to prevent the antenna from wobbling back and forth, which can cause the picture to fade in and out. There are several critical points in a TVRO Earth station which can affect reception and which do not apply (or at least not as much) to vhf reception of standard television broadcasts. Antenna stability is one of these critical points.

ENJOYMENT OF A TVRO EARTH STATION

A person does not get involved in personal satellite reception in order to go through an endless string of headaches. The previous information may have given the indication that microwave reception is fraught with all kinds of perils, which only an engineer could straighten out. This is true, but fortunately, the engineers who designed the receiving equipment which is presently available have worked out most of the difficulties and good basic construction techniques on your part will do the rest. TVRO Earth stations are fun to put together and open a whole new world of entertainment and information on the same television receiver that you have been watching your favorite shows on for the past fifteen years. The installation of the antenna and its mount requires the greatest amount of space, but this can be easily provided for in the front or backyard of most homes. Even apartment dwellers may be able to use the roof of their building for such an installation if adequate lot space is not available.

Due to the complexity and critical nature of the circuits used to detect satellite transmission and to make them usable by standard television receivers, the home construction of satellite receivers, LNAs, and other microwave portions of the system is not possible for the average experimenter. A satellite receiver kit is being offered by a manufacturer of completed TVRO products and more on that will be included in a later chapter. This is quite a different situation, in that all of the critical wiring is done at the factory and the builder simply assembles the parts into a working unit. Some persons who have had success in building simple transmitters, receivers, and hobby devices from the many project books available to the electronics experimenter today have often wondered why more microwave kits are not available. The reason for this is that it requires a lot of experience in microwave technology to properly build an LNA or receiver from the ground up. Sure, almost anyone can buy the parts needed to build one, but few people are qualified to perform the delicate and exacting steps required to arrive at a working unit. In order to build microwave cavities used at the feed horn, a knowledge of metalworking is required, along with extremely close tolerances.

For today, the best approach to building your own personal Earth station is to buy the premanufactured antenna, LNA, and receiver. It would certainly be possible to construct the modulator at home, but these devices are relatively inexpensive and can probably be purchased for about the same amount of money that it would cost to build one.

If you're an avid home builder, don't be dismayed. There is still an awful lot of assembly required in putting one of these stations together from the manufactured major components. Often, the dish antenna comes in several pieces and must be assembled at home. One antenna which is discussed in Chapter 6 requires a great deal of construction technique in order to assemble the parts. But to build the same antenna from discrete components bought at random rather than in the form of a kit with all struts and mounting hardware included and properly cut would be an enormous task. The assembly of these antennas reminds me of many of the building experiences I had in putting Heathkit equipment together. All of the parts are there, along with the instructions, but it still requires a lot of assembly.

For those of you who detest putting anything together, take heart. Many manufacturers offer packages where all components are delivered to your prepared site and set up by technicians. The result is a correctly operating and guaranteed Earth station with no physical labor on your part. The bill for these services tends to be a bit high, however.

QUESTIONS AND ANSWERS

There are a few basic questions which persons ask first about satellite reception and Earth stations in general. Most of these have already been answered by the previous discussion; however, a bit of a recap at this point is in order.

Question. Can I use my present television set to receive signals through the Earth station?

Answer. Yes, you can. The Earth station effectively serves to receive the satellite signals and to convert these microwave frequencies to those your television receiver was originally designed to pick up. If you are going to go to the expense of installing a TVRO Earth station, you will certainly want to have a good television receiver which is in good working condition. The Earth station will in no way improve the basic performance of your present set. In other words, it will not make up for any internal deficiencies in your TV receiver. Persons with older sets which do not offer uhf tuning will want to specify to the manufacturer the inclusion of a modulator which offers an output on vhf (channels 2 through 13). Normally, the modulator will relay the signal on channel 2 or 3, although some units offer the option of choosing any vacant channel desired.

Question. After the Earth station is installed, can I still receive standard television broadcasts?

Answer. Certainly. The satellite reception is accessed by tuning your television receiver to one prearranged vhf or uhf channel. When you change channels within the satellite system, this is accomplished by turning a separate control housed within the satellite receiver or remote receiver control unit. If you desire to receive standard broadcast channels, all that is necessary is to flip the channel control on your television receiver to the station desired. Your set will operate exactly as before, but you will have a special channel which is used for satellite reception only.

Question. Does my present television set require any modifications?

Answer. Usually not. Occasionally, interference from radio sources and power lines may be picked up by the TVRO equipment. When this occurs, the attachment of a small filter at the antenna terminal on the back of the set will remove or reduce this problem. No internal modifications whatsoever should be required.

Question. What kind of programs can I receive from the satellites?

Answer. To give a complete answer to this question would require enough pages to fill up a small book. A general answer would include first-run television programs, uncensored movies (without commercials), sports, stock market reports, all news stations, religious programs, live coverage of congressional meetings, children's programs, and possibly even Russian spy satellites (although the received information will be incomprehensible to the average person not related to James Bond). These are just a few of the programs you can expect with a properly operating TVRO Earth station.

Question. In what areas can I receive U.S. satellite signals?

Answer. United States satellite signals may be received in most of North America, Central America, and a portion of northern South America. In general, all areas of the Continental United States are within a window of at least some of the orbiting satellites. In certain specific locations, local obstructions, such as mountains, steel towers, and tall buildings may interfere with reception. In these instances, moving the installation site to a clearer point a short distance away will completely open the transmission window.

Question. How can I find out which satellites may be received in my area?

Answer. Chapter 8 is devoted to this subject alone. Most manufacturers of TVRO Earth station equipment will provide you with a computer printout showing you in which direction each satellite is located with regard to your area. The printout will indicate, based upon your latitude and longitude, the number of open windows which exist in your area. These windows are then matched with a list of satellite positions. From the two lists, you can determine which satellites are available to you and which are not. Many computer printouts will also provide an indication of roughly how strong a signal you can expect in your area. This will help in the determination of which receiving system you purchase, especially in regard to antenna size. These printouts are usually provided on a charge basis, with the money often deducted from the price of the equipment should you decide to purchase it from the manufacturer providing the printout.

Question. Do I need a license from the Federal Communications Commission to install my own TVRO Earth station?

Answer. If you intend to operate your Earth station for your own personal non-commercial purposes, you do not need a license from the Federal Communications Commission. The manufacturers of this equipment are required to have their designs type-approved by the FCC before offering them for sale. This assures that no interfering emissions from the receiver or modulator circuits are allowed to cause problems with other television users. Legally, the FCC states that you can own and operate a receiver-only satellite system and you are not required to have a license. However, if more than 50 dwellings are involved and they are not under common ownership or control, it will be necessary for you to register as a cable company. In either case, you will need to obtain permission from and in some cases pay a moderate fee to program owners.

Question. What kind of people install TVRO Earth stations?

Answer. The same people who drive automobiles, motorcycles, and airplanes. The same people who own houses, live in apartments, and own thousand-acre estates. The same people who are amateur radio operators, CB enthusiasts, and home experimenters. The same people who are housewives, blue collar workers, and corporate presidents. All types of people are installing their own TVRO Earth stations. The setups are not limited to the technically inclined, although these persons will probably have an easier job should they elect to do their own installations.

Question. Do I have to use special tuning methods in order to receive satellite signals through my Earth station?

Answer. Not really. Once the Earth station is properly installed and connected to your television set, all standard tuning is done at the television receiver in a normal manner. It will be necessary to rotate and change elevation on the antenna whenever you desire to receive signals from a different satellite. This may be accomplished by electric control with the master panel located in a unit which is placed on top of the television set. Alternately, hand controls are installed at the antenna mount which allow these same adjustments to be accomplished manually.

Question. Don't you have to change satellites each time you want to receive a different program?

Answer. Definitely not. One satellite may transmit twenty or more different channels. The different transmission frequencies or channels are called transponders. On transponder 1, you will pick up a different channel than on transponder 2. Each transponder transmits at a different frequency. To access the various transponders, you simply change channels on the satellite receiver control. This is exactly the same way you change channels on your home television at present.

Question. With all those satellites up there transmitting on the same frequencies, isn't there a lot of interference?

Answer. No. As has been previously stated, it is necessary to have your antenna in almost perfect alignment with (pointing directly at) the satellite whose signals you wish to receive. When this alignment is slightly off, you get no reception. While there are many satellites in orbit and all transmitting on the same frequencies, your antenna discriminates against unwanted signals by not being directed to an open reception window. Your antenna will tune signals through only one window at a time. This is shown in pictorial form in Fig. 3-11.

Question. Don't TVRO Earth stations require a lot of service?

Answer. No more so than does your present television receiver. As is the case in most industries, there are good products and there are bad. I have attempted to present and discuss only the products of reputable manufacturers who are well known in the TVRO Earth station field. If proper care is exercised in making the installation and the preventive maintenance instructions from the

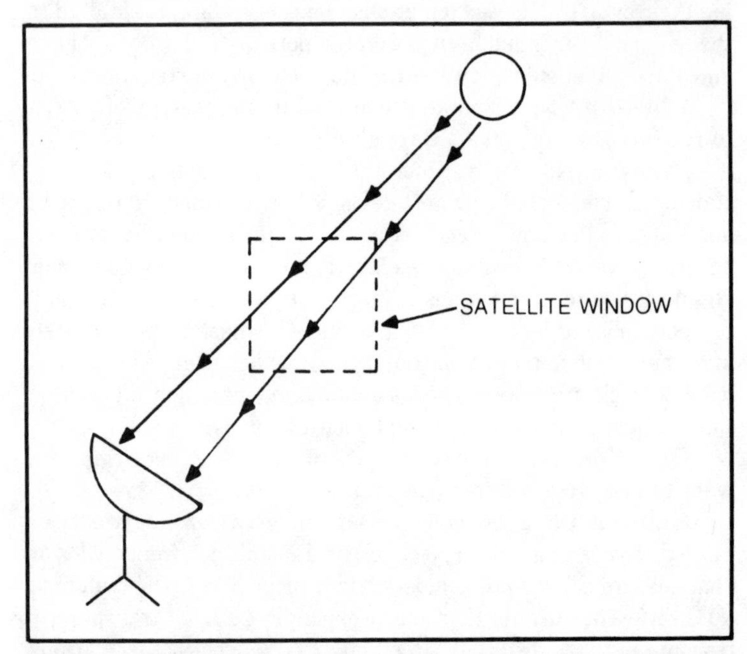

Fig. 3-11. A TVRO Earth station antenna will tune signals through one satellite window at a time.

manufacturer are followed closely, then your Earth station should provide you with many long years of excellent service. Naturally, if you are located in an area with severe climatic conditions, the wear and tear on the exposed components will be greater than if you were situated in a milder area. Special precautions may need to be taken in those areas which experience unusually high winds, hail, snow, or which are subjected to water-borne salt (sea coast areas). Of course, internal difficulties can develop with Earth station circuits. These will usually dictate the necessity of sending the defective unit back to the factory for repair. Your local television shop won't be able to handle this job for you.

Question. Can I connect more than one television in my home to the TVRO Earth station?

Answer. Definitely. All that is required is the attachment of a standard 70-ohm splitter to the incoming TVRO line. This allows for several televisions to be operated from a single input. The splitters may be purchased at your local Radio Shack or other dealer along with the cables needed. The splitter serves to divide the signal, so remember that you will have half the signal strength at each television when operating two sets from a single line. Splitters which offer outputs for three televisions provide a third of the normal signal strength at each set. If the signal coming from the main line is marginal, you may experience some reception difficulties. This would be true only if there was a very great distance between the modulator and the television sets. This will be encountered very rarely in most TVRO installations. Signal strength to the set is provided by the modulator. The strength of the satellite signals will not be directly affected by the inclusion of splitters.

Question. Can I use my video tape recorder with a TVRO Earth station?

Answer. Yes, you can. The recorder will be hooked up in the usual manner and will function in exactly the same manner as it does with standard television stations.

Question. Don't TVRO Earth stations present a lightning hazard?

Answer. No. As a matter of fact, your present television antenna is more of a hazard regarding lightning hits than would be a typical Earth station. Most Earth station antennas are located near the ground because of their size and weight. Standard television antennas are mounted high in the air. Also, the antennas used for satellite reception are usually mounted on metal platforms whose supports extend into the ground. Lightning hits should not be of undue concern to anyone contemplating the installation of a TVRO Earth station.

Question. Will the Earth station improve my standard television reception?

Answer. No. The two systems are completely different. If you have poor vhf and uhf reception due to your location (and not because of the condition of your television set), it will be just as poor after the Earth station is installed. Your reception of satellite signals, however, will be excellent.

Question. Will the quality of reception using a personal Earth station be as good as that which I obtain from my cable company which has its own Earth station?

Answer. This will depend upon a lot of different situations. If you are located in a marginal signal area, a very large dish antenna might be required to match the picture provided by your cable company. Chances are, they have a very large dish as well, which is typical of commercial installations. A computer printout of satellite windows and signal strengths in your area will help find the answer. For most areas, chances are very good that your personal Earth station, if constructed from good components by a reputable manufacturer, will compare favorably with the cable company. The received image may even be better.

Question. Are there any electrical hazards presented by a TVRO Earth station to small children and pets?

Answer. No more than would be presented by an outdoor lighting fixture. Power to the receiver and LNA is usually derived from the 115-volt house supply. No high voltages are involved in most installations. Owners would do well to restrict access to the Earth station area, as the science fiction look of the antenna is bound to attract the curious of all ages. Unfortunately, the Earth station will probably be a very appealing target of vandals who would take delight in reducing your expensive setup to rubble. For this reason I recommend backyard installations which are generally out of sight of most passersby.

Question. Is it necessary for the antenna to "track" the satellite?

Answer. No. This question is often asked by persons who have seen radar antennas in operation. The satellites we wish to receive are in geostationary orbits. This means that they are always at the same point in the sky. They rotate with the Earth and do not move as far as the antenna is concerned. Actually, the antenna automatically tracks the satellite. As the orbiter moves, the Earth which the antenna is mounted on moves with it. Therefore, you can receive the satellite signals at all times of the day, night, or year without ever moving the antenna in relationship to the Earth.

Question. Can weather affect reception?

Answer. Yes. Heavy moisture content in the air can interfere with satellite reception. During periods of heavy downpours of rain or snow, you will most likely notice a bit of signal fading. Freezing rain can accumulate on the dish and feed horn, interfering with reception and causing structural stresses. Areas which are plagued by these weather conditions often see TVRO installations where the antennas are protected by large non-conducting covers called radomes. This cover in no way interferes with normal reception. Heavy wind conditions can cause the antenna to vibrate. Sometimes, this will cause the signal which is received at the television to fade in or out or to flutter. Most antennas, when properly mounted, will provide normal reception in winds up to fifty miles per hour. They have survival ratings in winds of over 100 miles per hour in most cases.

Question. How long will it take to assemble and make operational a typical Earth station?

Answer. This will depend upon which system you purchase. Preparing the Earth station site can take a week, but most of this time is allotted for allowing the concrete footers for the base to set. The actual work involved in the site preparation can probably be completed in a day and certainly in a weekend in most locations. If you purchase a system which requires antenna assembly, this can take most of a day in some cases and possibly several days when larger or more complex antennas are chosen. Once the antenna is installed on its base, the rest is relatively simple. Interconnecting wiring between the low noise amplifier and the receiver is attached. Another cable is connected between the receiver and modulator (if these are two separate units). A multiconductor cable is run from the receiver to the television set when remote control tuning options are included. This usually takes very little time, especially when there is only a short distance between the antenna and the receiver. As a general rule, it would be fair to state that a full weekend of assembly and adjustment should produce a fully functioning Earth station.

Question. How many people are required to set up an Earth station?

Answer. Again, this will depend upon the system chosen. Generally, it will require three persons to lift a parabolic dish antenna onto its mounting platform and to make the proper attachments. While one man can prepare the site for the antenna mount, two persons can get the job done a lot faster. After the antenna is mounted in place, one person can usually handle the rest of the assembly.

Question. What are the chances of my personal Earth station becoming obsolete?

Answer. Certainly, improvements in TVRO Earth stations will follow the pattern of the entire electronics industry, but manufacturers are working within a range of frequencies which have been established for satellite transmissions. Present receivers are designed to detect signals within this preset range. It is conceivable, although not likely, that the frequency range could be extended, but certainly, modifications to present equipment would be made available to owners. In making a comparison of Earth stations to citizens band radio, we find that when the FCC expanded the CB frequency range to allow for 40 channels of operation instead of 23, the older 23-channel units were not obsolete in that they still provided the same effective communication within the frequency range they were designed for. Your present Earth station would still continue to do exactly what it was designed for even if the satellite frequency range were expanded in a like manner. Right now, these receivers cover the entire entertainment frequency range. If an extension occurred, they would still continue to cover the same range, but you would not have access to the new range without purchasing another receiver or modification kit. Your present TVRO Earth station should provide the same practical reception for many years to come.

Question. What kind of maintenance will need to be performed on my Earth station?

Answer. A properly designed TVRO Earth station requires a very minimal amount of maintenance. All cables should be checked periodically for wear and moisture infiltration. The connectors should be inspected for the same problems. Mounting hardware at the antenna will need to be tightened occasionally and should also be treated with a protective compound in areas near the ocean where salt can be a problem. Of course, you will want to remove any dust which can accumulate within the receiver housing if this unit is mounted out of doors. Weather conditions will play a large role in determining how much maintenance is necessary. After a heavy wind, for example, you might check for loose bolts at the antenna base. As was previously mentioned in this chapter, it's a good idea to remove snow which has accumulated on the dish and to protect the antenna as much as possible during icing conditions. Most maintenance is approached on a common sense basis. A properly installed TVRO Earth station should be easier to maintain than a power lawn mower.

Question. Can my Earth station be protected by my present homeowner's insurance policy?

Answer. Yes, in many instances it can if it is permanently attached to your property. The rates and coverage will vary from area to area and will depend upon such criteria as local weather conditions, crime and vandalism rates, and the amount of coverage you want. I talked with a local insurance agent about this matter and was advised that an Inland Marine policy might be best because of the broad coverage offered. Generally speaking, Inland Marine policies provide insurance on an all-risk basis. Some areas may require certain exclusions. It is difficult to provide a cost figure on this insurance which would be applicable to all areas of the country, but total cost of insuring a $6,000 Earth station on an all-risk basis should come to appreciably less than $200 per year. When contemplating the installation of a TVRO Earth station, a check with your local insurance agent is strongly advised. More than likely, he will want to inspect the installation to make certain that proper construction techniques have been used throughout.

Question. Suppose I move from my present home? Can I take my Earth station with me?

Answer. Certainly. The only thing you will have to leave behind is the concrete footers. The dish antenna is simply removed from the base, along with all electronic equipment. The base is unbolted from the concrete footers and the entire station can be moved in a small trailer. It may be necessary to disassemble the dish if one piece construction is not used on this component. It's a good idea to save the original packing carton in which your equipment arrived. When moving, these will help assure that no equipment damage occurs in transit.

Question. Suppose I move to an area which offers marginal signal reception? What can be done to make my station more sensitive?

Answer. There are many things that can be incorporated in a TVRO Earth station to improve reception in marginal areas. Most of these, however, are rather expensive. The addition of another LNA will bring about some improvement, but the best method is to use a larger antenna. Fortunately, some manufacturers who offer TVRO antennas also offer an extender kit. The Wilson WFD-11 is a good example. This is an 11-foot dish (3.35 meters) which offers a gain of 40.1 dB. If more gain is required, an optional enlargement kit can be ordered which includes extra panels that attach to the present antenna. When the project is completed, you will have a 13-foot dish (4 meters). This offers slightly greater gain which is amplified at the LNA. This is far less expensive than purchasing an entirely new and larger antenna.

Question. Can I mount a TVRO antenna on my roof instead of on the ground?

Answer. This will depend upon the structural strength of the mounting surface. There is absolutely no reason to mount a TVRO antenna high off the ground unless this is necessary to clear large obstructions. It would be wise to consult a local contractor and have him look closely at the intended mounting site and advise you as to what would be required for a safe and durable installation. A typical TVRO antenna and its base can weigh 500 pounds or more. When struck by high winds, the pressure exerted on the assembly can be six to ten pounds per square foot. Your contractor will have to take this into consideration. Again, while rooftop mounting is possible, it should be used only when it is impossible to provide good reception with an Earth station that is mounted on the ground.

Question. Does my property become more valuable when I install a TVRO Earth station?

Answer. Probably. An Earth station is something you don't find in every backyard. From an appraisal value standpoint, your property will probably not increase substantially in worth. But from a resale point of view, A TVRO Earth station could mean the difference between quick sale and no sale at all. Certainly, the price of the Earth station would be added to the asking price of the home and property if you intend to leave it in place when you move. An Earth station facility can be likened to a swimming pool. For those persons who like to swim, the pool will be a good incentive to buy a certain piece of property. Persons who are interested in receiving television signals directly from satellites (and this includes almost everybody) would probably find your house and property a more attractive buy. Looking at this situation from another angle, the Earth station certainly doesn't decrease your property value as it can be moved when you do.

Question. How about financing?

Answer. As always, financing will certainly be available to those individuals with good credit ratings. Some companies may offer their own financing program or you can arrange a loan at your local bank or lending institution. In these days high of interest rates, it's a wise move to shop around for the best loan rate. Some banks may wish to handle these transactions on a collateral basis, using the Earth station to back up the loan. This is very similar to purchasing a new automobile, with the bank technically owning your Earth station until the loan is completely paid off. It will be necessary for you to get a quote from the manufacturer you have chosen and then make all the arrangements in advance of the purchase date. Financial arrangements may differ from bank to bank and certainly from person to person, but in general, the lending institutions I have contacted have no qualms about loaning money to qualified individuals for the purchase of TVRO Earth stations.

The previous questions and answers are typical of the concern of persons interested in personal TVRO Earth stations. There are others, however, which will be dealt with in this chapter. Contech Antenna Corporation, 3100 Communications Road, St. Cloud, Florida, publishes an excellent primer on satellite receive-only Earth stations. Other questions and answers were derived from this manual and are reprinted here with their permission.

Question. Why use satellites for television transmission?

Answer. Satellites permit more economical communications (television, telephone, radio, telemetry, etc.) than ground stations or where hard cable connections are required. One satellite can provide 24 transponders with television entertainment to cover all of the United States. Normally, large United States cities have five or six television stations. The reception becomes marginal from the stations for distances in excess of fifty miles from the transmitting station. In telephony, satellites replace literally tons of copper cable that would normally be required to establish telephone communications across distances of a thousand miles or so.

Question. How do the satellites perform their function?

Answer. Satellites are placed in a geostationary Earth orbit. Geostationary implies that the satellites rotate at the same speed as the Earth. This phenomenon is achievable at an altitude of approximately 23,000 miles directly over the Equator. Once a satellite has been stabilized in its orbit, it is used as a a transponder. It receives a signal from an Earth station at 6 gigahertz, converts it to 4 gigahertz, and amplifies and retransmits a directional signal back to the Earth.

Question. Does that mean that everyone on the Earth can receive these signals?

Answer. No, the satellite from its position in orbit can see approximately 40% of the Earth's surface. In order to provide a sufficiently strong signal to desired locations, the satellite transmit beams are shaped. The U.S. domestic satellites, for example, point their beams at the center of the United States. The signal strength decreases as you go away from the area where the beam is peaked.

Question. Can I receive U.S. satellite signals outside the continental United States?

Answer. Yes, but an engineering evaluation should be conducted to determine the equipment that will be required for the reception of these weaker signals.

Question. Where are the U.S. satellites physically located?

Answer. All of the U.S. COMSATs are located on the Equator over the Pacific Ocean. The nearest U.S. COMSAT is WESTAR III located at 87° West longitude, which is approximately 600 miles west of Ecuador. The farthest U.S. COMSAT is SATCOM I located at 135° West longitude, which is approximately 3500 miles west of Ecuador.

Question. How many transponders are there?

Answer. The two SATCOMs and three COMSTARs each have 24 transponders. The three WESTARs and three ANIKs each have twelve transponders, which is a total of 192 transponders. Approximately 55 presently carry some type of video programming.

Question. Why do some satellites have 12 transponders while others have 24?

Answer. The ANIKs and WESTARs provide transmission only in horizontal polarization. They have only 12 transponders each. The COMSTARs and SATCOMs alternate vertical and horizontal polarization. Therefore, they can utilize the same total frequency band more efficiently.

Question. Will all of these additional satellites operate in the 4 gigahertz range?

Answer. No. Some of these will be used in the Ku band—12 GHz range.

Question. Will the higher frequency range make present equipment obsolete?

Answer. Not in the near future. The expected satellite life is presently eight years. With the advent of the space shuttle and platform, life may be extended by preventative maintenance.

Question. How does the antenna contribute to system performance?

Answer. The antenna must capture sufficient signal to overcome the receiving system noise. The weaker the satellite signal in your area, the larger the antenna needed. The signal strength is defined as a power density—watts/square. The larger the antenna, the larger the area or square. Therefore, more power is focused on the antenna feed for reception.

Question. What effect does improper shaping or surface irregularities have on signal received?

Answer. Both shaping and surface rms detract from received signal strength. For example, surface rms of 0.125 inches causes a signal loss of 1.5 dB, which corresponds to 20 percent power loss of received power. The net effect is that one could have used a physically smaller antenna to perform the same function, or the cause of picture problems may be the antenna.

Question. What if my antenna is spherical rather than parabolic?

Answer. The spherical shaped antennas do not provide the same focusing efficiency as a parabolic. Hence, you will need a larger spherical antenna to do the same job as a parabolic.

Question. Why do you need to perform a structural installation on an antenna?

Answer. There are two basic reasons: (1) Safety—to prevent wind forces from tearing away the antenna from the installation; and (2) The antennas have "pencil beams" which must be pointed directly at the satellite.

Question. What type of antenna mounts are available?

Answer. The two basic antenna mounts that exist are: (1) the elevation/azimuth, which requires independent azimuth and elevation adjustment; and (2) the polar mounts, which permit the "tracking" of the satellite arc through a single motion.

Question. Which one should I use?

Answer. If one does not plan to change satellites often, the elevation/azimuth mounts are less expensive. However, if one desires to change satellites often or to effect changes remotely with a programmed unit, then the polar mount should be chosen.

Question. What options are available for satellite Earth stations?

Answer. Just about every component of the Earth station has many options. It would be best to discuss each one individually.

Modulators. Depending on the manufacture, it can be a simple channel converter or selectable to any of the vhf channels on your television set. The most commonly used modulators have a switch to put your signal on either channel 3 or channel 4 frequencies.

Modulators become more expensive when utilized in commercial installations where they are required to drive a few hundred television sets.

Receivers. Two basic types of receivers are currently available. One type is the 36 MHz bandwidth unit designed for Cable TV. The large bandwidth produces good color quality and picture definition. The larger bandwidth accounts for a higher noise figure. This type is differentiated mainly by the "bells and whistles" on the final units. The second type is the one designed for the home market. The units do not meet any industry criteria other than the ones dictated by the marketplace. The home units in general are smaller bandwidth devices that give up picture definition but have better noise figures (better in marginal C/N areas).

The latest innovation in receivers affects the frequency conversion at the antenna. In this manner, you can use both less expensive and longer cables between the antenna and the receiver. The rf frequency being transmitted from the antenna to the receiver will be at the i-f (immediate frequency) rather than at 4,000 MHz.

LNA. The low noise amplifiers are described by a noise temperature. The most common LNA is the 100° K unit which is basically a 1.5 dB noise figure device. Most LNAs have a gain of 50 dB. This device in general establishes the signal to noise figure of the receiving system. Lower temperature LNAs are used to improve marginal system performances. Before spending money for a 90° LNA, it is suggested that signal strength in your area be determined from a computer printout. If it is indicated from this sheet that a lower noise figure is needed in the low noise amplifier, then the extra expenditures can be made.

Question. Who can help me decide on the options?

Answer. It is suggested that you rely on your dealer for the technical parts of the system, such as LNA temperature, size of antenna, etc. The remote control options are really a function of your personal desires and budget.

Question. Why do you need different size antennas for different parts of the United States?

Answer. As previously mentioned, the satellite antenna is pointed at the center of the United States. The signal decreases as you go away from the peak of the beam. A "footprint" is developed for signal strength.

Question. How do I know what size antenna I need?

Answer. Again, the company you buy your TVRO equipment from will be in the best position to aid you in your selection. Additional information can be garnered from computer printouts for satellite and signal losses in your area. In general, you will want to choose an antenna with the highest gain (largest dish diameter) which you can afford and which is reasonable for your area.

Question. Is an Earth station right for me and my needs?

Answer. You will have to answer this one for yourself. You have probably viewed some of the programs offered from satellites, possibly on your local cable system. Chances are, you have obtained program listings from several different satellite users. You must ask yourself, "Will the pleasure and use I get from my Earth station justify the expense of purchase and installation?" Let's assume your Earth station costs about $4,500. This is about 1 1/2 times the price of a good, large-screen television projection system, and about 6 times the price of a good color television set. Do you have small children in your home who could take advantage of the many educational programs offered? Would your occupation be benefitted by the 24-hour news and information channels available? Do you live in a rural area where first-run movies are shown at local theaters six or seven months after they are released? Do you watch a lot of television? Do you have cable television access or do you live in a fringe area where only a few stations can be marginally received? All of these and many other questions must be answered by the only person qualified to do so—you. I think that a thorough self-examination of your motives will greatly aid you in determining if now is the time to install your own personal TVRO Earth station.

SUMMARY

The expression, "There's a whole new world out there" does not adequately describe all that is to be had with a personal Earth station. The signals are beamed to you from far out in space. This alone opens up many, many possibilities. Experimenters will delight in receiving foreign and military communications satellites and when the experimenting stages are over with, you can sit back and relax with the sights and sounds of a first-run movie.

The installation of a properly functioning ground station is not all that difficult. It can be performed by anyone who has just a bit of mechanical or electronic aptitude. Those persons who have no inclinations in these areas but who desire to receive satellite pictures can contract with Earth station manufacturers for a turn-key operation.

Most owners of Earth stations will be surprised at how fast they adapt to that Martian-looking spectacle in the backyard and to its many uses. At first, the reception of satellite signals on your television receiver will be a sort of anti-climax to all of the planning and work that went into the Earth station installation. But as the user becomes more familiar with all that is available from those transponders far out in space, a universe of possibilities will be opened to you, and all from the comfort of your living room.

It is my firm belief that in one decade satellite television reception will be the rule rather than the exception. Broadcasters are intrigued with this system, which opens up an even larger number of possibilities for them. As development continues in this area, the public should be offered more and more services and satellite use and interest will grow more intense. Satellite television reception will most certainly be the most startling development which has ever occurred in the home entertainment field.

Chapter 4

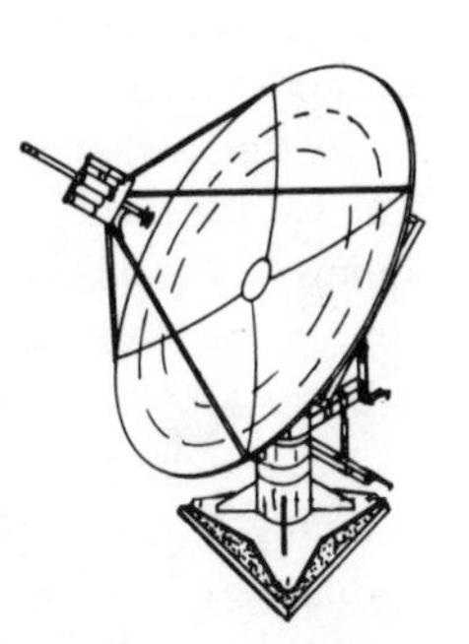

Obtaining Surplus Components

IT IS BEYOND THE SCOPE OF NEARLY EVERY TVRO ENTHUSIast, regardless of technical background, to construct a complete Earth station from the ground up using discrete components to build the many complex circuits. This is discussed elsewhere in this book and alludes to the highly critical nature of frequency-determining circuits at microwave frequencies. Today, when persons build an Earth station, they generally do so by assembling or interconnecting manufactured devices, such as receivers, LNAs, and modulators. Unless you elect to have the manufacturer's technicians come to your site and completely install the Earth station you have purchased from them, you will be doing a major part of the building, which involves antenna site preparation, mounting and interconnecting the devices needed for reception.

Chapter 7 overviews some of the few kits which are available to the TVRO enthusiast for building TVRO receivers, LNAs, etc. There is another way to go, however. You may be able to obtain some components and completed circuits through surplus channels which can be modified to serve as a part of an operating TVRO Earth station.

Admittedly, the surplus market at present is all but devoid of devices which can be directly applied to satellite Earth stations. One area where this does not especially apply, however, is in regard to dish antennas. You have probably already noticed that the parabolic antennas used for TVRO Earth station purposes closely resemble those which have been used for many years in government radar applications. While there are many different surplus radar antennas to choose from, only those with diameters of ten feet or more are practical for our purposes.

By consulting the catalogs offered by government surplus suppliers, you may become lucky and find one that is suited to modification and will save you many hundreds of dollars over the cost of a commercially manufactured TVRO dish. What we are mainly concerned with when selecting a surplus dish is the diameter of the reflecting surface. The feed horn and any waveguide which may come with it will probably have to be discarded and replaced with components designed specifically for TVRO purposes. These are not excessively priced from most outlets. I have seen feed horns advertised for $25.00 from TVRO companies.

Another important consideration is the mounting base atop which the parabolic dish is situated. You probably won't find too many of these, but it pays to check with suppliers to see if anything might be available. I have used many of the products offered by Fair Radio Sales, which is one of the leaders in electronic government surplus. While their catalogs list many thousands of items, a phone call to this outfit with a specific request for items not offered in their publications is often productive. With the ever increasing interest in satellite television reception, it would be a fair assumption that many surplus outlets will begin concentrating on this growing market.

There are many ancient radar receivers and auxiliary equipment that are offered for sale by government surplus outlets. Most of these are leftovers from World War II, although the Vietnam War has caused newer devices to begin appearing. The older circuits are nearly all of the tube type, but it may be possible to obtain some of the frequency-determining portions of them to serve as a basis for building a receiver for TVRO applications. Certainly, some experimenters will develop conversion designs for a few pieces of military equipment which can be applied to your TVRO earth station.

Another possible source of surplus parabolic dishes will be your local phone company. This source is not often thought of by most experimenters, but it can offer a wealth of discrete components and entire circuits which can serve many purposes. Did you ever stop to think what happens to all of that sophisticated telephone company equipment when it is replaced by newer models? Chances are, all identifying placards are removed and it's hauled to the local junkyard and tossed on the heap. Some obsolete equipment may be retained for parts salvage, but most is simply discarded.

Being an avid amateur radio enthusiast, I often check with my local phone company to see if they have any "junk" they want hauled away. It is necessary to check with a ranking supervisor in most cases. Some of the "junk" which has been carted to my home has included sophisticated radio teletype machines, high-speed line printers, innumerable printed circuit board cards, and other nameless devices for which there are presently no known uses. The circuit board cards alone often contained parts which would cost well over $100 if purchased separately. Many of these cards were spares that had never been used and were discarded when the equipment they were designed to fit became obsolete.

Telephone companies depend heavily upon microwave communications (including satellite applications) to make worldwide phone service possible. Surely, many of the components used in these applications will become outdated and will have to be discarded. For this reason, it pays to get in touch with an official at your local phone company and let him know of your interests in their obsolete microwave antennas and communications equipment. It may be some time, but you may eventually receive a phone call asking you to haul away some of their "junk" if you're still interested. This can be a two-way benefit. First of all, you can obtain a lot of exotic playthings; and secondly, the telephone company won't have to go to the expense of having their obsolete equipment trucked away.

At present, I have not been able to obtain any dish antennas through telephone company outlets, but I have been informed that some may be available in the near future. Since my desires have been made known to the equipment supervisor, there is a good chance that I may get first choice when equipment becomes available. If you're interested in pursuing the surplus conversion route for TVRO Earth station equipment, contacting your local telephone company should be your first step. If you don't know exactly who to get in touch with, the best place to start is the service department. One of their technicians will certainly be able to tell you the name of the person who can give the authorization for you to receive the obsolete equipment. Additionally, (and this is most important) try to establish a good relationship with this technician, because he could be of great help in informing you as to what might be available in the near future. The person who must give permission for you to receive the various pieces of equipment will undoubtedly be quite busy, and your request will not be one of his top priority items. However, when your local technician friend lets you know that a piece of equipment is being junked, you can wait a week or so and call the company official again. This may be enough to jar his memory and get you the equipment you want.

Your local cable company is another source of future surplus equipment. This topic is discussed in Chapter 9. Most of the equipment that could become available from these outlets should be directly applicable to home TVRO purposes. Chances are, the telephone company equipment that you receive will be in reasonably good working order, but this will probably not be true of any cable TV surplus which might be available now or in the near future. Cable companies tend to keep their equipment for a long time and have it repaired whenever a defect occurs. After several years of use, equipment repair of defective units may no longer be practical, so new replacements are purchased. Any surplus you would get through them would probably not be operational, but you could probably make it so by doing the work yourself and replacing any defective components. This could be very cost effective, but it will probably be some years before this equipment will be available as surplus in any appreciable numbers nationwide.

Follow the same inquiry procedure with the manager of your local cable company as was recommended in the discussion on telephone companies. It never hurts to ask to be put on a waiting list. The worst they can do is say no.

Another possible source of some limited TVRO components might be large corporations which are operative in your area. Near my location, such companies as DuPont, General Electric, Western Union, and many others maintain large manufacturing facilities. Some of these use microwave equipment for communications purposes. As old equipment is replaced with new, a possible surplus source exists. Again, it can pay you to make a contact with someone in authority to let them know that you would be interested in removing any surplus they may have. Large companies are often crowded for storage space and may be receptive to such a request. There is a good chance that you would have to pay a small sum for this equipment, but often it's simpler from a bookkeeping standpoint for the manufacturer to simply write it off. This policy will vary from company to company.

In all of these dealings, diplomacy will play a very large role. After all, anyone who grants your request is most likely doing you a favor which could mean a little extra paperwork on their part. I normally make a direct contact by telephone in order to find out the name of someone who might be in a position to help with my request. If possible, the general manager should be contacted. An appointment is then set up at the facility and an "eyeball-to-eyeball" conversation is the result. This is most important, as the person who is in a position to grant your request can learn alot about what you are planning to do with the equipment and may even become interested in the project himself. If you can get his full attention and interest, you have a much better chance of being thought of when something becomes available. This approach is often successful and was borne out some years ago when I contacted a company about some surplus equipment to be used in my amateur radio endeavors. The president of the company became so interested in the project that he paid a visit to my home to see just how his surplus equipment would be used. Not only did I get the equipment, but the president of the company helped me set it up and align it. Afterwards, this same individual expressed his interest in amateur radio and is now an active operator. This will not occur in most cases, but it does demonstrate the possibilities that can be opened by personal contact with individuals within a company that may be able to supply you with surplus equipment. If you are really interested in TVRO Earth stations, it should be relatively easy to get others to feel the same way.

Government surplus was mentioned earlier in this chapter and is available from commercial outlets. However, there is another way to get this type of equipment which involves eliminating the middleman—the supplier. The United States Government regularly holds auctions of surplus equipment which can be purchased by sealed bid. A contact with your local government representative will aid you in obtaining these lists. If you make the highest bid on a piece of equipment, then it's all yours and all that needs to be done is to pick it up and haul it away. Sometimes, it's possible to view their offerings, checking for specifications, condition, etc. Be careful when taking this route, because once your bid is accepted, you are obligated to pay for it and to remove it from government property within a specified period of time. If you don't show up, the Federal Government begins charging storage fees in some instances. Most of the time, however, if you do not pay the balance due, the equipment goes back on the market again. This consumes time and additional taxpayer dollars, so don't bid on anything unless you're sure you really want it. I point this out because as a lad of sixteen, I received a bid sheet that had a piece of equipment listed that I desired. There were other interesting goodies on there too, so just for the heck of it, I penciled in fifty cents as my bid on each one. The piece of equipment I wanted was obtained by a higher bidder, but my fifty-cent bid on one item was a high bid. I was informed by the U.S. Department of Agriculture that I had purchased a giant egg incubator (probably weighing several tons), and would I please pick it up by a certain date or be charged for warehouse storage. The intervention of angry (but understanding) parents and the fact that I was under age saved me in this fiasco.

There are additional ways of at least obtaining information on equipment which may become available. An ad in a large newspaper may produce some results from people who have been collecting surplus components for many years. These have a habit of building up into mammoth proportions and filling needed space. You might even try purchasing an ad in an electronics magazine, but this can be rather costly. Normally, the companies which have surplus equipment are advertising their products in the same magazines.

If you are interested in a particular piece of U.S. government surplus, you can probably get complete information on it by writing the U.S. Government Bookstore in Washington, D.C. for a list of the many publications they offer for sale (and sometimes for free) at reasonable prices. Their military publications are often designed for training technicians and will include a wealth of technical data as well as practical operating instructions.

There are many possible outlets for surplus equipment which may be accessed by the TVRO Earth station enthusiast. By keeping in constant contact with these many potential resources, you can be advised and updated as to equipment which may be offered in the future. It might be a good idea to begin hoarding different types of equipment as it appears in order to one day arrive at enough to build an Earth station or improve a present one. Right now, surplus channels are mainly used to obtain obsolete reflectors from large parabolic dish antennas originally incorporated into military radar systems. Some of these may even be available with motorized controls. These alone may be more than worth the low asking price.

Check around to see what's available in your area. Order all of the surplus catalogs that you can. These are advertised in the popular experimenters magazines sold at most newsstands. Certainly, you would have to wait a long time to collect enough surplus to build a complete TVRO Earth station, but you may find a device or two which can be used in place of newly purchased and manufactured Earth station components. If you can buy a surplus dish antenna for $100 or so, this can mean a savings of well over $1,000 when it is used to replace a parabolic dish offered by TVRO equipment manufacturers.

By keeping yourself up to date as to the offerings of the surplus market as a whole, you will be in the best position to take advantage of attractive deals as soon as they are made available. In future years, the resourceful TVRO enthusiast has much to look forward to in the surplus market.

Chapter 5

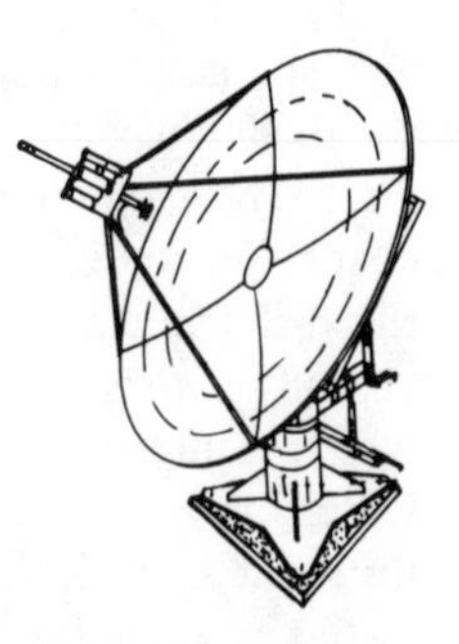

Principles of TVRO Antennas

THE MOST PROMINANT FEATURE OF A TVRO EARTH STATION is the antenna. This is always a large device, usually at least ten feet in diameter. Most Earth station antennas are generally circular in nature and require sturdy mounts, owing to their relatively high weight. At microwave frequencies, the transmitter radiates power in the form of electromagnetic waves. In satellite television applications, these radio waves are beamed toward a satellite located in a geostationary orbit 22,000 miles above the equator. These transmissions are intercepted by a receiver at the satellite and retransmitted back to Earth again. At the receiving station, the electromagnetic waves must be intercepted and fed to a receiver whose electronic circuits have the ability to remove the video and audio information and pass it along to the television receiver.

An antenna serves to link the receiver with the rest of the universe. Simply put, it is a matching network which couples a transmission line to free space with as much efficiency as possible. Efficiency of an antenna can be thought of as its ability to gather microwave energy and to pass it along to the remaining circuitry. An inefficient antenna system will reflect much of this energy. As efficiency improves, more and more of the energy which strikes the antenna is passed along to the receiver. Of course, there are losses within the receiver system, as well as at the antenna. For TVRO ground station purposes, antenna efficiency would be a comparison of the amount of energy which strikes the reflector and the amount of energy which is delivered to the input of the low noise amplifier.

TVRO antennas are shaped to propagate the electromagnetic waves in a particular direction and more precisely, to a fine focal point. The energy which strikes the reflector is broad in nature or spread out. The dish serves to gather this energy from a large physical area and then combine it in an area which is physically small. This basic principle has been previously discussed.

The most popular type of TVRO Earth station antenna is the parabolic dish. It serves as a reflector at microwave frequencies due to the geometric properties which are obtained from its physical shape. We know that most of the microwave energy which strikes this dish will be reflected to the focal point. This is where the energy actually enters the receiving antenna proper. The reflected wave which is intercepted at the focal point is made up of parallel rays which are all in phase. The antenna beam can be shaped in many different ways. This will depend upon the shape of the dish. Larger dishes produce far narrower beams, while the smaller dishes have broader beam widths.

WAVEGUIDES

The high frequencies employed by satellite television transmissions often necessitate waveguides in certain portions of the receiving system rather than standard dipole or vertical antennas. Waveguides also take the place of standard transmission lines. A *waveguide* is a metallic pipe which is used to transfer high frequency electromagnetic energy.

Basically, there are two methods of transferring electromagnetic energy. One is by means of current flow through wires. This is the most common method known today, but is far more practical at lower frequencies. The ribbon or coaxial cable which is presently attached between your television receiver and antenna is an example of this type of energy transfer.

The second method of transfer is by movement of electromagnetic fields. The transfer of energy by field motion and by current flow through conductors may not appear at first to be related. However, by considering two-wire lines as elements which guide electromagnetic fields, the current flowing through the conductors may also be considered the result of electromagnetic field motion.

At the microwave frequencies used for satellite reception, a two-wire transmission line is a poor means of transferring electromagnetic energy because it does not confine electromagnetic fields. This results in energy escaping by radiation. Electromagnetic fields must be completely confined. This is where coaxial cable comes into the picture. A coaxial cable consists of a center conductor which is surrounded by an outer conductor. This is shown in Fig. 5-1. The outer conductor serves to contain the energy and greatly reduce or prevent radiation. The two-wire conductor can be thought of as the standard 300-ohm ribbon cable which is often used between a television set and antenna. The coaxial cable is presently more popular than ribbon line, although at standard television frequencies, the ribbon line may be more efficient. Even the coaxial cable is not very efficient at the microwave frequencies we will be dealing with in TVRO Earth stations. Long cable runs can quickly render the entire system useless.

Energy in the form of electromagnetic fields may be transferred very efficiently through a line that does not have a center conductor. A waveguide can be thought of as a hollow transmission line without a center conductor. The configuration of the energy field in a waveguide is different from that in a coaxial cable due to this missing conductor. Waveguides may be rectangular, circular, or elliptical in cross-section.

There are three types of losses in radio frequency transmission lines. These are copper losses, dielectric losses, and radiation losses. The copper loss is due to resistance within the copper conductor. This is often referred to as I^2R losses. I^2R is part of Ohm's Law for computing power. Power losses are equal to the resistance of a conductor (R) times the current flow (I^2). The conducting area of a transmission line will determine the amount of copper loss.

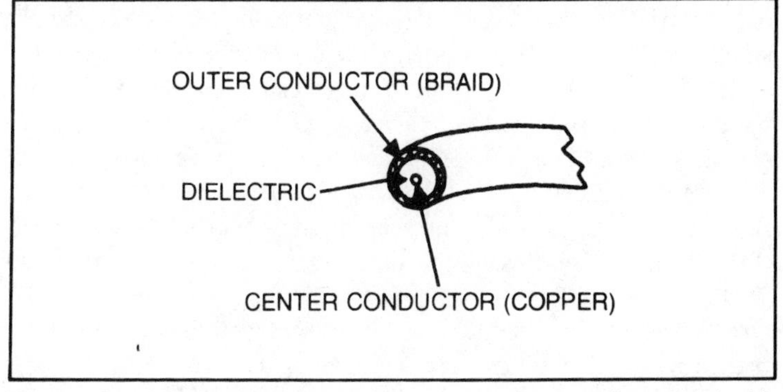

Fig. 5-1. Coaxial cable is made up of a center conductor and an outer braid.

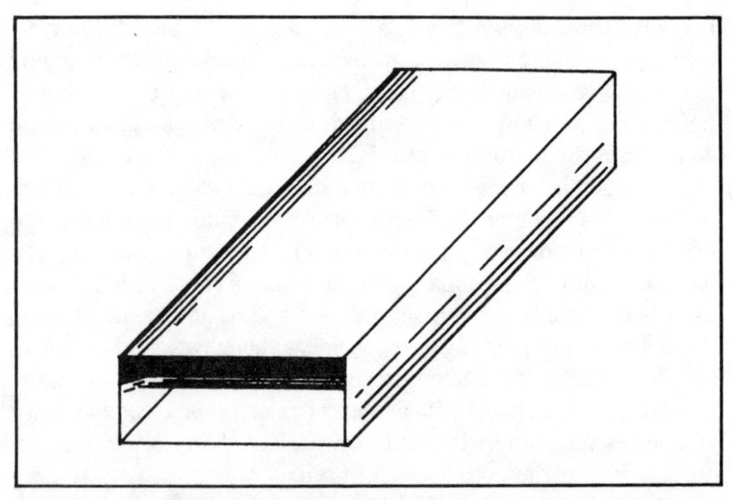

Fig. 5-2. A rectangular waveguide has the advantage of a larger surface area.

The larger the cross-sectional diameter, the lower the losses. Dielectric losses are due to the heating of the insulation between the conductors of a transmission line. This can occur when a transmission line which is simply too small is used to transmit high amounts of power or when the dielectric does not provide sufficient insulation between the center conductor and the outer one. A dielectric which is a good insulator at 30 megahertz may act like a conductor at 3,000 megahertz (3 gigahertz). Radiation losses are due to the radiation of energy from the line. Coaxial cable usually prevents most of this radiation.

With these facts in mind, the advantages of waveguides over a two-wire and coaxial transmission line can be better understood. A rectangular waveguide with a large surface area is shown in Fig. 5-2. A two-wire line consists of a pair of conductors with relatively small surface areas. The total surface area of the outer conductor of the coaxial cable is large, but the inner conductor is small in comparison. At microwave frequencies, the conductor becomes electrically smaller. This is due to a phenomenon known as *skin effect*. Put simply, skin effect is a tendency for high frequency energy to flow only on the outside surfaces of the conductor rather than through the entire cross-section. So, coaxial cable used in TVRO Earth stations will restrict a certain amount of energy flow. The size of the energy field is limited by current flow and is thus restricted to the current-carrying area of the conductor.

With this in mind, you can see that a waveguide will have the least copper loss of the three types of transmission lines under discussion because it has no center conductor. Dielectric losses are very low too. This is due to the fact that no solid supports are needed for a center conductor. The dielectric in a waveguide is air. This has a very low dielectric loss at microwave frequencies compared with solid conductors. Radiation losses are negligible in a waveguide as well, since the electromagnetic energy is wholly contained within the structure.

With all of these many advantages, one might ask why waveguides are not used throughout the entire TVRO Earth station facility. We know at this point that coaxial cable is incorporated between the LNA output and the receiver. While a waveguide is used at the LNA/feed horn connection in TVRO Earth station, its physical size and structural requirements make it expensive and difficult to work with. Therefore, we accept the losses presented by coaxial cable in order to save some money and to have a system which is easier to set up and adjust.

The dimensions of a waveguide are critical. They must be approximately a half-wavelength at the frequency of operation. At four gigahertz, the waveguide width would be about 1 1/2 inches. If these dimensions are not adhered to, energy at four gigahertz, for example, and all frequencies below it would not travel down the guide whose opposite end would be connected at the receiver. There is also an upper limit to the frequency which may be transported by a waveguide. Therefore, the frequency range of any system utilizing a waveguide is limited. This is not the case in standard transmission lines, such as coaxial cable. Coax will transport electromagnetic energy over a tremendously wide range of frequencies. But as the frequency increases, losses become more and more abundant.

The travel of energy down a waveguide is similar but not exactly identical to that of the propagation of electromagnetic waves in free space. The difference is found in the fact that energy in a waveguide is confined to the physical limits of the guide itself.

HORN ANTENNAS

A horn antenna is shown in Fig. 5-3 and provides a method of matching the input impedance of a waveguide to free space. Normally, the receiving pattern of an open-ended waveguide is broad in both the vertical and horizontal planes. The evolution of the horn antenna came about as a result of the effort to minimize the reflec-

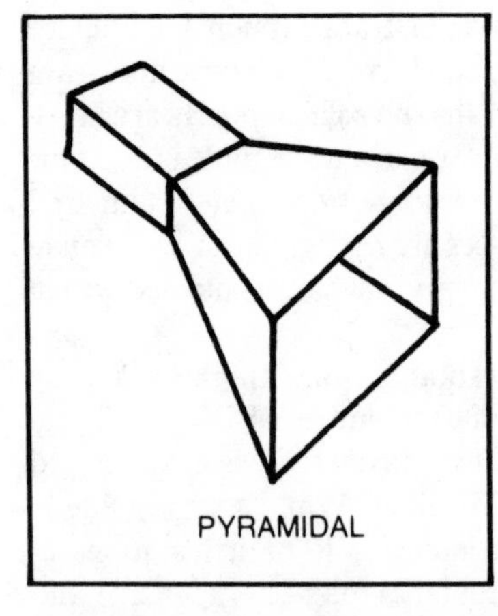

Fig. 5-3. A horn antenna serves to match the input impedance of a waveguide to free space.

tions that occur when a straight piece of waveguide receives microwave energy. In other words, a length of waveguide connected to a receiver could serve as an antenna, but it would be a very inefficient antenna because of its mismatch to free space. The horn antenna is actually a piece of waveguide which has been flared at the open end. Figure 5-4 illustrates this.

When a waveguide is flared out, a horn antenna is obtained; but the flare must be very gradual so as to permit a better match between antenna and free space. The horn is very practical at microwave frequencies, since its physical size is not prohibitive. Since a resonant element is not used, this type of antenna is capable of wideband operation.

Another type of horn antenna is shown in Fig. 5-5. This is called a conical horn. It is sometimes seen in TVRO Earth station uses. The previous type is called a pyramidal horn and has equal directivity in both the vertical and horizontal plane.

The reader may be wondering at this point why the discussion

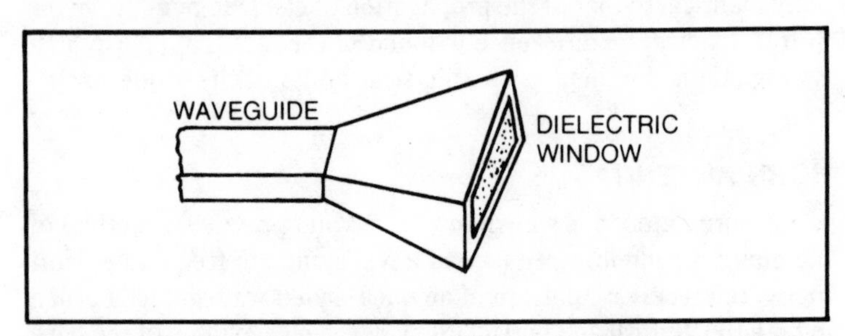

Fig. 5-4. In actuality, a horn antenna is really a piece of waveguide which has been flared at the open end.

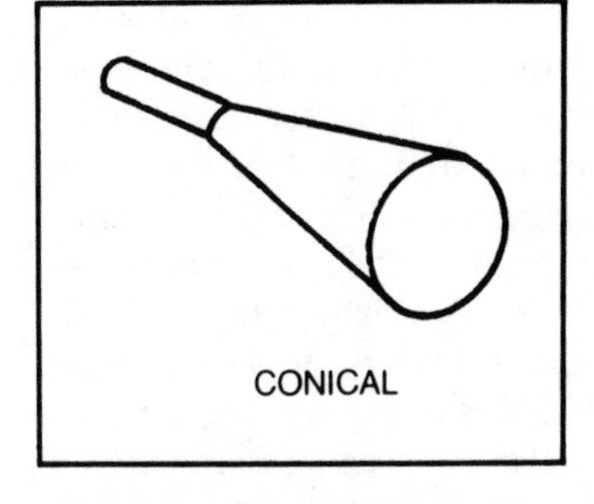

Fig. 5-5. A horn antenna serves to match the input impedance of a waveguide to free space.

has become so dry and technical. The reason is that it is necessary to understand a few basic principles in order to perceive the operation of Earth station antenna systems. The horn antenna is normally used at the focal point of the dish. It is connected to a waveguide at the input of the low noise amplifier. This is a critical point in the antenna circuit, because no amplification of the received energy has taken place yet. We use a very short length of waveguide between the horn and the LNA, because the use of coaxial cable or some other type of solid transmission line at this point would sharply reduce the amount of signal delivered to the amplifier circuitry. The feed horn/waveguide input to the LNA is extremely efficient and very little of the energy at the focal point is reflected away.

The magnitude of the signal input to the input of the amplifier is increased within the LNA circuitry. Now, we have much more signal to work with, having gotten through the free space portions of the system. Coaxial cable, in short lengths, may now be used to transfer the higher magnitude signal to the receiver circuitry without incurring losses so great as to input a deficient signal to the receiver. Sure, it would be far better to use waveguide between the LNA output and the receiver input, but due to the short distances involved, the price and complexities are simply not worth it.

PARABOLOID REFLECTORS

The paraboloid or parabolic reflector serves as a collecting and focusing device and is an integral part of an Earth station antenna. Since the feed horn dimensions are small, the reflector is large in terms of wavelength. A reflector surface is chosen which will provide a constant phase. This means that all of the gathered energy striking the dish will travel the same distance. For our purposes, this also means that all of the transmissions at any one instant which were sent by the satellite will strike the feed horn at the same time.

But how can this be? The feed horn is obviously closer to the center of the disk when using a parabolic reflector than it is to the outer edges. It would seem that the signals which strike the outer edges would have to travel farther than those which strike at dead center. The answer is that they do. Signals which strike at the center of the disk have a shorter distance to traverse to the feed horn than those which strike at the edges. This is shown in Fig. 5-6. For the sake of discussion, let's assume that the distance from the feed horn to the center of the disk is 8 feet. Also assume that the distance from either edge is 10 feet. There is no doubt about it; ten feet is a longer distance than is eight feet. But do the signals really travel farther?

Now that you're completely confused, let's attempt to clear this matter up by using Fig. 5-7. To be in phase, the signals must travel the same distance. But this does not mean the same distance from the dish which collects the transmissions to the feed horn, but from the satellite to the feed horn.

Figure 5-7 shows the satellite, the parabolic dish, and the feed horn. Again, the feed horn is located at a distance of eight feet from the center of the reflector and ten feet from either edge. Let's assume that the satellite is located exactly 40,000 miles fro the back of the feed horn. The incoming signal from the orbiter must travel 40,000 miles to get to this point. But it must travel an additional

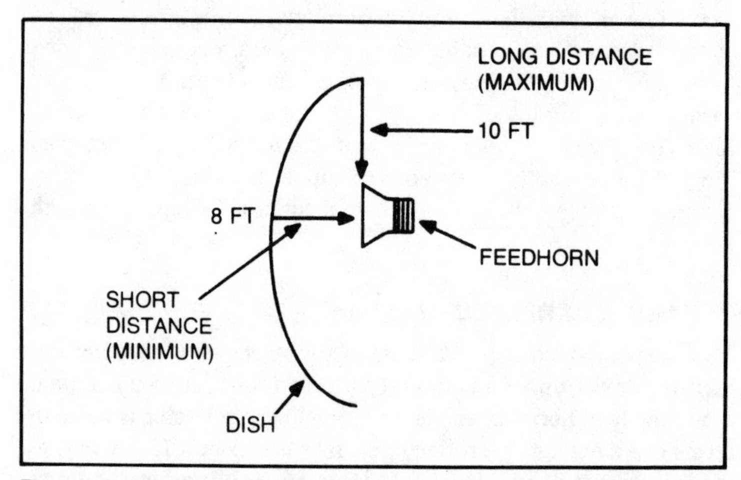

Fig. 5-6. Signals from the satellite travel varying distances from points on the dish to the feed horn.

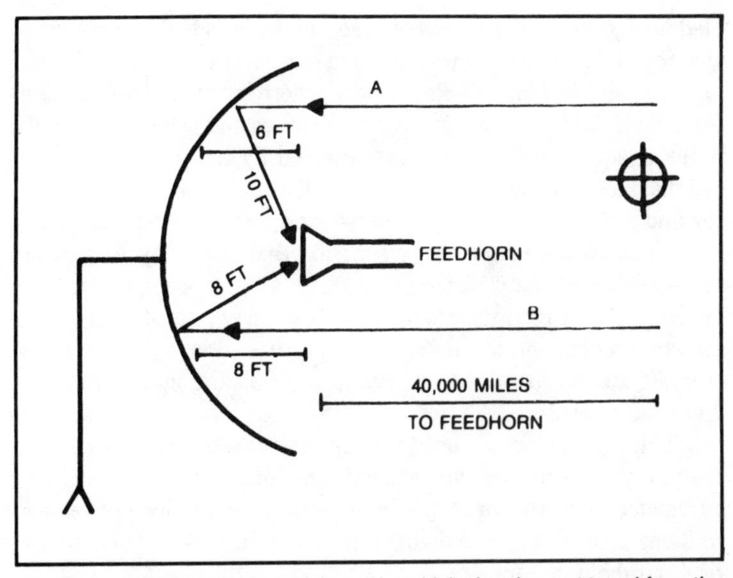

Fig. 5-7. A simplified diagram of the paths which signals must travel from the satellite to the feed horn.

eight feet to get to the center of the dish and then eight feet more after being reflected to enter the aperture of the feed horn. Now, we must remember that the parabola is a curve. Therefore, the center of the dish which lies directly in a line with the satellite and the feed horn is the farthest point from the orbiting transmitter in the whole antenna system. The outer edges of the dish are farther from the feed horn but closer to the satellite. By studying the parabola in great detail, we would learn if the center of the dish is 40,000 miles plus eight feet from the satellite, then (in this example) the outer edges must be 40,000 miles plus six feet from the satellite. Let's repeat this again from clarification purposes. To get to the center of the dish, the satellite signal must travel 40,000 miles and eight feet. The same transmission will travel only 40,000 miles and six feet to get to the outer edge of the dish. Forget the feed horn for the moment. the outer dish edges are physically closer to the satellite than is the dish center. The transmission from the satellite, then, does not strike the different areas of the dish at the same time.

Refer to Fig. 5-7 again. Transmission ray A from the satellite strikes the outer edge of the dish at a distance of 40,000 miles and six feet. This ray is reflected toward the feed horn, where it travels an additional ten feet to reach the aperture. Total distance traveled is 40,000 miles and sixteen feet. Now, let's look at transmission ray B. It travels 40,000 miles and eight feet to strike the center of the dish. It is then reflected toward where it travels another eight feet. Total distance traveled is 40,000 miles and sixteen feet.This is the exact distance which was traveled by ray A. Again, there is a distance of eight feet between the dish and aperture at the center and a distance of ten feet between aperture and outer edge.

It can be seen from this discussion that while ray A is being reflected by the outer dish edge, ray B is still traveling toward the center. When this ray is finally reflected, ray A has already traveled two feet of the distance to the aperture. It is eight feet away from its target. At this same instant, ray B is also eight feet away from the aperture.

This discussion has used two-dimensional drawings to depict the parabolic reflector, which is a three-dimensional device. If you can conceive of this principle being applied to millions of rays at millions of points on the dish, then you understand the operation of a parabolic antenna.

The operation of a parabolic antenna helps to explain why alignment of the antenna with the satellite is so critical. If the dish is

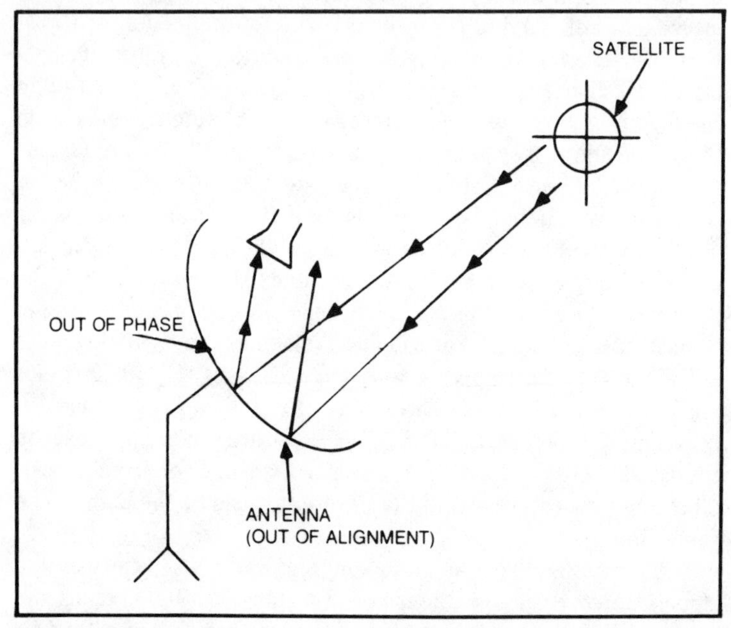

Fig. 5-8. If the antenna is not properly aligned with the satellite, signals will be out of phase.

only slightly off the beam, so to speak, the transmitted rays would strike the dish in such a manner that the signals reflected to the aperture would be out of phase. This is shown in Fig. 5-8. Here, the leading edge of the antenna would be the first area struck. The trailing edge would be the last to receive the signal. The original transmission would be reflected to the feed horn over a time span which does not coincide with the time required for the original transmission. For proper reception, a one-second burst of transmitted energy and the signal rays contained therein must strike the receiving antenna (the feed horn) over a span of exactly one second. If this latter period is 1.5 seconds, most of the transmitted information is lost due to the out of phase condition. For this reason, a TVRO Earth station antenna must be sighted with the same precision marksmen use in shooting at small targets except that your target is thousands upon thousands of miles away.

SPHERICAL ANTENNAS

Another type of antenna which is becoming popular for TVRO ground station applications is called the spherical antenna. Like the parabolic antenna, this design focuses the microwave energy reflected from its surface to a point in front of the antenna. The parabolic antenna has one focal point which lies directly in front of and perpendicular to the center of its reflector. For this reason, it must be directly aimed at the satellite which is to be received. This method is often referred to as *boresighting.* This term was probably garnered from the fact that the parabolic dish is aimed at the satellite in the same manner that a gun is aimed at a target.

There are major differences, however, in the operation of a spherical antenna. This design focuses microwave signals received by as much as plus or minus 20 ° off boresight (perpendicular to the center) with a negligible loss of efficiency. This means that a spherical antenna does not have to be exactly aligned with satellite, but more importantly, a spherical antenna positioned to reflect and focus the signal from one satellite will simultaneously reflect the signals of others. These must be adjacent satellites within plus or minus 20 ° of boresight. The spherical antenna reflects these adjacent satellite signals most efficiently but at different points in front of the antenna. In other words, when the spherical dish is boresighted on one satellite, it efficiently reflects these signals to a specific focal point. At the same time, the signals from an adjacent satellite are also being reflected, but to a different focal point. Both points still lie in front of the antenna.

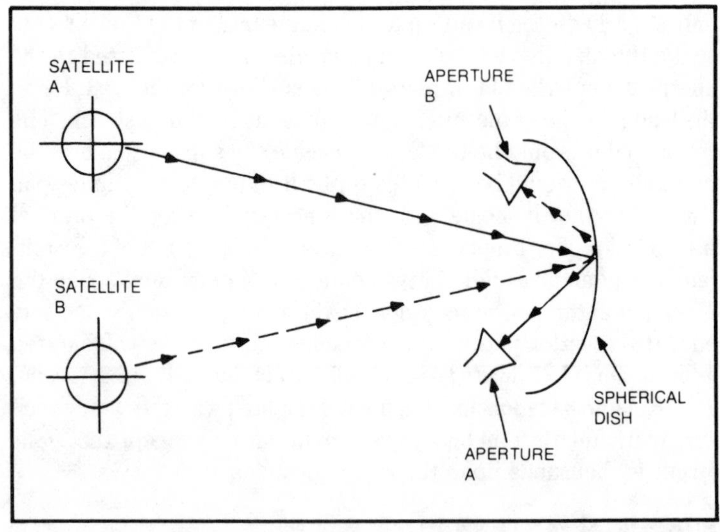

Fig. 5-9. In a spherical antenna, signals from two satellites may be received without moving the antenna.

Figure 5-9 shows a graphic representation of the spherical antenna using signals from satellite A and satellite B. It can be seen that the signals from the first satellite are reflected to aperture A which lies more to one side of the reflector than does aperture B. Notice that this is accomplished without moving the reflector. A feed horn placed at aperture A would receive the transmission of satellite A. If you took this same feed horn and moved it to focal point B, satellite B would be received.

A low noise amplifier and feed horn combination properly positioned at either of these focal points will amplify the maximum amount of signal that this antenna is capable of reflecting from a given satellite. This occurs provided that the signal is received at the reflector at less than a 20° angle from the boresight. By moving the LNA/feed horn from one focal point to the other, several satellite signals are received with about the same efficiency and no repositioning of the large dish is required. It should be noted that while the reflector's efficiency is generally constant within the 20° receiving arc, highest efficiency is approached at boresight and to angles either side of this point.

Figure 5-10 shows a 12-foot spherical dish that is put together from a kit. We will use this design as an example for the continuing discussion. Since the spherical antenna can reflect and focus the signals from more than one satellite simultaneously and in an efficient manner, two variations of antenna azimuth positioning may be considered.

If you desire to receive only one satellite, the antenna will be set up in the normal manner with its center being directed toward the azimuth heading for that satellite. Once this is done, the LNA and feed horn will be positioned at the focal point which lies directly in front of the antenna's center.

For multiple satellite reception, the antenna will be directed toward the azimuth heading of the center of the group of satellites whose signals can be properly focused by the reflector. This is shown in Fig.5-11. The LNA/ feed horn is then positioned at any focal point for reception of a given satellite in the group. When reception of another satellite is desired, the feed horn is moved to another focal point. This, of course, is done manually. The feed horn/LNA is not physically attached to the spheroid reflector but is mounted separately some distance away. The antenna uses a concrete base to support the feed horn. The base is heavy enough to hold its assembly in place but can be moved by hand to different focal points.

Alternately, two or more LNA/feed horn assemblies may be positioned for reception of multiple satellites simultaneously. A

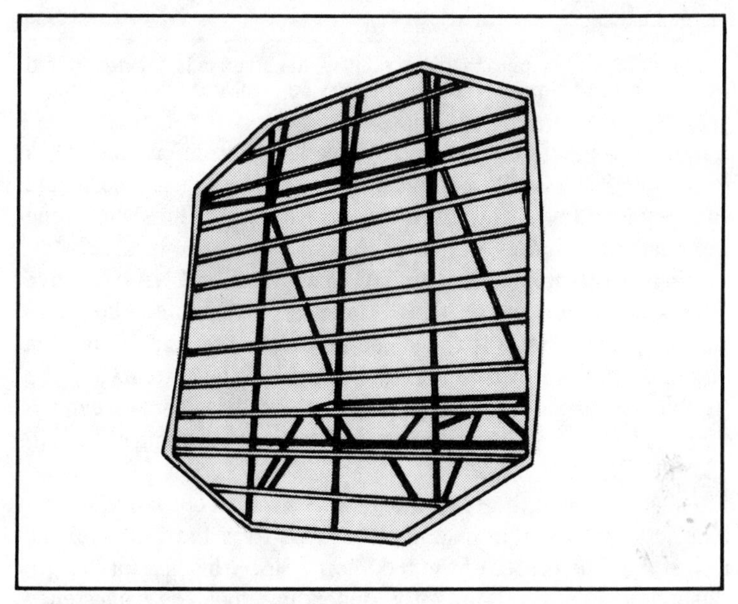

Fig. 5-10. The Skyview I antenna is a spherical design. Courtesy of Downlink, Inc.

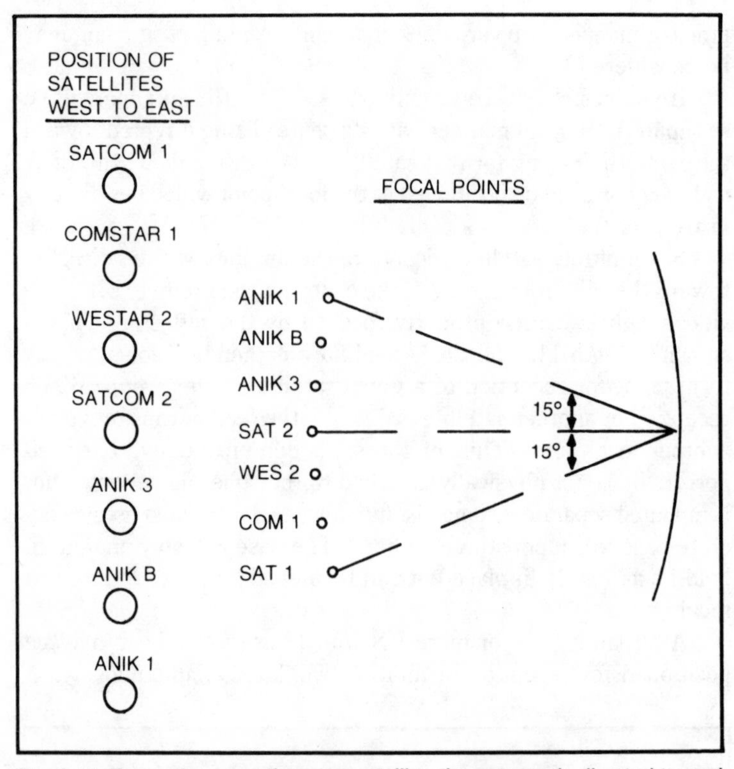

Fig. 5-11. To receive more than one satellite, the antenna is directed toward a centered azimuth heading of the group to be received.

remote switching unit can be used to change from one feed horn to another. Technically, it would probably be possible to incorporate as many feed horns and LNAs as there were satellites within the 20° receiving angle.

In order to properly position the feed horn and LNA to receive the maximum amount of signal, three factors must be taken into consideration. First is the focal length of the antenna; second, the angular difference between the inclination of the antenna and the satellite elevation; and finally, the angular difference between the azimuth heading of the antenna and the azimuth heading of the satellite.

The focal length of the model antenna (12-foot model) is fifteen feet. This is the distance measured from the center of the reflector to the center of the feed horn. Since this is a kit design, this distance may vary slightly, depending upon the consistency of the reflector's curvature. Knowing this, we also know that the feed horn will always be fifteen feet from the center of the antenna, and the mouth of the horn will always aim directly at the antenna center, no matter which focal point is used. This is shown in Fig. 5-12 as an arc along which the feed horn may be positioned.

The height of the LNA/feed horn assembly will vary, however. This will depend upon the elevation of the satellites and the antenna inclination. A signal coming into the reflector at an angle which is measured from a line drawn perpendicular from the center will be reflected at an equal angle in the opposite direction. When the feed horn is repositioned left or right in front of the reflector to receive another satellite, its height above the ground must be changed. Figure 5-13 shows how the focal point varies in height for each satellite elevation above the horizon. This height can approach impractical limits with some inclinations, so this must be worked out beforehand.

Positioning of the LNA/feed horn to a focal point in the azimuth direction requires approximately one foot of movement (left or right) for every 4° of change in azimuth. The azimuth focal point of a specific satellite in relationship to the center of the antenna is found by using the formula: feed horn movement from center (in feet) = the satellite azimuth − the antenna azimuth, with the

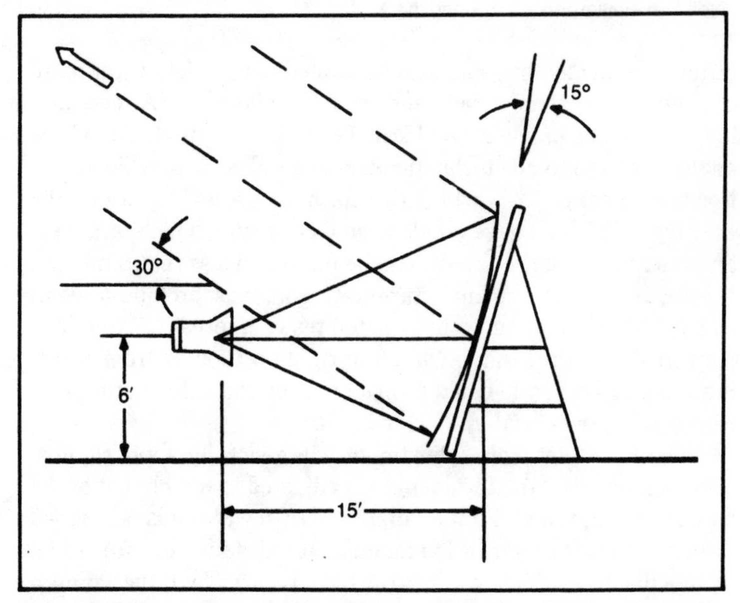

Fig. 5-12. The arc along which the feed horn may be positioned to receive the maximum amount of signal.

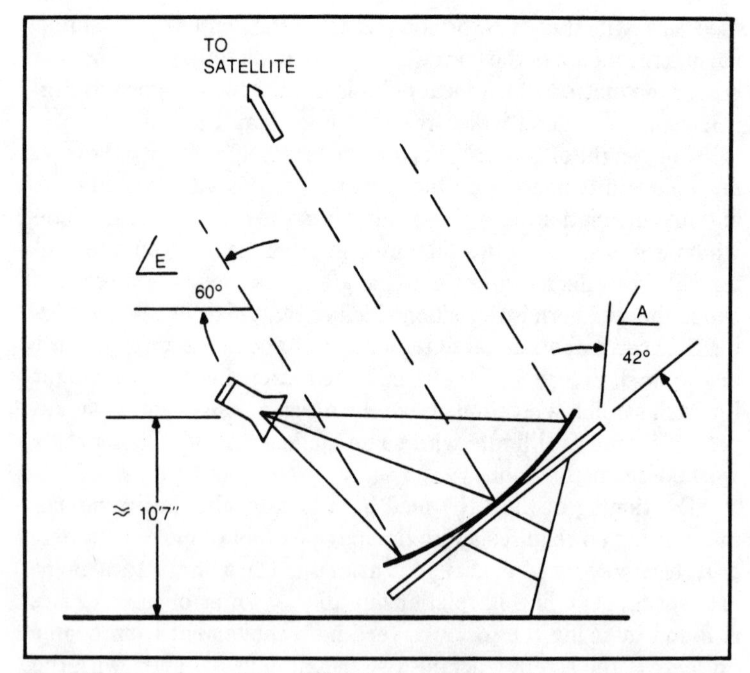

Fig. 5-13. Here, it can be seen that the focal point varies in height for each satellite elevation above the horizon.

difference divided by four. The LNA/feed horn height is far more complicated mathematically and reads as follows: LNA/feed horn height = 6 cos (angle A) + 15 sin (2 angle A − angle E), where angle A is the antenna inclination and angle E is the satellite elevation. For most of us, working this formula is a little scary on the first try, but if you have an electronic calculator which offers sin and cos functions, the answer can be figured in less than a minute. Fortunately, most manufacturers of antennas provide a lot of calculated information on most satellites of interest to American viewers. Since the formula for left or right positioning from center is simple to work, you could probably set up the antenna for proper receiving on a trial and error basis.

The LNA/feed horn adjustment will generally apply to most spherical dishes if the focal length is adjusted to match that of the antenna being tuned. First of all, the LNA/feed horn is positioned directly in front center of the antenna at a distance of fifteen feet (or at a distance which matches the focal point). Next, the azimuth position is adjusted. This is calculated by moving the LNA/feed horn left or right of center by the appropriate distance figured from the formula. Remember to maintain the focal length (15 feet in this case) at all times. Calculate the required height of the feed horn and aim it toward the center of the antenna. The feed horn will now be within a few inches of the exact focal point. This will depend upon the accuracy of the antenna's curvature and the inclination/azimuth setting.

Now, while watching a monitor, move the feed horn slightly and systematically until a picture is obtained. You will want to make certain that the LNA/feed horn is set up for the proper polarity (vertical or horizontal) for the satellite you are trying to receive. Once a picture has been obtained, it's an easy job to make minute positional adjustments for maximum signal strength.

PARABOLIC/SPHERICAL COMPARISON

Now that the two main types of antennas designed for TVRO Earth station applications have been discussed, it is appropriate to compare the two. This is a difficult task in some ways because they each have equal advantages and disadvantages. In the end, the specific operating needs of the Earth station owner will most likely be the determining factor in deciding which design will be best for his particular needs.

The spherical antenna has an advantage in simplicity of mounting hardware. The antenna is set and then forgotten. Only the feed horn/LNA assembly is moved for tuning. The parabolic dish antennas must be aimed each time a different satellite is to be received. This aiming requires the movement of the entire antenna structure instead of just the feed horn. But from a positive standpoint, the parabolic dish can be controlled from the television receiver location when optional equipment is purchased. Most manufacturers offer motorized drives which are fitted to the parabolic dish base mount. These motors are activated by a control box which is mounted at the television receiver. This certainly adds to the convenience of TVRO Earth station operation. The spherical antenna, on the other hand, requires the physical moving of the LNA/feed horn assembly each time a different satellite is to be accessed. The dish is not physically attached to the feed horn, so rotation is neither possible nor desirable. While two LNA/feed horn assemblies may be used with a single spherical reflector, this becomes prohibitive from a cost standpoint.

In all fairness, it is proper to mention at this point that most TVRO Earth station owners will not be switching from satellite to satellite all that frequently, regardless of what type of antenna system they have. Sure, in the first few weeks of operating the Earth station, every satellite with an open window will certainly be accessed. This would be part of a normal breaking-in period for the Earth station as well as for the excited owner. But after the novelty wears off, the Earth station's operation becomes more in line with what is was originally purchased for—the reception of entertainment channels in a convenient and readily available manner.

Most satellites which carry entertainment programming have many other channels as well. For example, the video services might include the following: Nickelodeon, PTL, WGN, The Movie Channel, WTBS, ESPN, CBN, USA Network, Showtime West, Showtime East, CNN, WOR, Reuter/Galavision, Cinemax East, HTN, Home Box Office West, Cinemax West, Home Box Office East, and several spare channels. That's at least nineteen major programming services. Everything from children's educational programs to religion to first-run movies is included on this one satellite. So assuming you don't like the movie that's playing on The Movie Channel, which is on transponder 5, all you do is switch to transponder 24, where Home Box Office is airing something you consider to be decent, or vice versa. If you get tired of first-run movies, you can always go to CNN, the all news network. Satcom I and other such satellites individually carry a wide enough range of programming to suit almost anyone. The result of all this is to make it unnecessary to go to more than one satellite. This is called competition between programming services.

So, don't place more importance on picking up every satellite in the sky instead of just one or two than this practice merits. Of course, it is very convenient to be able to alter the position of the antenna in order to change satellites. Most TVRO Earth station owners keep track of all the services that are being offered through monthly publications. This gives them the opportunity to check out each program almost as soon as it becomes available. Many TVRO Earth station owners like to experiment with their installations. There are many, many satellites of all types and from all nations in orbits which can be intercepted in the United States. The appendices of this book contain information of many of them. By owning an antenna which can be rotated, many of these satellites can be accessed, although the information they transmit may not be intelligible (this would especially apply to the Russian spy satellites). Persons why try to access any satellite they have a window to can be directly compared with shortwave listeners who are not so much looking for information as they are seeking any kind of transmission which can definitely be attributed to that particular orbiter. They may even keep lists on which satellites they have been able to access and those that are still on the want list. QSL cards which are verifiers sent out by amateur radio operators and other transmitting stations are never seen in TVRO service, but keep trying. All of these activities would not be possible without an antenna system which can be directed to many different satellites.

The choice of a spherical or parabolic dish will most likely depend on the Earth station owner's personal desires. The spherical antenna described in this chapter would receive along an arc plus or minus 20° from boresight. By obtaining a computer printout of satellite windows in your area, you can quickly determine where the satellites you are interested in will be found. If all or most of these could be hit with a plus or minus 20° reception area, then the spherical dish might be a good choice. On the other hand, if you want constant reception over a larger portion of the sky and don't like the idea of going outside to change satellites by readjusting the feed horn, then the parabolic dish would be the way to go. Most parabolic dishes, however, are not fitted with motorized controls. At present, these are the exception rather than the rule in most TVRO Earth stations. So, even with the parabolic dish, you will probably have to go outside to change satellites, although the readjustment may be simpler than that which is necessary for spherical designs.

Another comparison which is necessary is that of cost. This is another difficult parameter to measure, because there are great fluctuations within the market. The spherical antenna is usually offered as a kit of parts and for this reason, it may be cheaper than most parabolic dishes of the same diameter. Parabolic antennas are also offered as kits, but there are just a few parts involved in assembling the latter, whereas the former is a major project. The struts and base assembly for the parabolic dish will be more costly than those for the spherical design, which will most likely be simply placed with the lower edge on the ground and a few simple supports from behind. This again tends to make most spherical antenna designs less expensive than comparable parabolics when considered as a whole system (dish, feed horn, mounts).

There is an aesthetic consideration as well. In my opinion the parabolic dish is neater in appearance. Most structures are rather striking and massive, but the seemingly one-piece design of parabolics provides less of a chicken-wire effect than does the spherical. The spherical antenna usually consists of a fine mesh screen mounted on a support lattice which is visible from front and back. Fiberglass construction is pretty much universal with TVRO Earth station dishes designed for personal applications when the parabolic design is considered. Fiberglass is not a conductor of electricity, but serves only as a smooth-surfaced container for the screen which is sandwiched between two molded sections of the fiberglass.

Another problem with spherical antennas is positioning of the feed horn, especially when the dish is tilted to intercept satellites which lie many degrees from the horizon. When the upper edge of the dish is tilted back, the LNA must be elevated higher from the ground. This can present structural problems, in that it is necessary to provide a mast support for the feed horn which may have to be fitted with guy wires for stability. This tends to add a lot of permanence to the feed horn installation, and changing satellites is much more difficult, in that the guy stakes which anchor the wires to the earth must be removed and the entire assembly set up again at a different focal point.

It can be seen that there are many criteria to be considered when deciding between a spherical and parabolic antenna design. Most companies that sell spherical antennas also offer the more standard parabolic types, and once you explain your location and intended use of the completed Earth station, they should be in an excelled position to give you further advice on your selection. I have found in talking to many representatives of companies which offer equipment for personal TVRO Earth stations that they are more than willing to provide information and assistance, even if you're not sure that an Earth station setup is for you. They must feel (and rightly so) that the more they disseminate information about Earth station operation, the more interest will be generated about the field as a whole. This helps the entire industry.

OTHER TYPES OF ANTENNAS

While the parabolic and spherical antenna designs are used almost universally for TVRO Earth stations, there are other types of antennas that have microwave applications. A mutli-element array can be used, but its design is very complex when compared to the relative simplicity of microwave dishes. Figure 5-14 shows a multi-element array which consists of many elements or individual antennas that are combined to provide a single output. The length of each horizontal element section would be a quarter wave at the operating frequency. Two of these sections are combined in the same plane to produce a half-wave dipole. Each element is fed at the center. Though often massive is size, the use of thin wall aluminum tubing makes the overall design very light in weight.

A major problem is encountered with this type of design. When the elements are in the horizontal plane, the entire system responds efficiently to only horizontally polarized transmissions. To become vertically polarized, the system would have to be canted 90° to place each element in a vertical position in respect to the Earth. With a structure this large, this type of rotation is not practical. Some designs may incorporate both vertical and horizontal elements fed from the same line. This is shown in Fig. 5-15. Now, the antenna system is both vertically and horizontally polarized and need not be canted on end for vertical reception. This, of course, doubles the complexity of the design and makes the structure about twice as heavy. Wind storms tend to play havoc with these types of designs.

Another type of antenna which has seen some limited usage at microwave frequencies is the quad array shown in Fig. 5-16. This antenna consists of a large number of square elements whose

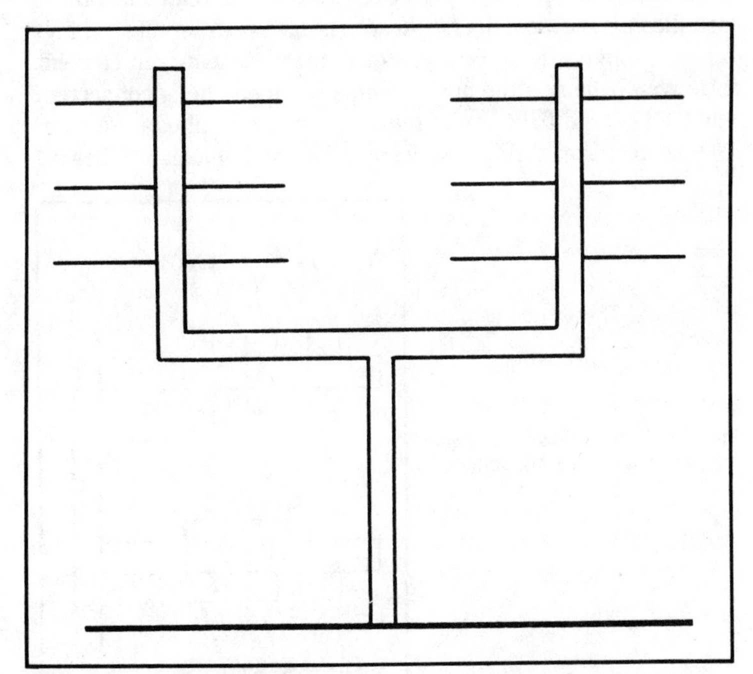

Fig. 5-14. In a multi-element antenna system, individual elements are combined to provide a single output.

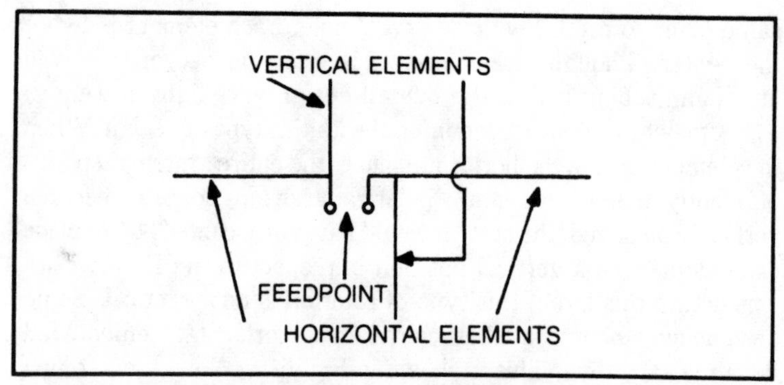

Fig. 5-15. In this design, vertical and horizontal elements are fed from the same line.

parameters equal a full wavelength each. A single quad antenna element is shown in Fig. 5-17. The feed point is at the bottom horizontal portion of the element. A quad is a mixture of both vertical and horizontal components, so it will respond reasonably well to both horizontal and vertical polarized transmissions. In the quad design, only one element is attached to the feed line. The others are called parasitic elements and each is cut to a different size. Figure 5-18 shows one section of a quad array. The driven element or active element is the one to which the feed line is connected. The elements to the back of the active one are called reflectors. The perimeter of the first reflector (the one immediately behind

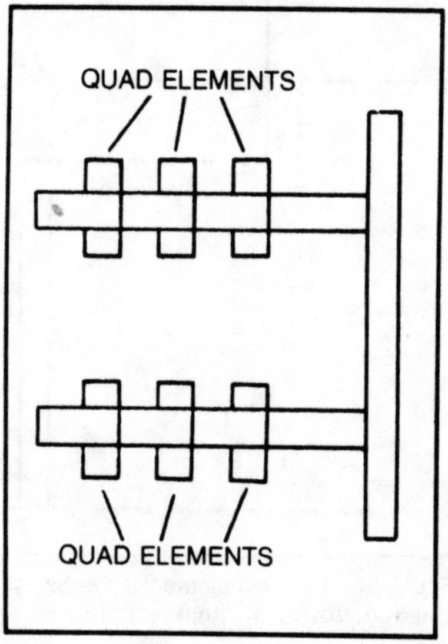

Fig. 5-16. Quad arrays have limited usage at microwave frequencies.

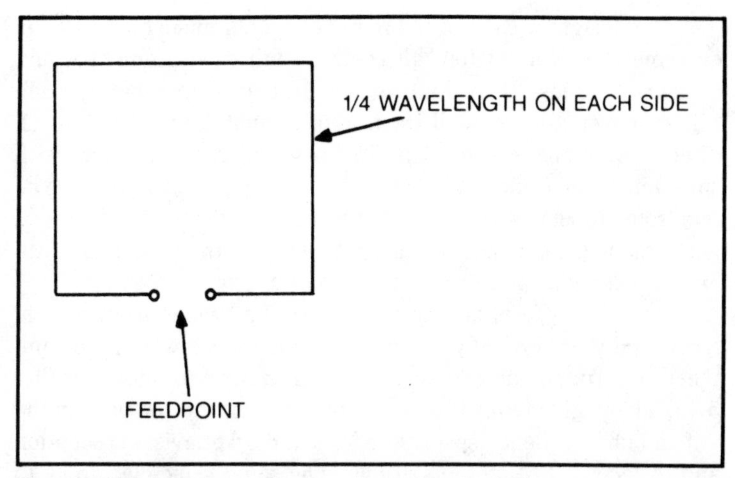

Fig. 5-17. This single quad-antenna element has the feed point at the bottom horizontal portion.

the driven element) is slightly larger than the one to which the feed line is attached. The second reflector has a perimeter which is larger than the first. The reflectors increase in size as their numbers increase.

The elements in front of the driven element are called directors. The first director is slightly smaller in perimeter dimensions than the driven element. Each successive director on out to the front end of the antenna is smaller in perimeter size. Several of these sections are combined to form a quad array.

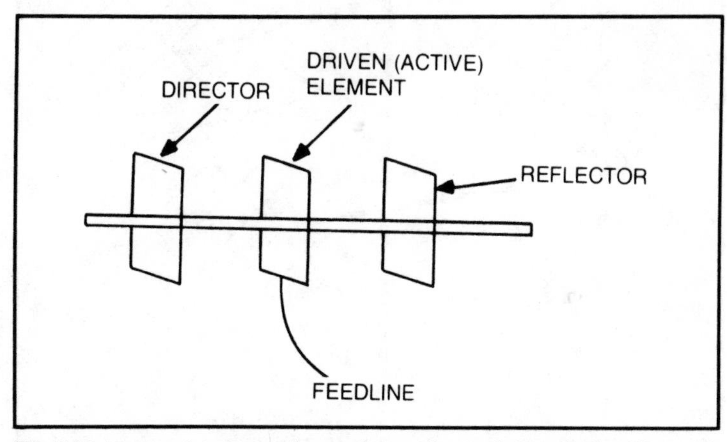

Fig. 5-18. In a quad design, only one element is attached to the feed line, with the others called parasitic elements.

The design of each of the many elements in such an array is extremely critical. At four gigahertz, a full wave element would be a fraction less than three inches around its perimeter. Each reflector would be a small fraction of an inch larger, while each director would be an equivalent fraction smaller. A quad array with substantial gain could easily have fifty or more elements. Any deviation from proper design length at any of these will result in degraded performance. Obviously, attention to detail with this and any other multi-element microwave antenna is crucial.

A similar type of antenna array may be formed from several horizontally or vertically polarized yagi antennas which are combined to form an antenna system. Such an array is shown in Fig. 5-19. The yagi antenna is a design which is most often used in the vhf and uhf frequency spectrums for standard television reception and to receive FM stereo stations. The same principles apply to the yagi as were discussed in the section of this chapter dealing with quad antennas. Each yagi element is about a half wavelength at the microwave frequency. As before, each major section is composed of a driven element, reflectors, and directors. A closeup of

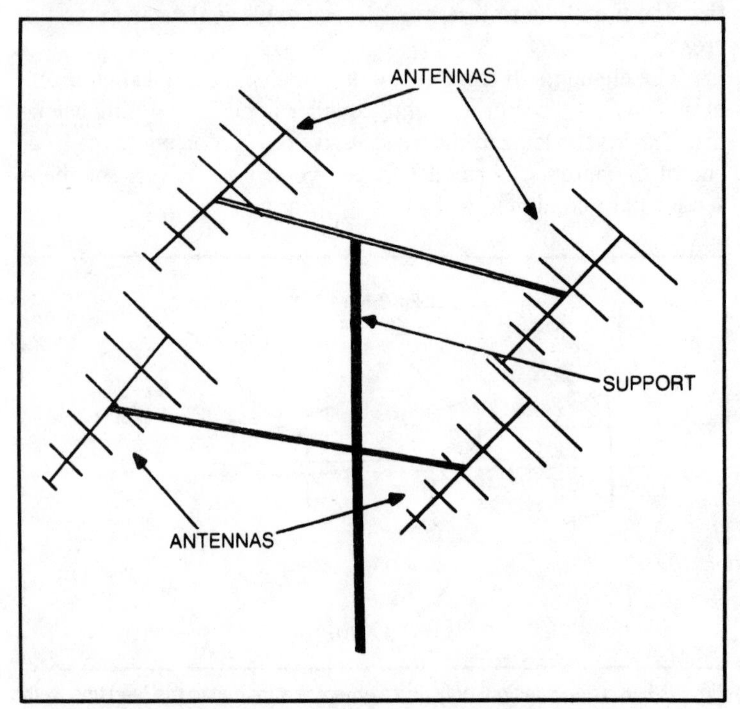

Fig. 5-19. Here, several yagi antennas are combined to form an array.

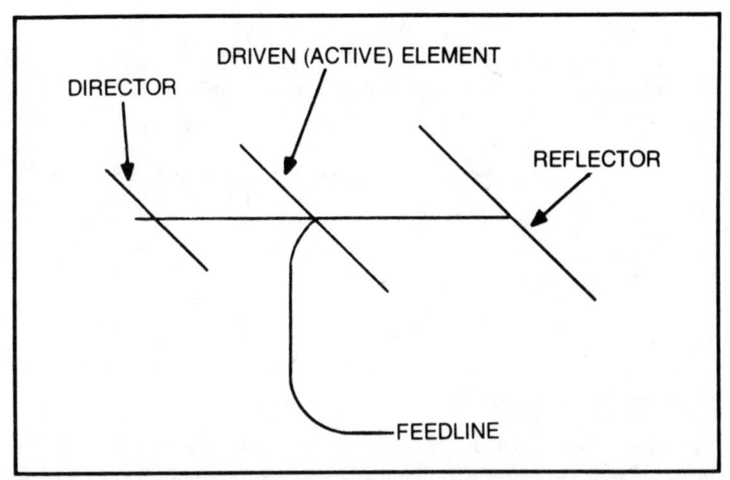

Fig. 5-20. In this closeup view of one of the sections of a yagi array, the driven element is fed at its center.

one of these sections is shown in Fig. 5-20. Here, the driven element is fed at its center. It is exactly one-half wavelength at the operating frequency. Each reflector is slightly longer, while each director is decreasingly smaller. As shown, the antenna is horizontally polarized. The entire system would have to be canted 90° to place the elements in a vertical plane in order to be used in the reception of vertically polarized waves.

Still another type of antenna which has seen some usage at microwave frequencies is the helix shown in Fig. 5-21. The helix consists of circular turns of copper conductor mounted in front of a reflector. Each turn is one wavelength long and the spacing between turns is approximately 0.25 wavelength. The turns are supported by a dielectric which gives good ratings at microwave frequencies. The reflecting screen is one wavelength square (about 3 inches by 3 inches or four gigahertz).

Several helix antennas may be combined in parallel to form a helix array. This is shown in Fig. 5-22. The reflector would be four wavelengths square if four antennas were used, six wavelengths square when using six antennas, etc. The total number of turns used to make each of the major antenna elements will determine the overall gain of this design.

The helix antenna is neither horizontally nor vertically polarized. At the same time, it will respond with fairly good efficiency to transmissions which are either horizontal or vertical in nature. The helix antenna is said to be circularly polarized. This means

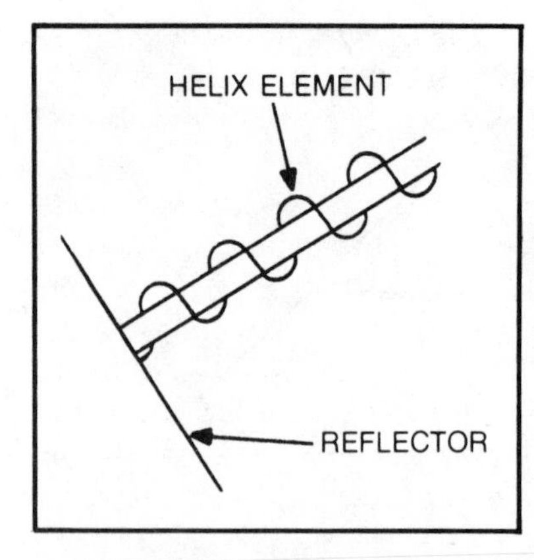

Fig. 5-21. The helix antenna is made up of circular turns of copper conductor mounted in front of a reflector.

it contains both vertical and horizontal portions. Actually, the helix antenna will be most efficient in receiving transmissions which originated from circularly polarized systems. The helix antenna has been used experimentally for uhf and vhf communications, and at one time, it was quite popular with many amateur radio operators who were attempting to build equipment to put them on the microwave frequencies.

All of the alternate antennas discussed in this chapter are presented mainly for academic purposes. Many entire books could be and have been written about the subject of quads, yagis, and

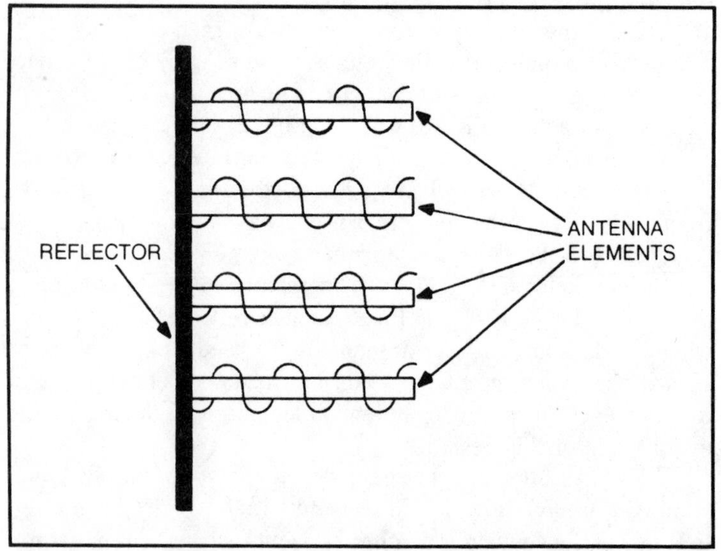

Fig. 5-22. Helix antennas can be combined in parallel to form a helix array.

helix antennas alone. Those of you who have had experience as amateur radio operators know that it's quite a simple job to put together a half-wave dipole. As the frequency increases, the physical length of the dipole decreases. But at microwave frequencies, other factors become significant and are relatively unknown for antenna designs in the vhf and hf bands. We have already mentioned the critical nature of element dimensions. It becomes very difficult to design many individual elements which have error factors of no more than 1/64th of an inch or even less. Also, different materials used for these elements will necessitate changes in the overall element length. If you use copper wire to make a quad element, for example, its length would be different than if aluminum wire were substituted. At these high frequencies, even the diameter of the wire used will affect the required physical length. Additionally, if you mount your elements on a wooden boom (the section of material which supports the elements) each will have a different length than if an aluminum boom were used. There are many, many variables which must be taken into account and a great deal of experimentation will be required.

During a special antenna test outing that was held for amateur radio operators, I witnessed one ham who had brought a very elaborate yagi array designed to operate at 1.2 gigahertz. This was not a TVRO antenna by any means, but this fellow's experience will help to properly relay the critical nature of antenna design at these frequencies. By working out the calculations, this antenna should have produced a gain of about 32 dB. When tests were run at this outing, the first results were unbelievable, so a retest was made. The same results were again obtained, so a third test was conducted. Finally, we had to accept what the tests continued to show us. Although the antenna should have produced a gain of about 32 dB at 1.2 gigahertz, it produced a gain figure of minus 1.3 dB. These gain measurements were in relationship to a single-element dipole. What the negative gain figure indicated was that this amateur operator's elaborate home-built array produced less gain than a single element design. His array consisted of over fifty of these single elements. What happened?

When the results were made known to the antenna's owner, he did some quick recalculating and realized that the element diameters were figured for a system which used wooden booms for element support. Later, he had decided to use aluminum booms and had not made the proper corrections. A hasty pruning effort after these facts were known brought the antenna close to specs.

I believe the final test showed a gain of about 26 dB, which is more in line with the theoretical figure. The amazing part is that the pieces of the antenna elements that were removed during the pruning process were just small slivers. This antenna was designed to operate at 1.2 gigahertz. Each element was about three times the length of the same type of elements which would be used for reception in the TVRO frequency spectrum. If pruning were required of a TVRO antenna's elements, it might be wiser to use sandpaper to remove fractional bits of metal than to resort to wire clippers. This example points out the highly critical nature of antenna element design.

The reflector antennas discussed earlier in this chapter have only one antenna element instead of fifty. The element is actually a flared piece of waveguide called a feed horn. It is far easier to accurately design a single element than it is to make many, all of which have to be a different physical size. This is one major reason why the dish antennas are so popular at microwave frequencies. The reader is urged to experiment with all types of antenna designs; but if you have not had a lot of antenna experience, especially in the vhf and uhf frequency areas, your move up to microwaves will most likely be a traumatic one.

SUMMARY

It is difficult to say that the antenna is the most important part of a TVRO Earth station. Certainly, without it, no signal will be received at the television set. But this can also be said of the LNA, satellite receiver, modulator, and even the lowly coaxial cables. Each and every part of an Earth station is a necessity, as is each component's efficiency. The largest antenna available coupled with the finest commercial LNAs and receivers will fail miserably if a short length of coaxial cable is defective. TVRO enthusiasts must look at their systems as a whole in order to achieve good results.

The choice of an antenna should be made after considerable research into your personal needs, future wants and the site availability. If you're like most of us, cost will be a prime determining factor as well. You can spend as much as $50,000 or $60,000 for a good commercial antenna (and much more than this without trying too hard), while others will cost less than $1,000. A computer printout of losses in your area will help determine how much gain you need from your antenna; and from that point on, the choice is yours.

If you are interested in building your own antenna, I recommend that you obtain the assembly manual for a commercial kit, of which there are many offered. This will give you a thorough understanding of all that is involved. The manual will probably cost a few dollars, but most manufacturers are happy to supply them, because after all the facts are considered, it is usually about as cheap to purchase the kit than to buy individual components. Perhaps the builder would like to supply some of the parts which are readily available and order the specialty items from the manufacturer. Some companies will probably offer partial kits (mounting kits, azimuth-elevation control kits, feed horn/LNA mount, etc.) which will allow the builder to construct what he can and then complete the project with a few additional kit items. New industries are entering the TVRO field every month, so it will pay to shop around.

Chapter 6

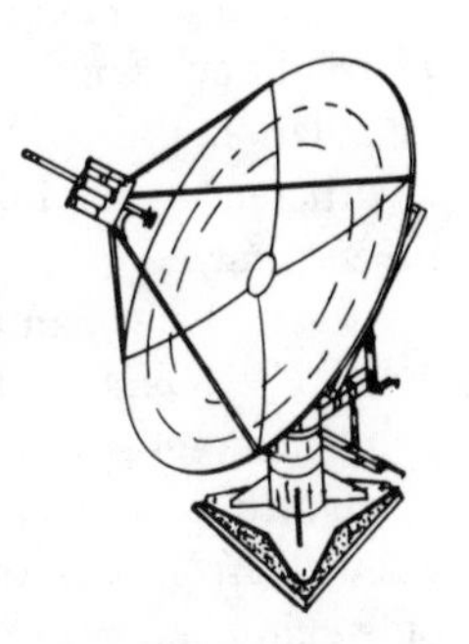

Commercially Available Antennas

THE MOST DISTINGUISHING FEATURE OF A SATELLITE EARTH station is the dish antenna, which at one time drew wondering stares from passers-by. Today, however, these antennas have become a common part of the landscape and are not so much the status symbol they once were. The space-age appearance of these devices belies the fact that they are rather simple structures from a purely electronic standpoint. On the other hand, they are highly complex mechanically and their design is quite crucial. A deviation in dish curvature of only a fraction of an inch can mean the difference between good reception and poor or nonexistent pictures at the receiving end.

While many TVRO Earth station owners will elect to purchase and install packaged systems that include the antenna, LNA/down converter, receiver, etc., others will elect to build a customized system by choosing an antenna from one manufacturer, a receiver from another, and proceed in this manner until the project is complete. This chapter will overview the TVRO antenna offerings. Many of the products discussed are available separately or as part of a packaged system.

The antennas used to receive signals from satellites have encountered no drastic operational changes since the first edition of this book was written. Rather, many of the changes have occurred in the physical structure of the antenna proper. These changes are aimed at decreasing overall weight, wind resistance, and in making for an easier installation. Today, it is common to see parabolic dish antennas made from a wire mesh and supported by a skeletal structure, whereas a few years ago, these were rare or nonexistent. The solid fiberglass dish is still the most common type of antenna seen, but this solid construction naturally increases the weight, and more importantly, the wind resistance. This places more rigid requirements on the supporting and mounting structures, but these antennas are less expensive than their mesh counterparts. If weight and wind resistance is not a critical factor in a particular installation, the solid antennas make very good buys. They are just as efficient as the mesh antennas, assuming both are of the same physical size.

Due to new antenna construction techniques that use wire mesh, TVRO Earth station antennas are now economically available in much larger sizes. It is not uncommon today to see a 12- or 14-foot mesh dish mounted atop a roof, something that was extremely rare a few years ago. In the early days of Earth station technology, the only suitable place to set a parabolic dish was on the ground unless you wanted to resort to spending thousands of dollars on specialized mounting.

Another major change in satellite antennas does not involve the antenna itself, but the method by which it is aimed at a satellite. At one time, most personal Earth stations were fixed to a single satellite and left in place 90% of the time. Redirecting the antenna was done manually or by means of a series of manually-operated levers attached to the antenna mount. Today, a number of companies offer motorized computer-controlled devices called trackers that can redirect the antenna to a large number of different satellites simply be pressing a button at the television receiver. One such device, the Tracker III Plus, shown in Fig. 6-1, contains its own on-board microprocessor that allows the user to preprogram the settings for 72 different satellites. Additionally, there are no wires connecting the programming box with the motorized antenna adjuster. This is all done by uhf wireless remote control. The Tracker III Plus from Houston Tracker Systems in Houston, Texas boasts a low-profile design and involatile memory. This means that once the satellite locations are programmed, they will not be lost when the unit is disconnected. It has a range of approximately 100 feet, which should be adequate for most personal Earth station installations.

Another tracking unit is the MTI 4100 from PenTec Enterprises, Inc., in Salt Lake City, Utah. Shown in Fig. 6-2, this device incorporates the latest in microcomputer technology and offers

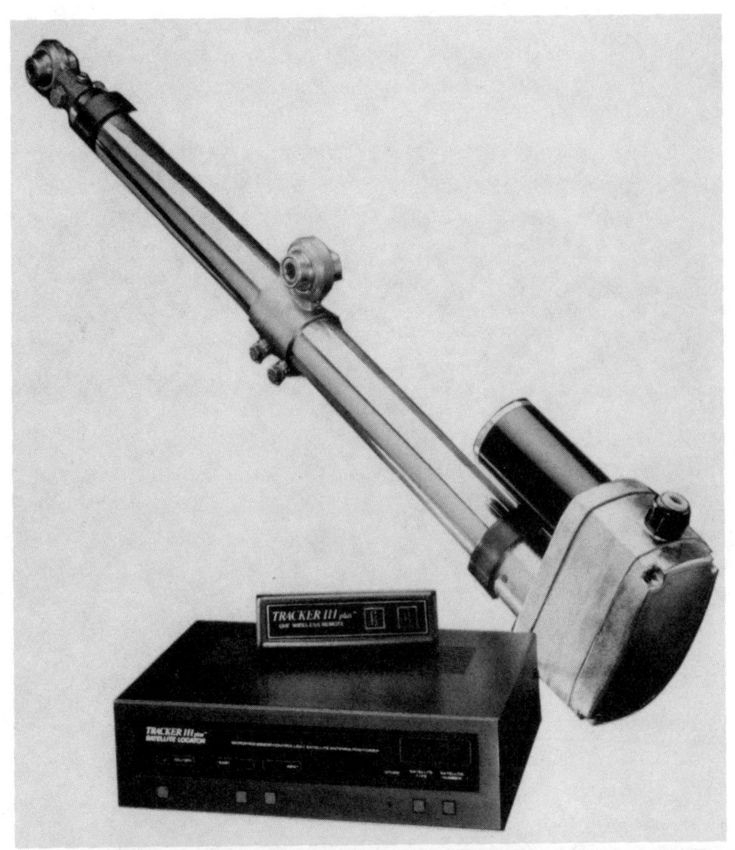

Fig. 6-1. The Tracker III Plus Antenna Positioner allows 72 different satellite positions to be preprogrammed.

single-chip operation. The system includes a battery-free memory, infra-red remote capabilities, display, and motor drive. It uses infra-red command reception and decoding and offers position sensing, polarotor pulse generation, display, and motor control in the same microchip. As is the trend these days, nonvolatile memory is used rather than a battery backup to hold the preprogrammed satellite location. Two mechanical actuators are offered with this unit. One is designed for a rated load of 500 pounds, while the other is a super heavy-duty model rated at 1500 pounds. Mechanical limit switches simplify installation and eliminate over-travel. Also, a power switch is built in that prevents unintentional and unintended motor activity. This unit sells in the $800 price range. This same company also offers less expensive antenna positioning units that are designated the MTI 2100 and 2800.

Fig. 6-2. The MTI 4100 antenna positioner offers single-chip operation and uses a remote control.

CHANNEL MASTER ANTENNAS

Channel Master is now offering an antenna line that uses fiberglass in a new construction process for improved performance along with a faster setup time. It also offers a more aesthetic appearance. Shown in Fig. 6-3, this model uses the new sheet-molding-composition (SMC) fiberglass manufacturing method. Using the SMC method, better uniformity and accuracy of the parabolic shape are obtained. Also, the shape remains consistent regardless of weather conditions. This was not always true of earlier fiberglass dishes, which tended to warp out of shape as temperatures changed significantly. Conventional fiberglass-clad metal dishes, the industry's standard antenna material, expand and contract according to temperature. This changes the unit's parabolic shape, thus affecting reception. Standard dishes also have more problems with ice build-up in the winter and reflected heat in the summer.

Channel Master's SMC design and more uniform shape translates into significant performance improvements. For example, with a 10-foot antenna, there is a 25 percent increase in signal gain over .3 F/D hand lay-up models and a 10 percent gain increase over .4 F/D. This, along with 3 dB more rejection in side lobe patterns and reduced noise temperatures, means these dishes have the additional margin necessary for good reception from satellites spaced 2° apart.

Channel Master calls this antenna their Ground Star model, and it comes with a written guarantee stating that it is certified to perform without interference from adjacent satellites within the degree spacing just mentioned. This conforms to the new FCC decision regarding certification of 8-foot systems. What this means to the present owner of a Ground Star is that when 2° spacing becomes a reality, this antenna will not become obsolete. Adjacent

Fig. 6-3. The Channel Master Sheet-Molding composition antenna. Courtesy of Channel Master.

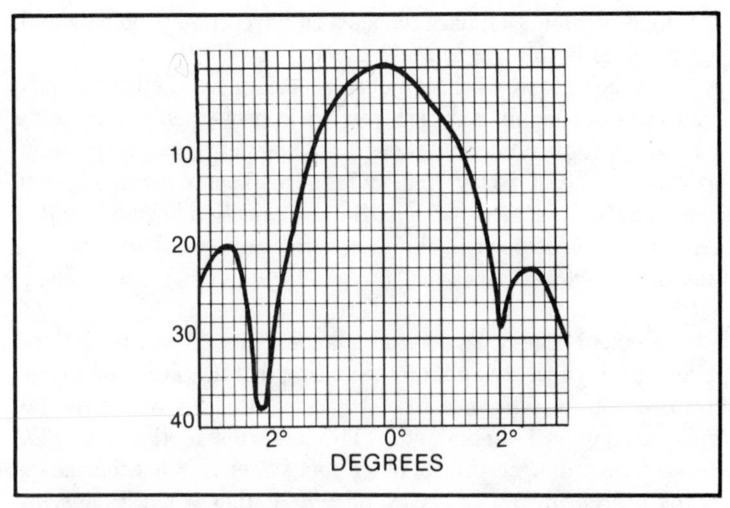

Fig. 6-4. Response curve of the Channel Master Ground Star antenna. Courtesy of Channel Master.

satellite interference is becoming more and more critical today, because all satellites that are received by today's Earth stations transmit on the same frequency, making it impossible for a satellite receiver to distinguish one signal from another. It's up to the antenna to capture the maximum signal from the desired satellite while rejecting all other transmissions. Figure 6-4 shows a typical Ground Star response curve. Notice how strong the antenna's response is to the signal in the center and how it falls off at 2° on either side. Again, this high efficiency is possible due to the extremely accurate surface and shape of these dishes, which use SMC construction. As a matter of fact, the Ground Star dishes conform to military and broadcast telecommunications specifications.

This is a well-constructed antenna that is quite easy to assemble and install. For instance, all bolt holes are machine set and drilled to ensure a perfect alignment during assembly. The back of the dish displays the molded support ribs which are installed during production. This means that no extra bolt-on supports or oversized back frames are required. The Ground Star dishes come in four completely interchangeable sections. This means the sections do not have to be worked into place. Even the surface of the dish is a bit different. This is a textured surface which is said to be scratch- and mar-resistant. Such a surface also diffuses reflective heat from the sun. In solid white antennas, this heat can damage the feed assembly. The textured surface is also quite attractive.

Channel Master rates this antenna to withstand hurricane-force winds in excess of 100 mph.

Channel Master also offers a mesh dish antenna in response to a growing consumer preference for this relatively new type of design. Shown from several different angles in Fig. 6-5, this 10-foot, 4-petal dish is constructed from heavy-duty perforated aluminum. This four-quadrant design allows for production of dishes with surface accuracy that conventional mesh types can't match. Performance is substantially improved by closer manufacturing tolerances in the forming of the parabolic shape as well as in the assembly by the dealer. With only four petals to match, maintaining the factor-produced parabola is much easier for the installer than when assembling most mesh dishes, some with as many as sixteen sections. Because of these manufacturing techniques, Channel Master is certifying in writing that the antenna will perform without interference from adjacent satellites at the 2° orbital spacing.

This is the first departure from fiberglass design on the part of Channel Master, and it would appear that their reasons for offering it were primarily consumer-oriented. It is true that today, mesh antennas have become somewhat of a fad, and if that's what people want, then that's what companies must offer. Channel Master apparently took a long hard look at the market and certainly must have done a great deal of research to see if they could develop a mesh antenna that would closely match the performance of their SMC fiberglass model. The Vice President of Channel Master still maintains that their Ground Star dish is the industry's best performing antenna, so this apparently means that this mesh antenna runs a close second. In advertisements promoting their perforated aluminum antenna, Channel Master states, "This satellite antenna is designed to meet the requirements of those who want the see-through properties of a mesh screen and who need the high-performance capabilities of the SMC-type fiberglass dish." I interpret this to mean that this mesh antenna is touted to perform better than a lot of other mesh antenna designs, but not quite on a par with their SMC fiberglass dish.

Channel Master resisted the temptation to rush into a mesh design and felt that a high-performance parabolic antenna could not be designed to their specifications using a conventional mesh material. Finally, Channel Master elected to use .125 perforated aluminum, whose strength allows for all the good features of a see-through antenna without sacrificing important gain and side-lobe characteristics essential for clear pictures.

Fig. 6-5. The Channel Master mesh dish antenna design. Courtesy of Channel Master.

Performance is also aided by the use of a four-leg feed support which assures correct positioning of the feed horn in relation to the dish. Channel Master recommends this dish for customers who want a less noticeable appearance. The black color absorbs light rather than reflecting it, as is the case with other colors. Black mesh seems to blend better with most environments.

Like the SMC dish, the Channel Master mesh version is certified for 2° spacing. This is accomplished by adhering to closer manufacturing tolerances in forming the parabolic shape. With only four petals to match, maintaining the factory-produced parabola is much easier than when assembling most mesh dishes.

Channel Master has recently introduced two new motor drives to position TVRO Earth Station antennas. As these join the current line of antenna positioners from Channel Master, the company has stressed that the new models are built for simplicity of operation.

The Model 6253, shown in Fig. 6-6, is a basic unit that features easy-to-use pushbutton controls for simple up/down operation. The second unit is basically the same as the first, but has an infrared, hand-held remote control. This is the Model 6252. Operation of either unit by the consumer is learned quickly. The dish is moved by pushing one of two buttons. The consumer compares the LED readout with the satellite locations recorded during installation. The latter are contained on a permanent satellite locator card kept near the remote control unit. Both units use low, 36-Vdc operation, feature nonvolatile memory, heavy-duty ball screw linear actuator drives, and are protected by a weatherproof expandable jack bellows and motor boot. The basic unit retails for $395; the Model 6252, which is equipped with the remote control option, sells for $495.

Both units are intended for discrete operation as opposed to some versions that are designed to serve as a unit in a central-controlled console that gives access to all TVRO Earth station functions, such as receiver tuning, polarization, and other attributes in addition to antenna positioning. All of these functions are handled from one unit. These discrete controllers are often installed on systems that were originally equipped with manual antenna positioners. These controllers, then, find many uses in conversion of current TVRO Earth Stations.

PARACLIPSE ANTENNAS

A very popular and quite striking antenna that is often seen

Fig. 6-6. The Channel Master model 6253 TVRO antenna positioner. Courtesy of Channel Master.

as a major part of TVRO Earth station designs today is called the Paraclipse, made by Paradigm Manufacturing, Inc. Several models are available and choice depends on signal strength in a given area. The 9-foot model is ideal for open areas where satellite signals are quite strong. In other parts of the country where signals are somewhat weaker or in an installation where some signal shadowing may take place, the 12-foot dish is recommended.

The Paraclipse antennas are quite futuristic in appearance. They remind one of a government radar installation or even a radio astronomy observatory. Shown in Fig. 6-7, the Paraclipse antenna

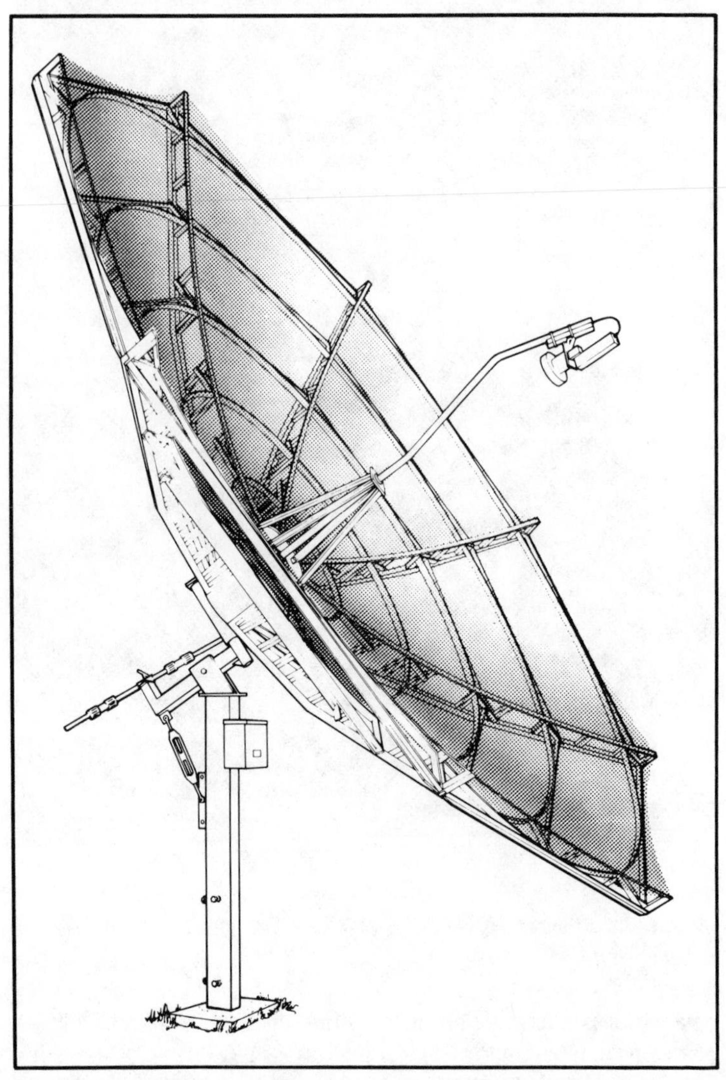

Fig. 6-7. The Paraclipse antenna is a mesh design with a very distinct appearance. Courtesy of Paradigm Inc.

line uses a rib and ring truss system that supports an aluminum mesh. The antenna is even available in "designer" colors. The supporting system is made of aluminum alloy for lightweight strength and corrosion resistance. This extremely rigid framework shapes the heavy expanded aluminum mesh reflector into a true parabolic dish. Concentric rings assure true parabolic shape in two directions.

The standard mount is a single-post polar type as shown. This mount has been around for a long time; it is simple, strong, and offers adjustable declination. It is also very easy to motorize. The 9-foot model offers a gain of 39.3 dB and it has a 2.1° beam width and a 65 percent aperture efficiency. Carrier-to-noise ratio is 10.1 dB. The 12-foot model offers a 42.3 dB gain. Figure 6-7 shows this lightweight mesh dish mounted atop a residential roof. More and more, mesh antennas are filling the needs of individuals who have no suitable mounting location on the ground and must resort to a roof or tower mount.

This is not to say that solid fiberglass antennas cannot be successfully mounted in the same manner. The problem is not so much with the weight of the antenna as it is with the wind resistance. Many mesh antennas weigh just as much as their fiberglass counterparts (some even more), but the wind resistance of a mesh antenna is far less than any comparable solid fiberglass antenna.

ASSEMBLING YOUR ANTENNA

Most commercial antenna kits require quite a bit of assembly before they can be utilized. The following section describes an assembly the author did several years ago using an antenna kit from Wilson Microwave Systems. Perhaps this recount will give you a good idea as to whether or not you will wish to make your own installation as opposed to hiring an installer to put up the antenna proper, allowing you the opportunity to attach your own receiving components. Again, this describes the installation of a specific antenna. Other TVRO antennas may require slightly different procedures, but this discussion will give you a general idea.

Construction involves the selection of a mounting site for the antenna that will permit it to be laid flat on the ground during assembly. It must then be elevated to the operating position with wide clearance from all obstacles. This advice particularly includes power lines, telephone lines, and their associated poles, towers, guy lines, and related structures. The antenna must be firmly anchored to the base, and the exact requirements will depend upon your local soil and weather conditions and any other particulars at the site selected. Wilson strongly recommends the obtaining of advice from a local engineer or contractor before starting any work. While these individuals are probably not directly involved with TVRO antenna installations, they will be able to supply you with some sound advice on engineering principles that should be easily understood. In some instances, a surveyor might be needed to provide the proper magnetic headings required to obtain optimum satellite acquisition.

Wilson states that one of the best anchors for the base plate is concrete, with the anchor bolts cast in place. If an alternate method is chosen, you must be certain that it is equally strong and effective. This is most important because in an 8 mph wind, a 100-square foot antenna would be subjected to a force of approximately 2,500 pounds. This amount of force can easily upset the antenna should it not be properly mounted. The anchor must be able to withstand this or any greater force that may reasonably be expected to develop. You can probably obtain a record of normal and unusual wind conditions for your particular area from a local radio station or weather recording service.

If the soil in your area is considered to be within normal limits, a reinforced concrete base four feet deep with the anchor bolts cast in the concrete should be adequate. Normal soil, stated by E.I.A. Standard RS-22-C, is "cohesive type soil with an allowable net vertical bearing capacity of 4000 pounds per square foot and an allowable net horizontal pressure of 400 pounds per square foot per lineal foot of depth to a maximum of 4000 pounds per square foot. Rock, noncohesive soils, or saturated or submerged soils are not to be considered as normal." If your soil does not meet these requirements, you may need a differently constructed base.

In order to construct a base in soil that is considered to be within the above stated limits, four holes are dug four feet deep using a conventional post hole digger. Wood forms measuring 2″ by 4″ are then secured at the tip, and the holes are interconnected with an 8″ deep trench, as shown in Fig. 6-8. Antenna orientation will be discussed later in this chapter, but it is important that this is done properly with regard to the fixed base in order to allow the sweep of the rotating base to acquire all of the satellites of interest.

Preassemble the four anchor bolts to a wooden template with a nut on each side. Allow about 1/2 inch of the bolts to protrude above the upper nuts, as shown in Fig. 6-8. The template is centered on the form and temporarily nailed into place. The concrete can then be carefully poured into the hole, making sure the reinforc-

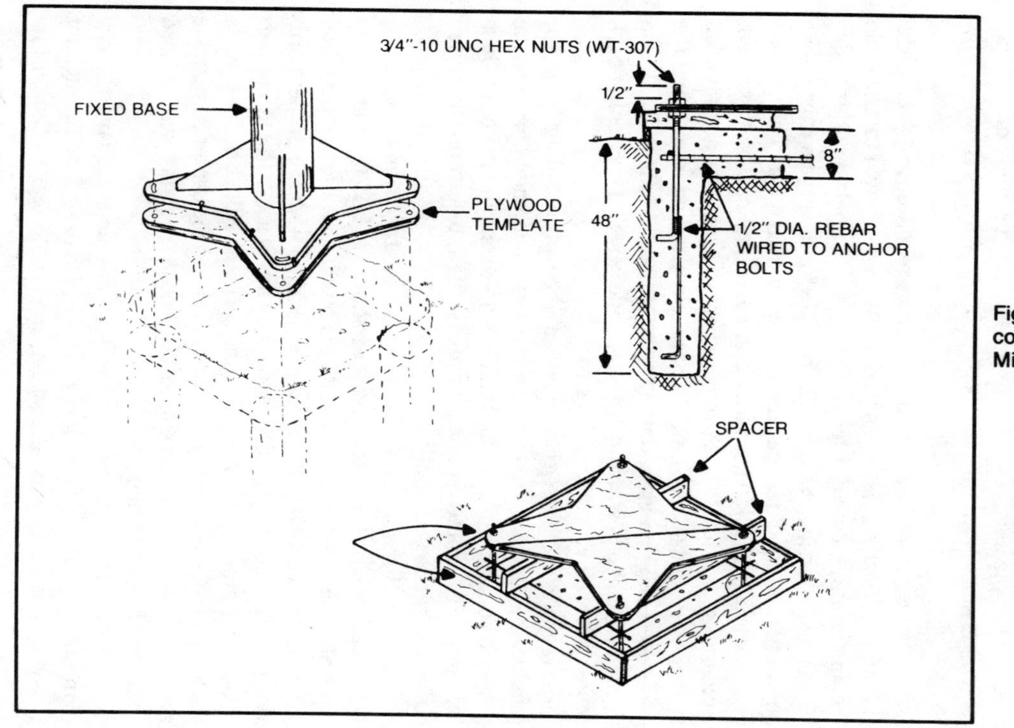

Fig. 6-8. Detailed instructions for the concrete base. Courtesy of Wilson Microwave Systems, Inc.

ing framework is not disturbed in any way. Once the concrete has been allowed to set the proper amount of time (until it is still workable), the top nuts and the template are removed, making sure that the anchor bolts are not disturbed. It's a good idea to leave the form in place for a seven-day period, since the edges of the concrete will be especially prone to damage during this period of time. Allow the concrete to cure for at least seven days. At this point, the form may be removed by tapping lightly with a hammer to break the form loose. The base can then be installed. This is secured with four additional nuts on top. Using a carpenter's level, make sure the base is straight and true, making any necessary adjustments by means of the nuts. These can be tightened securely once it has been determined that the base is completely level, as shown in Fig. 6-9.

After the base has been securely fastened down, attach the antenna support frame, as shown in Fig. 6-10. Begin to assemble the antenna by placing three of the quarter panels on a level surface and supporting the centers about 19″ above the surface. The bolts and nuts should be installed loosely at this point. The remaining panel is then added, and all bolts and nuts can be tightened. Note in Fig. 6-11 that the third bolt from the center is not installed at this stage of the assembly. These are installed later when the completed antenna is mounted to the support frame. Also, begin tightening all bolts starting at the outside row and tightening each row sequentially, working inward.

Figure 6-12 shows the proper method of installing the rotor mounting plate to its four support struts. Also shown here is the proper installation of the rotor and low noise amplifier (LNA). Once this step is completed, the assembly is lifted, placed on the antenna, and attached securely.

To install the antenna, lower the mounting frame to a near-horizontal position and support it as shown in Fig. 6-12, or in any equivalent manner. Place the antenna on the four upright ears, and when all holes are aligned, it may be secured with the proper hardware. At this point, all bolts should be snugly secured but not permanently tightened, as it will be necessary to adjust the antenna's contour. First, the antenna support struts and the center turnbuckle are installed, as shown in Fig. 6-13. These elements will be used to adjust the contour.

The secret to obtaining a perfect contour is to have the circumference of the dish perfect to within 1/8 of an inch throughout the whole dish. This is almost an impossible feat in the normal

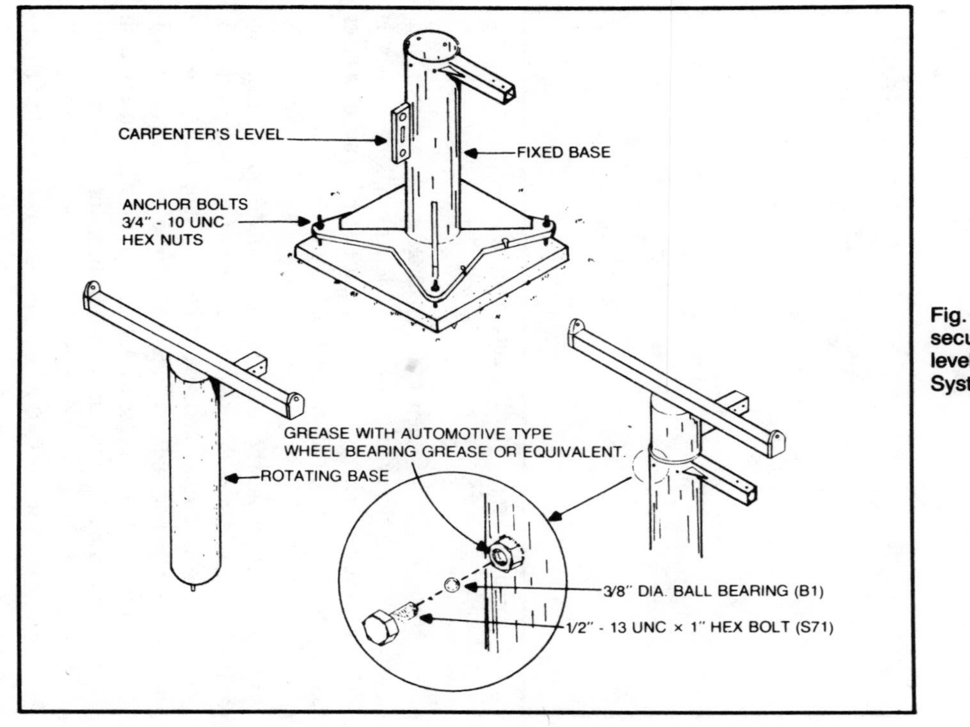

Fig. 6-9. The base is fastened securely once it is measured for level. Courtesy of Wilson Microwave Systems, Inc.

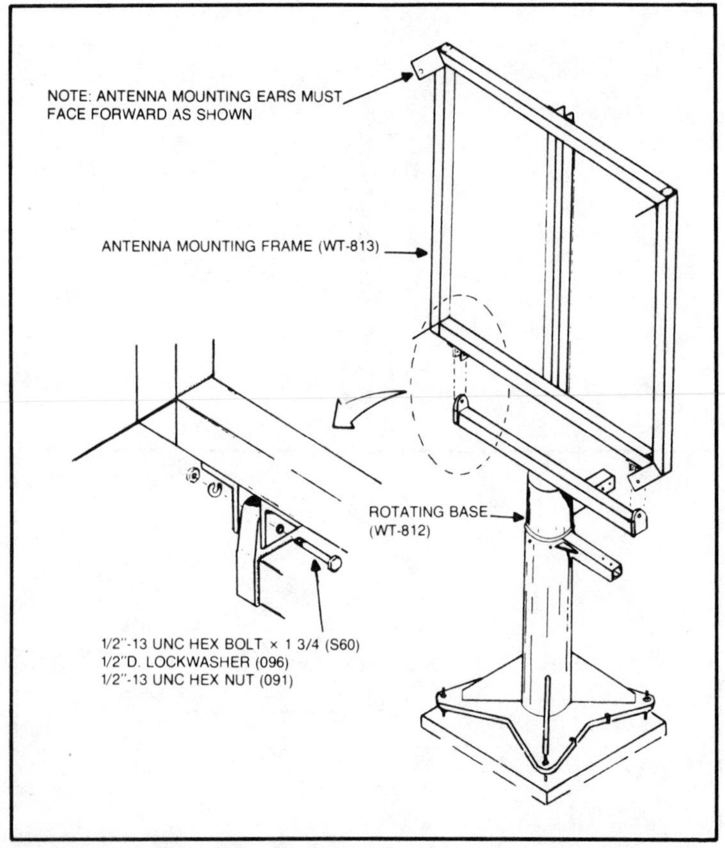

Fig. 6-10. The antenna support frame is attached to the base. Courtesy of Wilson Microwave Systems, Inc.

manufacturing process when four separate panels are involved. However, Wilson has overcome this problem by making the panels slightly smaller and making up the difference with the aid of washer spacers, thus allowing the installer to vary the circumference as necessary.

In order to adjust the antenna's contour to obtain a perfect parabolic shape, three different adjusting systems are utilized: spacers, support struts, and the center turnbuckle. If the circumference is proper, the turnbuckle should be in the neutral position (no force applying pressure either up or down) on the dish center. The turnbuckle may, however, be used for fine adjustment of the contour.

The first step in contour adjustment involves making a contour template which must be perfectly accurate. Wilson includes with their antenna complete instructions on how to make this template, and even states that the large box that the fiberglass quarter panels arrived in will suit quite adequately. The reason that the template must be absolutely accurate is that even an error of 1/4 of an inch will cause a considerable loss of gain. Once you are satisfied that the template is perfectly accurate, and with the antenna installed on the antenna support frame in the horizontal position, the template is placed in the dish, as shown in Fig. 6-15. The contour is checked across each panel seam and at 45° from the seams, as shown in Fig. 6-16.

Figures 6-17, 6-18, and 6-19 show some of the variations that may be noted when checking the contour of dish. Figure 6-17 provides the correct contour. If this is obtained with the template placed across both seams of the dish, no further adjustments are necessary. However, if a perfect contour is obtained in one position with the

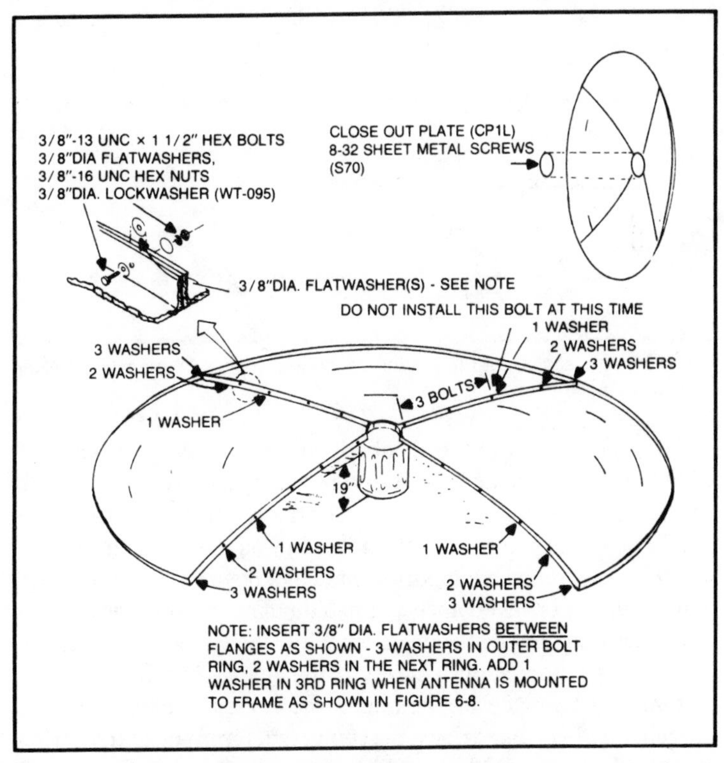

Fig. 6-11. All bolts are tightened sequentially working from the outside of each section. Courtesy of Wilson Microwave Systems, Inc.

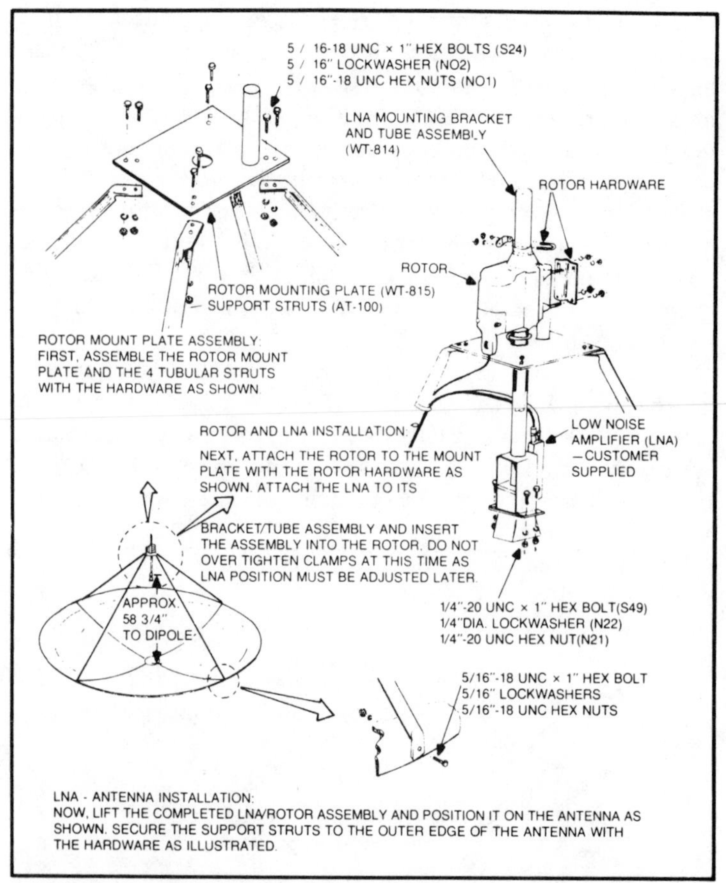

Fig. 6-12. Installation of the rotor mounting plate. Courtesy of Wilson Microwave Systems, Inc.

contour pictured in Fig. 6-18 obtained across the other seams, it will be necessary to adjust the antenna support struts outward. It should now be apparent why none of the securing hardware was tightly bolted in place when the antenna was mounted on its base.

Once the antenna support struts have been adjusted, the contour should be checked once again. You may find that the contour in one direction has been corrected, while the opposite contour has been changed to a slight degree. If the change is minor, the correction may be made by tightening the center turnbuckle to add tension, thereby pulling down the center and returning to the perfect contour. If, however, the change is somewhat pronounced, it will be necessary to remove one washer from each of the first and sec-

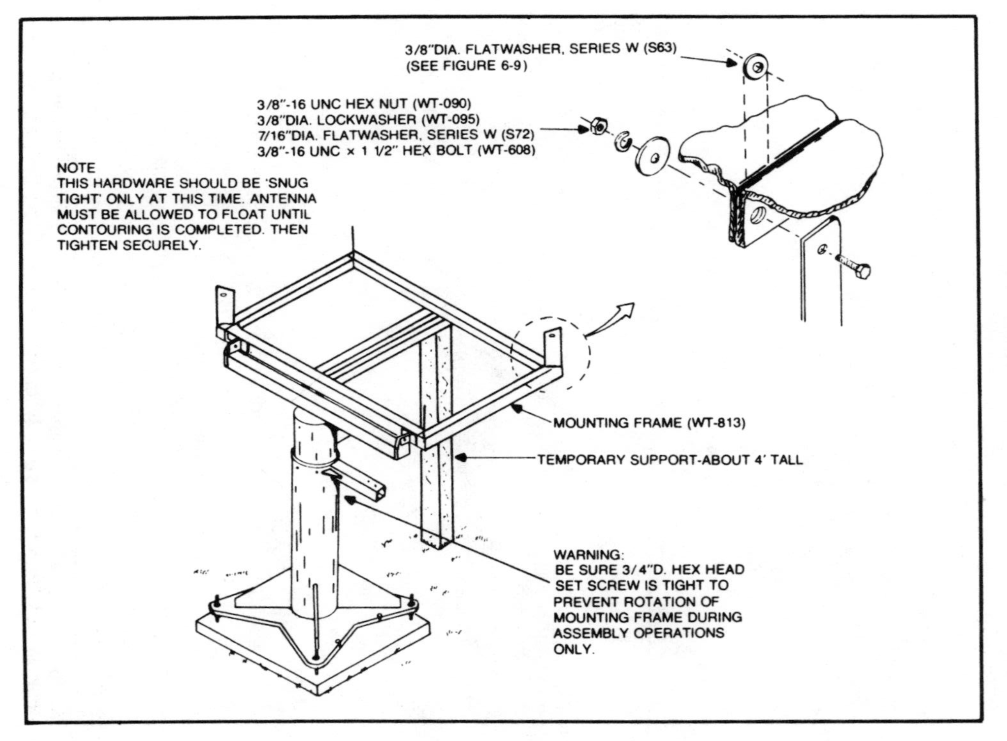

Fig. 6-13. The antenna is installed by lowering the frame as shown here. Courtesy of Wilson Microwave Systems, Inc.

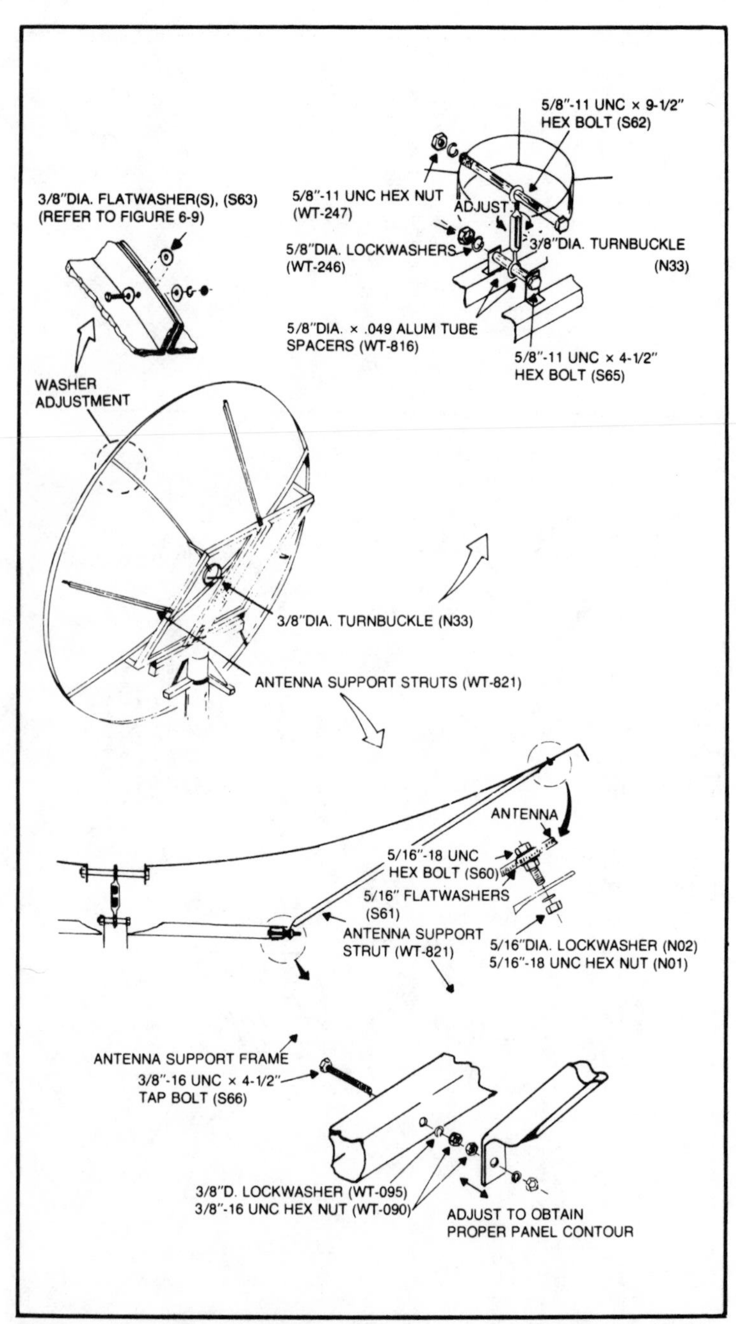

Fig. 6-14. Installation of the antenna support struts and center turnbuckle. Courtesy of Wilson Microwave Systems, Inc.

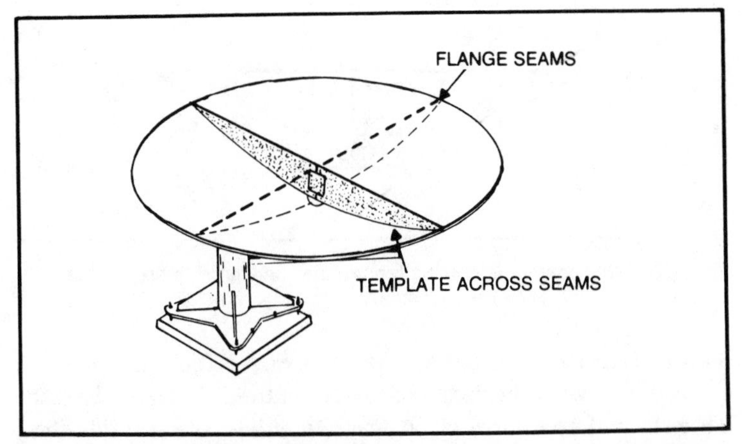

Fig. 6-15. The perfectly accurate template is placed in the dish. Courtesy of Wilson Microwave Systems, Inc.

ond holes at all four seams. Refer back to Fig. 6-14 for placement of this hardware.

The contour variation in Fig. 6-18 indicates that the circumference is too large. It is now necessary to make the circumference smaller by removing one washer at the first and second holes at all four points. It is important to note here that this particular contour will have the highest amount of loss, since errors at the tips of the dish cause more gain loss than errors at the center of the dish.

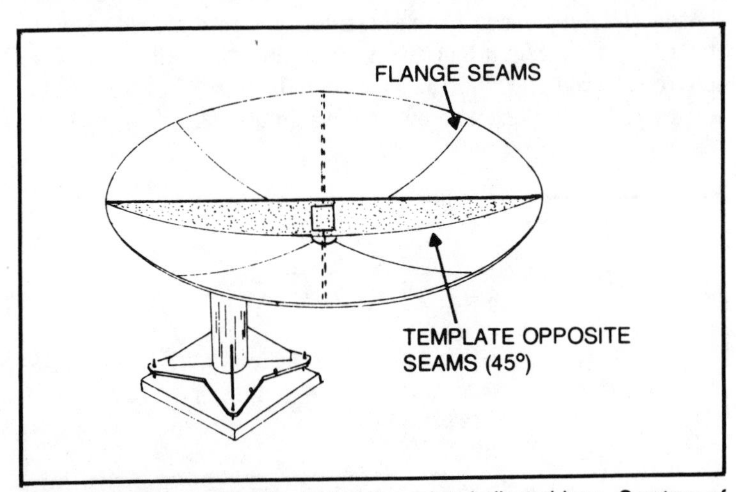

Fig. 6-16. The contour is checked at the points indicated here. Courtesy of Wilson Microwave Systems, Inc.

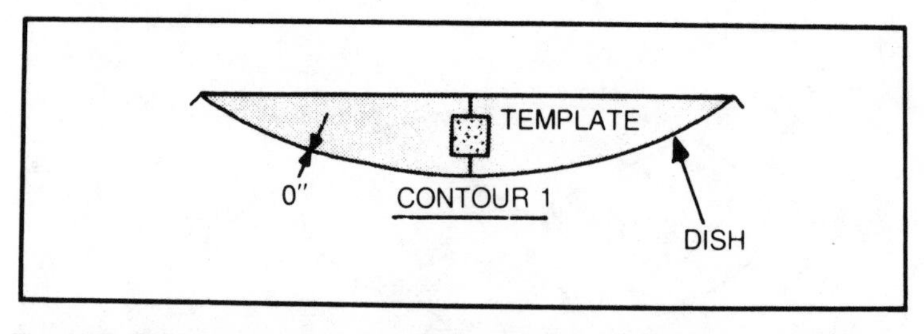

Fig. 6-17. If this contour is obtained, no further adjustments are necessary. Courtesy of Wilson Microwave Systems, Inc.

Figure 6-19 shows yet another type of variation that may be encountered when checking the contour of the dish. Here, the dish has a circumference that is too small. In order to correct this condition, it is necessary to add one washer at the first and second holes at all four points.

Once the contour has been adjusted properly, cover the flange seams with the vinyl tape and the antenna can be raised sufficiently to install the elevation and azimuth actuators. This is shown in Fig. 6-20. A temporary support should be used for this purpose. The first step is to install the elevation actuator support arms and spacer and the azimuth actuator attach brackets with the proper hardware. The next nuts are not tightened until the studs on the elevation actuator have been inserted into their mounting holes. The 1/2" diameter holes in the antenna support frame angles that best suit your particular satellite elevation requirements should be chosen. The upper end of the actuator is then attached. The installation of the azimuth actuator will follow this same basic procedure. The rod-end is secured between the attach brackets. Also, be sure to loosen the set screw on the azimuth before any azimuth rotation is attempted.

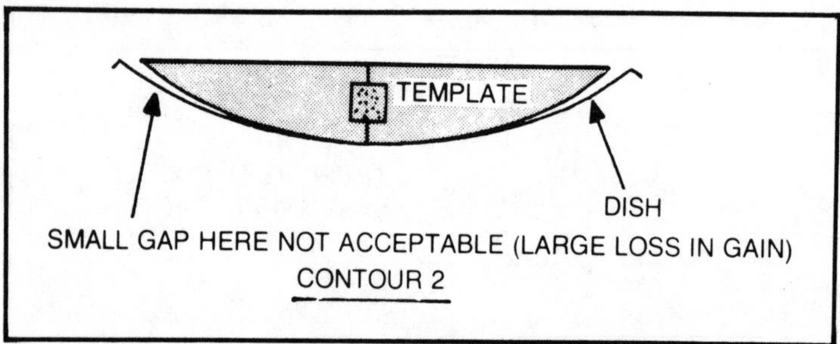

Fig. 6-18. If the contour appears like this, the circumference is too large. Courtesy of Wilson Microwave Systems, Inc.

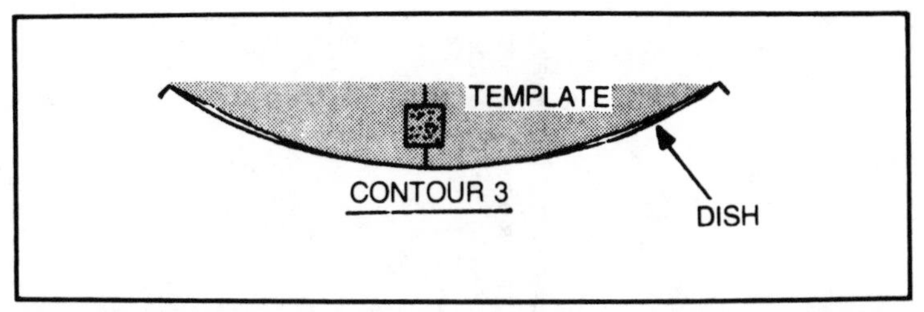

Fig. 6-19. If this variation occurs, the circumference is too small. Courtesy of Wilson Microwave Systems, Inc.

Figure 6-21 shows the proper method for installing the azimuth and elevation scales, as well as a caution label. Satellite azimuth and elevation logging scales are provided so that once a particular satellite's position has been obtained, its location can be recorded. This provides a method of easy access for relocating the satellite at a later date.

To install the elevation scale, peel off the protective backing from the label and press it into the elevation hinge ear. The pointer label is placed directly above it on the hinge angle with the pointer on or near the scale. This is accomplished by folding the lower edge around the edge of the hinge angle. Now, locate the azimuth scale on the top edge of the fixed-base tube with the "0-degrees" mark under the pointer and the azimuth actuator in its compressed position.

The coax and rotor cable are now installed following proper instructions as provided. After all connections to the receiver and the television receiver have been completed, orient the antenna to the satellite you have chosen and adjust the LNA by sliding it up and down until a clear, sharp picture is obtained. At this point, the installation is complete and all clamps can be tightened securely. Figure 6-22 illustrates what the completed antenna will look like once it has been installed and wired to receive satellite transmissions.

Admittedly, this was a project that was assembled a few years ago, and in recent years, TVRO Earth station antennas have become simpler to install and align. I have found it desirable to enlist the aid of two or three helpers in making such an installation. While this can be done with only two persons, the extra persons can cause the project to move more rapidly, and brainstorming the difficult portions is handled a bit more efficiently. Assembling the antenna can be a good family project, especially when a husky teenager or two is available.

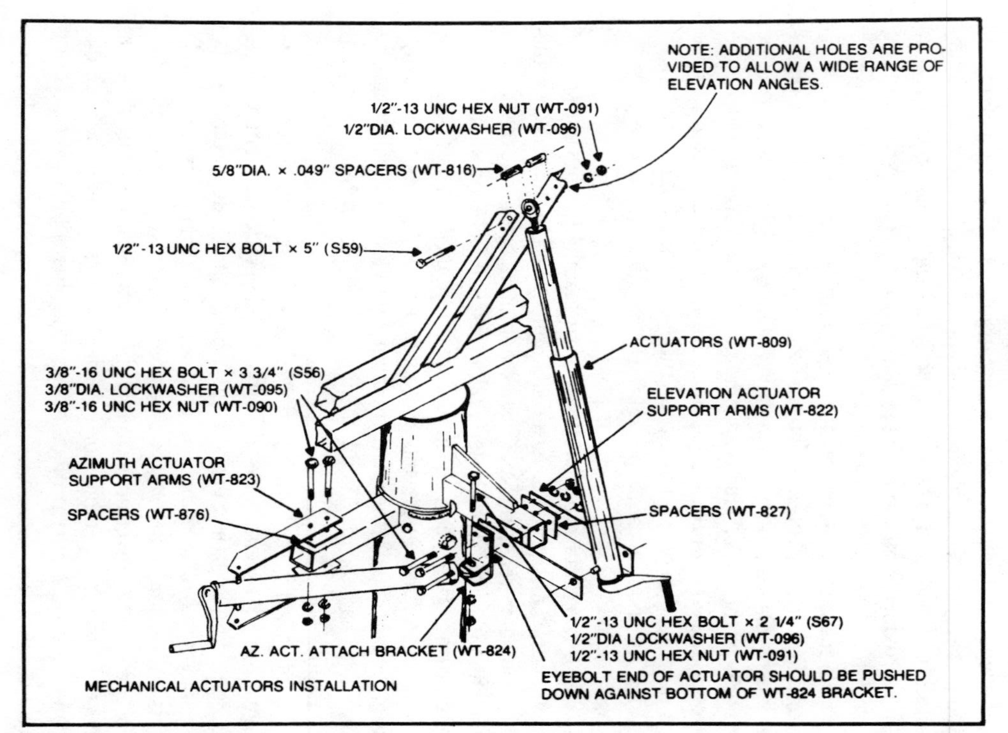

Fig. 6-20. Installation of the elevation and azimuth actuators. Courtesy of Wilson Microwave Systems, Inc.

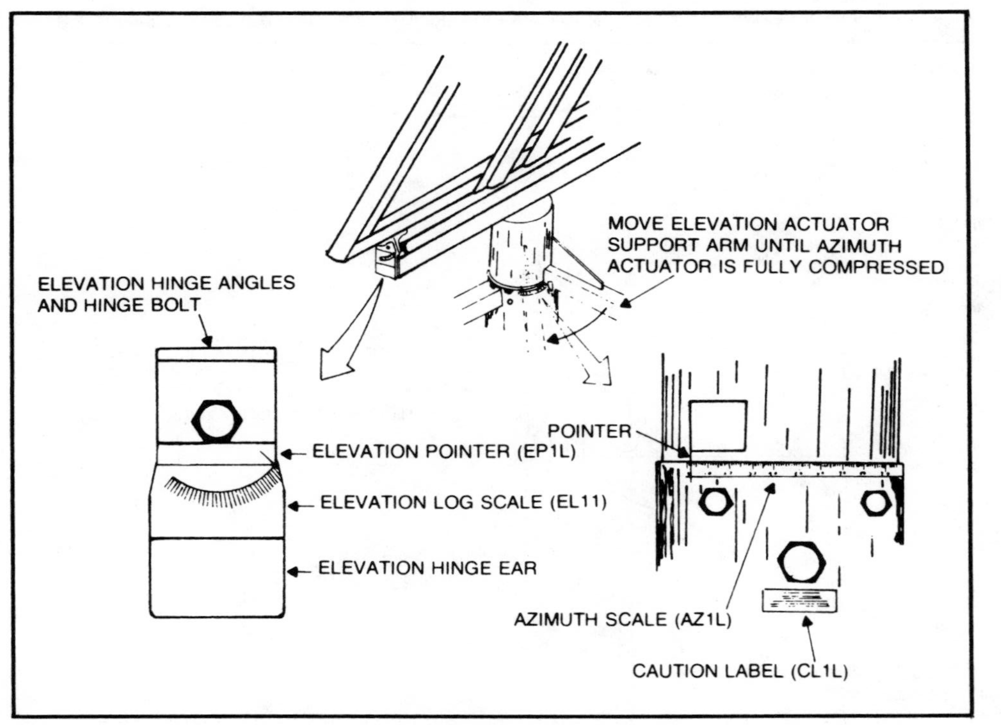

Fig. 6-21. Installation of the azimuth and elevation scales. Courtesy to Wilson Microwave Systems, Inc.

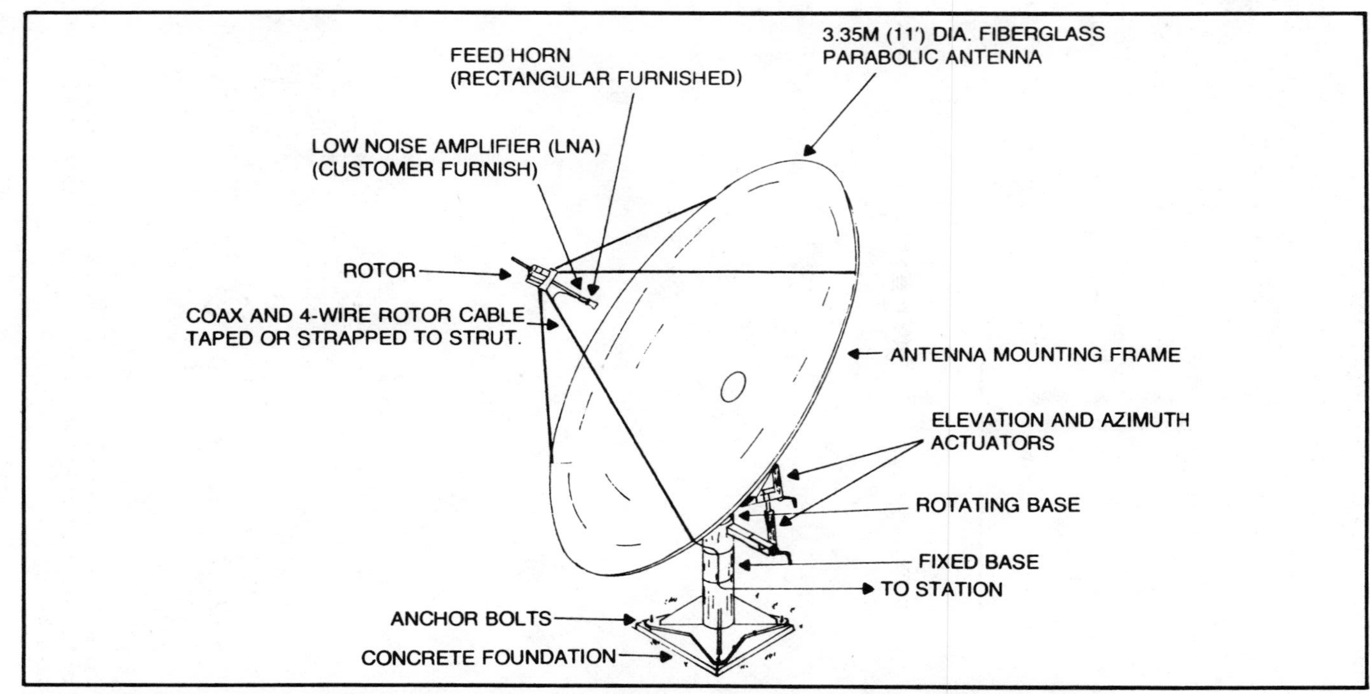

Fig. 6-22. This is what the completed Wilson antenna will look like. Courtesy of Wilson Microwave Systems, Inc.

ANTENNA ACTUATOR INSTALLATION

Many persons installed TVRO Earth stations a few years ago that included antennas of the manual-adjust type. Recently, however, motorized antenna positioners have become quite popular, and the price of these remote control units has dropped considerably. Many different models can be had for as little as $200. For those TVRO Earth station owners who must access more than one satellite on a regular basis, these control systems save a lot of time and the bother of trekking through deep winter snow to manually adjust the antenna.

Due to the popularity of satellite antenna control systems, many Earth station owners are electing to install these devices on already existing systems. In many instances, the installation is done by the owner, and it's not a terribly difficult job, although it's not ridiculously easy either. This section will describe the installation and alignment of the Gillaspie Model GCI 8200 Antenna Control System. The complete arrangement is shown in Fig. 6-23. As is customary, this is a two-part system that includes the motorized

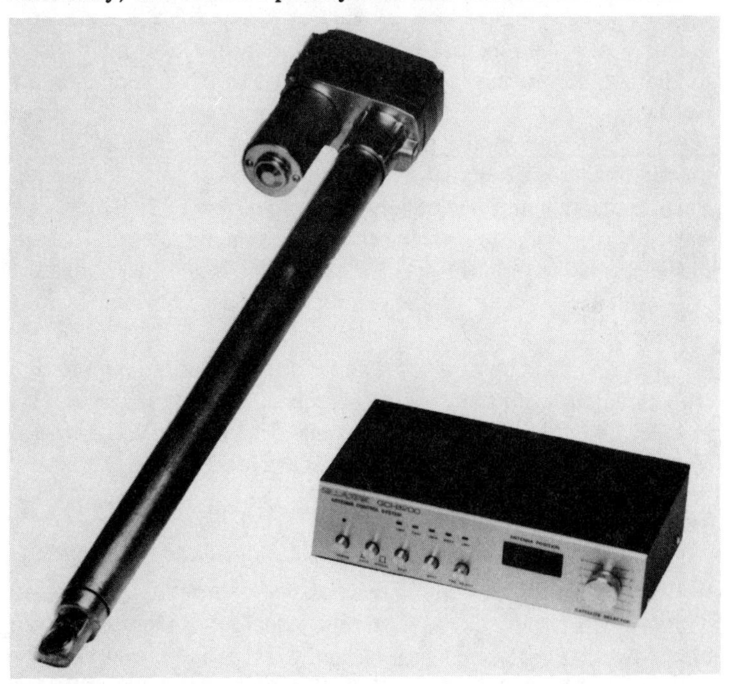

Fig. 6-23. The Gillaspie GCI 8200 Antenna Control System showing the motorized actuator and the remote control console. Courtesy of Gillaspie Communications, Inc.

actuator (left) and its remote control console (right). Figure 6-24 shows a close-up view of the remote control stacked with two other Gillaspie products in matching cases. This last picture might show a typical configuration of a complete TVRO Earth station control system. These three units would be mounted at or near the television receiver.

The Gillaspie GCI 8200 is an antenna actuator system that is ruggedly designed and should provide many years of satisfactory service. The system features twelve selectable satellite positions, heavy-duty gear-driven actuator motor drive, selectable safety limits, direction indicators, and large, easy-to-read digital position indicators. The control console is installed at or near the satellite receiver. It consists of the controls to operate the system and provides the power to operate the motor drive.

The second portion of the system is the actuator motor drive, which mounts on the TVRO antenna and, in response to the control console, provides force to move the antenna. The motor drive is a ball screw type that is powered by 36 Vdc (standard for TVRO actuators). The motor can provide an 18-inch stroke, and the mechanical system is protected by a clutch that prevents damage to the actuator when torque limits of 18 to 22 foot pounds are reached.

The actuator-motor drive-gear housing cover must be removed during the installation in order to make electrical connections. Since this unit is outdoors and susceptible to moisture, care must be taken when replacing the cover to make certain that the gasket is in place and the seal is tight. If this is not done, moisture will quickly invade the housing and perhaps necessitate replacement.

Figure 6-25 shows the front and rear panels of the remote control unit, along with the various controls, jacks, and indicators that are defined. You may refer to this drawing for clarification throughout the remainder of this discussion.

Wire Selection and Connections

Measure the distance of the proposed wire path from the control console to the antenna. This distance determines the size, or gauge, of the two wires required to provide 36 Vdc to the actuator motor (Rear Panel Connectors 4 and 5). It is important that the recommended wire size be used. A wire that is too large will damage relay contacts in the control console and a wire that is too small will cause the motor to perform poorly.

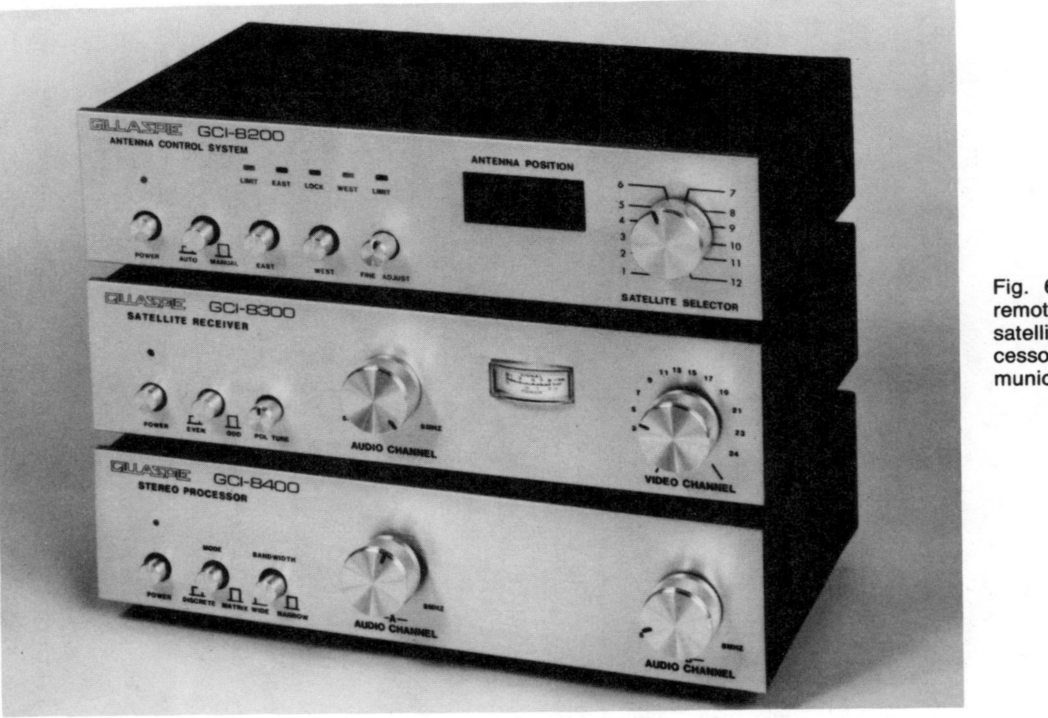

Fig. 6-24. Close-up view of the remote control unit stacked with a satellite receiver and stereo processor. Courtesy of Gillaspie Communications, Inc.

CAUTION:

Do not attempt to make electrical connections until you have read and understood this section.

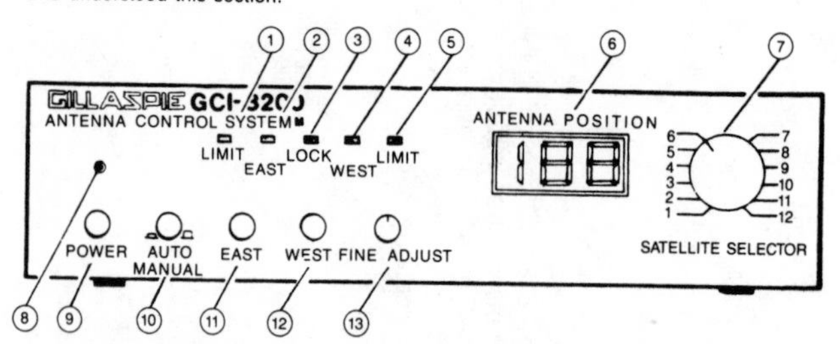

FRONT PANEL CONTROLS

① *East LIMIT position indicator light* - will come on when the antenna position has reached it's east limit position. Limits have not been pre-set and must be set during installation.
② *EAST movement indicator light* - will come on while the antenna is moving in an eastward direction.
③ *LOCK INDICATOR LIGHT* - WILL COME ON WHEN "tuning pot" ⑭ on back panel is synchronized with SATELLITE SELECTOR SWITCH ⑦.
④ *WEST movement indicator light* - will come on when antenna is moving in a westward direction.
⑤ *West LIMIT position indicator light* - will come on when antenna has reached it's west limit. Limits have not been pre-set and must be set during installation.
⑥ *ANTENNA POSITION digital display* - indicates present relative position of antenna.
⑦ *SATELLITE SELECTOR switch* - for automatic antenna positioning used to select individual satellites. SATELLITE SELECTOR switch has not been preset and must be set during installation.
⑧ *POWER indicator light* - will come on when power is applied to system.
⑨ *POWER on/off switch* - used to activate antenna control system.
⑩ *AUTO/MANUAL select switch* - used to place system in AUTO or MANUAL mode -

a. *AUTO* mode - In the *AUTO* Mode position the actuator will move in response to the position selected by the SATELLITE SELECTOR switch ⑦

CAUTION!

DO NOT PLACE AUTO/MANUAL SELECT SWITCH IN AUTO MODE UNTIL EAST-WEST LIMITS HAVE BEEN SET.

b. MANUAL Mode - In the MANUAL mode position the actuator will move in response to EAST ⑪ or WEST ⑫ movement controls.

Fig. 6-25. Front and rear panels of the Gillaspie remote control unit. Courtesy of Gillaspie Communications, Inc.

The GCI 8200 system has been designed to operate with .25 to .50 ohms at 3 amps in the combined length of both power leads to the motor. Installers who are bench testing this unit must provide minimum line resistance to prevent damage to the relays.

Recommended Copper Wire Size for Motor Power Leads

Maximum wire sizes to be used to provide correct resistance:

Under 50 Ft.	50 foot leads of 14 gauge AWG
50 Ft. to 75 Ft.	14-gauge AWG

(11) *EAST Movement Control* - used to rotate antenna in Eastward direction when *AUTO-MANUAL* switch (10) is in *MANUAL MODE.*
(12) *WEST Movement Control* - used to rotate antenna in Westward direction when *AUTO-MANUAL* Switch (10) is in *MANUAL MODE.*
(13) *AUTO FINE TUNE CONTROL* - used to make small antenna position adjustments when *AUTO-MANUAL* Switch (10) is in *AUTO* Mode only.

REAR PANEL CONTROLS & CONNECTIONS - Figure 2

(14) Satellite Position Adjustment "Pots" correspond to SATELLITE SELECTOR switch, position 1 thru 12 on front panel.
(15) EAST LIMIT adjustment. Establishes how far actuator will rotate antenna in east direction.
(16) WEST LIMIT adjustment. Establishes how far actuator will rotate antenna in west direction.
(17) SENSensitivity adjustment. (Counter clockwise increase - Clockwise decrease) Section G.
(18) DC FUSE - Replace with 5 AMP slow blow only.
(19) AC FUSE - Replace with 2 AMP fast blow only.
(20) Sensor Power High
(21) Sensor Signal Input - Wiper
(22) Sensor Power low
(23) & (24) Motor Power Connector - Provides 36VDC power to motor drive (Section B).
(25) AC Cord - Connected to 100 volt AC grounded outlet.

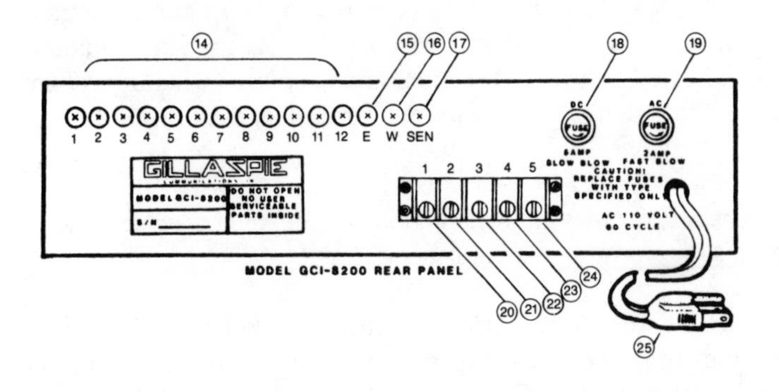

75 Ft. to 150 Ft.	12-gauge AWG
150 Ft. to 250 Ft.	10-gauge AWG
250 Ft. to 400 Ft.	8-gauge AWG
400 Ft. to 500 Ft.	6-gauge AWG

This wire should be suitable for underground burial and be resistant to ultraviolet light. Shielded cable is recommended.

In addition, you will need the same length of 3-conductor wire (24-gauge AWG or heavier) for connecting the console to the actuator potentiometer sensors. (Rear Panel Connections 1, 2 and 3.) Special 5-conductor wire with two power leads and 3 control leads is available from electronic supply house. For distances under 50 feet, it is not necessary to use a 50-foot minimum length of 3-conductor wire.

It is recommended that the wire be buried at least 6 inches deep. (Local codes may require deeper installation.) In installations where less than 50 feet of wire is required, it is important that a minimum of 50 feet of 14-gauge wire be used to power the motor. If less than 50 feet is used, the relay contacts in the control console could weld themselves shut due to an excess flow of electricity. The surplus wire may be coiled and buried.

Wiring Scheme for Actuator Travel Direction

Caution. Do not mount actuator motor drive to antenna until all electrical connections are made and system validated for correct operation.

The control console should be wired to the motor drive unit in a manner so that the digital readout increases when the motor drive is moved in a west direction, and decreases when moved in and east direction. This is important for proper setting and operation of the EAST/WEST limits.

Note. Do not alter connection on the motor drive. Make changes only on the control console.

In Fig. 6-26, "Motor Drive/Control Console Interface Wiring Diagram," the motor connections shown (4 and 5) are made to move the actuator inward for East direction and outward for West direction. The potentiometer connections 1, 2, & 3 are made to increase the digital readout numbers when antenna is rotated west and decrease when antenna is rotated east.

If motor control connections 4 and 5 on Control Console are reversed to produce inward for West direction movement and outward for east direction movement then connections 1 and 3 on the control console should be reversed as well. This will maintain proper integrity between digital readout movement and east/west limit controls.

Note. Control Console Connection Position number 2 is always connected to Actuator connector position number 2.

For the Western U.S. and Western Canada, antenna actuators

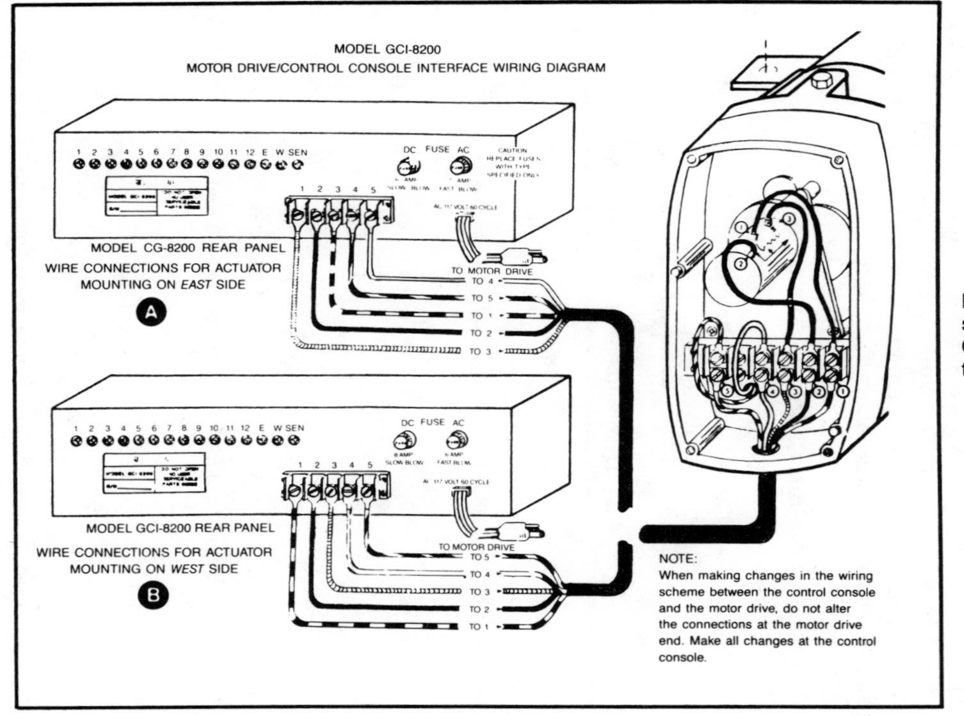

Fig. 6-26. Motor Drive/Control console interface wiring diagram. Courtesy of Gillaspie Communications, Inc.

are usually mounted on the east side of the antenna pivot point. In this configuration, to move the antenna in an east movement, the actuator shaft must move inward; a west movement will require the shaft to move outward.

When the actuator is mounted on the east side of the pivot point, wire connections should be as shown in Table 6-1. For Eastern U.S. and Eastern Canada locations, the actuator is usually mounted on the west side of the antenna pivot point. In this configuration, the direction of the shaft travel produces opposite results. East movement of antenna would require the shaft to move outward, while west antenna movement requires shaft travel inward. For actuator mounting on west side of the pivot, the control console connections should be as shown in Table 6-1.

By adhering to these procedures, east-west movement controls and limits controls will function properly.

Actuator Operation Check

The actuator motor drive (Fig. 6-27) should not be mounted to the antenna at this time. The actuator motor drive must be observed during the following step.

Upon completing the electrical and control connections to the

Table 6-1. Wire Connections for Actuator.

	Mounting on East Side (See Fig. 6-26A.)
Motor Drive End	Control Console End
1	3
2	2
3	1
4	5
5	4

	Mounting on West Side (See Fig. 6-26A.)
Motor Drive End	Control Console End
1	1
2	2
3	3
4	4
5	5

Note. Connection #2 is the same on both ends.

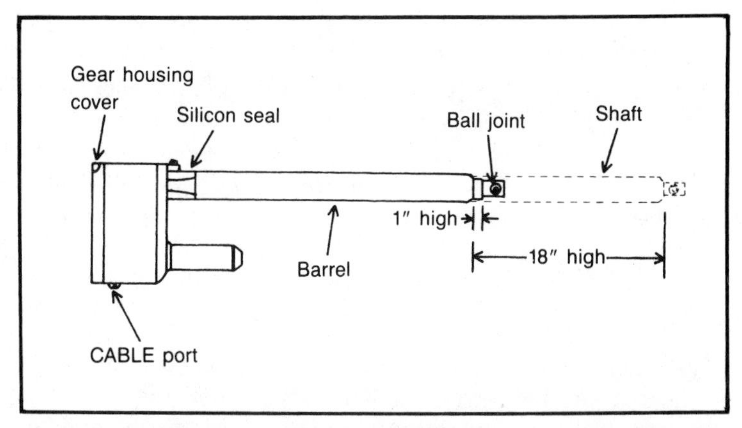

Fig. 6-27. The Gillaspie Actuator. Courtesy of Gillaspie Communications, Inc.

actuator motor drive unit and to the control console, test the system for the proper operation. Prior to connecting the console to power source, make sure AUTO/MANUAL switch 10 is in MANUAL mode. Activate system by connecting control console to a 110 volt ac grounded outlet and turning on the console with POWER switch 9 on the front panel. The power indicator 8 light should be on.

Keep the AUTO-MANUAL selector switch 10 in MANUAL mode. Push EAST movement control 11. The actuator motor drive should move. The digital readout should decrease. The direction of actuator movement will depend upon the position of the wires on connector block pins 4 and 5 as described earlier. When the EAST movement button is released the motor drive unit should stop. Push in the WEST movement button 12. The actuator motor drive should move in opposite direction and should stop when the button is released. The digital readout should increase while actuator is moving WEST. If motor drive fails to operate check all connections and fuses. If digital readout does not react as described review the Actuator Wiring section.

Mechanical East-West Limits

It is essential that operation of the motor drive unit not be attempted beyond its mechanical travel limits. Although a clutch device within the motor drive will prevent damage to the actuator, damage to your antenna may result.

The Gillaspie GCI-8200 system is equipped with electronic devices that, when set and properly adjusted, will automatically prevent operation beyond mechanical travel limits. Electronic limits are set with adjustments 15 & 16, located on the back panel of the control console.

The mechanical travel limits of the actuator are minimum inward travel of one inch and a maximum outward travel of 18 inches between the end of the outer barrel and the end of the shaft.

The extreme east and west electronic limits should be set a short distance beyond the most east and west satellites. This procedure will be explained next.

Installation and Setting Limits

Activate the motor drive until the shaft is compressed to within 3 inches of end of barrel. Install the actuator motor drive mounting clamp to the antenna mount. Connect the actuator motor drive by sliding the actuator barrel into the clamp (See Fig. 6-28). Do not tighten clamp to barrel at this time. Do not attach ball joint connector to the antenna at this time. Manually turn the antenna to the most east or west satellite. With the antenna held in position on the satellite, slide actuator barrel into the mounting clamp until the ball-joint end connector can be attached to antenna. Tighten the connecting bolt in the shaft ball-joint connector. Tighten

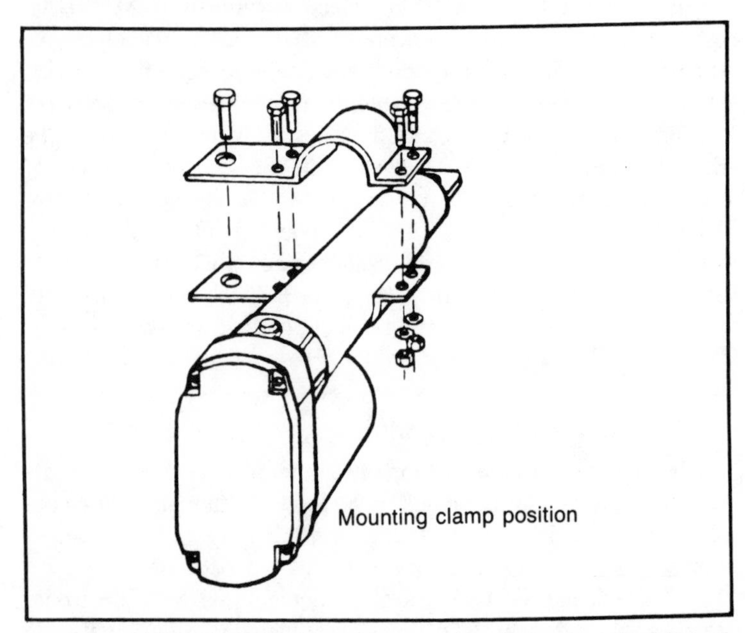

Fig. 6-28. Mounting clamp position for the Gillaspie Actuator. Courtesy of Gillaspie Communications, Inc.

the mounting clamp around barrel securely. The actuator should be positioned so that the cable entrance to the gear housing faces down.

Activate the movement control to extend the shaft while observing the actuator for any binding or other restrictions. List for motor stress or overload. Check for operation of east-west movement indicator lights and proper digital number sequence. Activate the actuator to slightly beyond the most east satellite position, but not within one inch or beyond 18 inches of the end of the barrel.

Locate the EAST LIMIT adjustment 15 on the back panel of the control console. With a small screwdriver, adjust EAST LIMIT until the EAST LIMIT indicator light 1 on front panel comes on. The east electronic limit is now set. Note and record the number displayed on the digital readout for future reference.

SUMMARY

The installation of the TVRO Earth station and the motor actuator (if desired) constitutes the most difficult mechanical portion of an Earth station, but most antennas and actuators are designed to be installed by the owner. Therefore, helpful instructions are always included. If you have reasonable mechanical abilities, the installation should go rather smoothly. If, however, you tend to be a "klutz" when it comes to such things, you might wish to have someone make the installation for you. Alternately, you might be in a position to watch the installation of someone else's antenna and get your pointers from their experiences.

Either way, antennas and their associated components are obviously quite rugged, so you don't run any real risk of damaging anything. At the very worst, you might get your antenna up and find that you need some professional assistance in realigning some of the parts. I have not yet run into anyone who attempted their own installation and encountered difficulties to severe that they could not continue.

Chapter 7

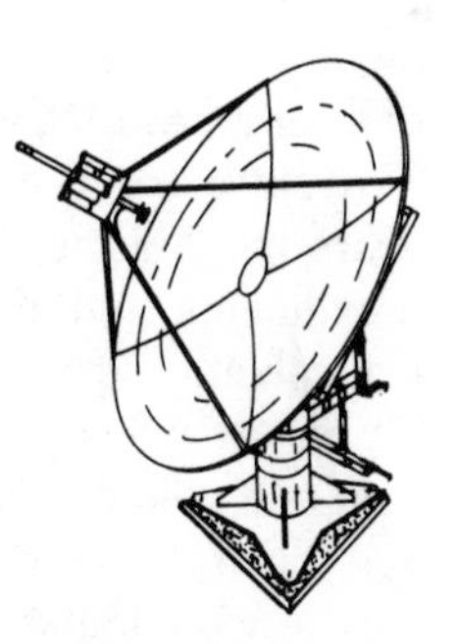

Commercially Available Receiving Equipment

THE ELECTRONIC CHAIN THAT MAKES UP THE RECEIVER section of the TVRO Earth station consists of the low-noise amplifier, which boosts the signal from the antenna; the down converter, which changes the microwave signal into a lower frequency equivalent; and the receiver itself, which demodulates the signal from the down converter. This demodulated signal may then be passed directly to a television receiver or to another component known as a modulator. Many receivers now have built-in modulators that can feed a television set directly, but some still output what is known as a composite video signal. This must be passed directly to a modulator that imposes the composite video signal onto a radio frequency carrier. The output from the modulator is connected to the antenna terminals of the television receiver.

In the first edition of this book, an entire chapter was devoted to electronic kits that were designed for home assembly. These kits were used to build LNAs, down converters, and even receivers. At the time the first edition was written, a kit offered substantial savings over many of the assembled units that were available. Today, however, most of these kits have fallen by the wayside as the price of all Earth station components has dropped considerably. Simply put, you can purchase an assembled and tested piece of Earth station equipment today for about the same price (or less) as any of the kits discussed in the first edition.

This is not to say that build-it-yourself kits are not available anywhere today. Occasionally, one will be offered. This can be an

LNA kit, a receiver kit, a down-converter kit, or a specialized kit that addresses TVRO accessories. None of these are mentioned in this book because they tend to have limited sales appeal, and many of the companies offering them eventually drop the product or move on to something else. Indeed, some go out of business entirely.

Two very large manufacturers of TVRO Earth station equipment, KLM Electronics and Gillaspie Communications, Inc., used to offer an assortment of kits. However, both of these companies have since dropped this line. These kits were quite popular a few years ago, but they have been made obsolete by technological improvements and the far lower cost of assembled and tested equivalents. Both of these companies are still at the forefront of TVRO technology, but all of their equipment is sold on an assembled and tested basis today.

When buying Earth station equipment, you will soon find that only a few companies offer a specific type of product or even a limited range of products. In most cases, a company will offer complete TVRO stations, but you have the option of purchasing one or more components that make up the system. For instance, a company's main product may be TVRO receivers, but they will also offer LNAs, antennas, and most other components. Possibly, the receiver is the only component the company actually manufactures. The remaining equipment they offer is often supplied by other manufacturers and may contain only the name of the company marketing the complete system. This is commonplace today, and it's not unusual to buy a TVRO Earth station from a company that makes only the receiver while the antenna and LNA are supplied by two different companies that specialize in these products.

In looking for the best deal, you may find that these packaged systems offer the best savings. The only way to make sure is to obtain specifications on an entire system and compare each component in the system with specifications provided for those components from other manufacturers. You may find that mixing and matching components will provide you with cost savings, but this is not true in every case.

This chapter will review a sampling of the receiving systems currently available. In general, this will include everything from the output of the antenna to the television receiver. Again, this is only a sampling of what is currently available. One must remember that satellite television is a fast-moving, high-technology field. It's not uncommon for specific models of components to be quickly replaced by upgrades. Therefore, this chapter will give you an idea of what is available, although you may find a specific model discussed here has been replaced by an upgraded model by the time you read these pages. Therefore, if you are planning to order a specific product discussed in this chapter, it would be very wise to contact the manufacturer to make sure the product is still their most current. The only way to offer a book on satellite television products that is completely up-to-date is to write a new book every other month. It should be stressed that any upgrades of the products discussed here should not alter the concept of satellite receiving equipment. These upgrades usually represent slight improvements designed to make operation a bit simpler. Such improvements are often the direct result of customer surveys conducted by the manufacturers themselves. Therefore, an upgrade to a specific receiver might be the same receiver with an extra switch or two added. The general specifications should apply for quite some time.

GILLASPIE MODEL GCI-8300 SATELLITE RECEIVER

Gillaspie Communications, Inc., in Milpitas, California, has been a leader in the TVRO Earth station field since the beginning of satellite television. Originally known as Gillaspie and Associates, this company began by offering a number of interesting kits for TVRO experimentation and has since moved into a sophisticated line of receivers, LNAs, and satellite antenna control systems. I have found the personnel of this company to be very supportive. If you run into a problem, you don't have to wait days or weeks for service or installation advice. This is available by telephone. I originally made contact with this company several years ago when their product line consisted primarily of kits. Any company that offers kits for home wiring must also be geared up to handle a great deal of customer inquiries. Their tradition of service has carried over into their advanced, fully assembled and wired line.

Their operation and installation booklets are quite easy to understand, and for the advanced customer, a great deal of technical information is provided. Installing their equipment is not a complex procedure because of their excellent documentation.

The GCI-8300 Satellite Receiver is shown in Fig. 7-1. It offers continuous audio and video channel tuning, video invert switch, automatic frequency control, easy-to-read signal strength indicator, polarizer control, and selectable channel 3 or 4 modulator. The modulator is built into the receiver cabinet.

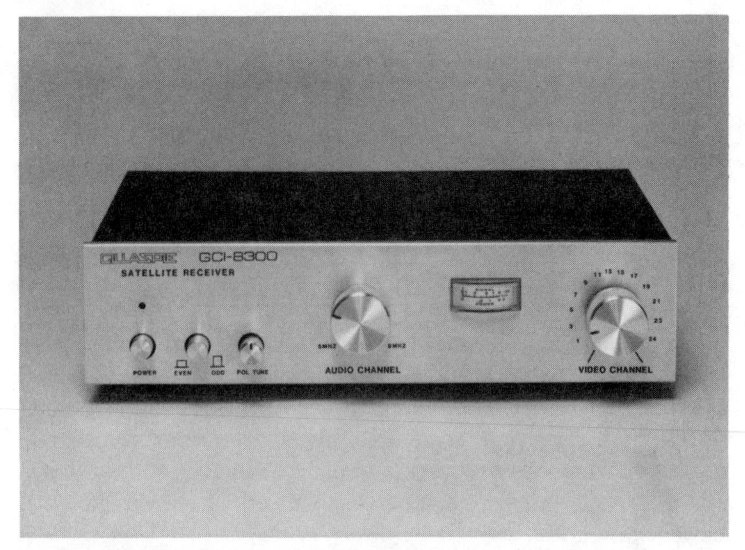

Fig. 7-1. Gillaspie GCI 8300 Satellite receiver. Courtesy of Gillaspie Communications, Inc.

This receiver consists of two separate units. First, there is the Image Reject Mixer (also known as the down converter). The second unit is the Control Console. This is the device that sits on top of the television set.

The down converter mounts at the antenna and receives its signal directly from the low-noise amplifier. The down converter converts the microwave signals from the satellite frequency (3.7-4.2 GHz) to a frequency compatible with the receiver section (70 MHz). This allows for the economical transmission of the signals from the down converter to the control console by means of inexpensive RG-59 coaxial cable. Some early TVRO receivers use this type of cable to transfer the microwave frequencies directly from the low noise amplifier, which resulted in a substantial loss. The sooner we convert from the microwave to vhf frequencies, the more efficient the system is. This is the main reason for a down converter mounted at the LNA.

The control console, installed at the television receiver, contains all of the controls for operating the satellite receiving system. This console also supplies the operating voltages for the image reject mixer, low noise amplifier, and polarizer. A built-in modulator provides either channel 3 or 4 output to the television receiver.

The GCI-8300 receiver system is completely aligned and calibrated at the factory. However, due to variations in cable length between the control console and the down converter, a minor touch-up of calibration controls may be necessary. Depending on the ambient temperature, the down converter will require 30 to 45 minutes of operation to stabilize the frequency. Proper channel calibration can be accomplished only after the system has reached frequency and temperature stability. This does not mean that you must wait 30 to 45 minutes before you can receive a good signal on your television receiver. It simply means you must let the system "cook" for this period of time before going through the initial alignment procedures. All alignment procedures are conducted with the controls provided on the receiving unit located at the television receiver.

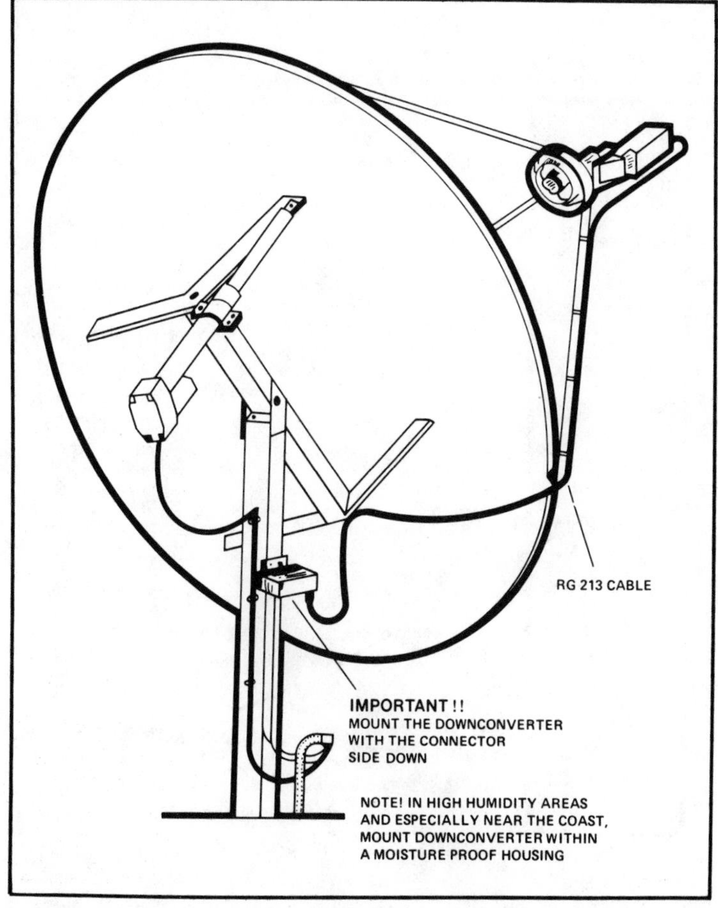

Fig. 7-2. The down converter is attached to the antenna pedestal. Courtesy of Gillaspie Communications, Inc.

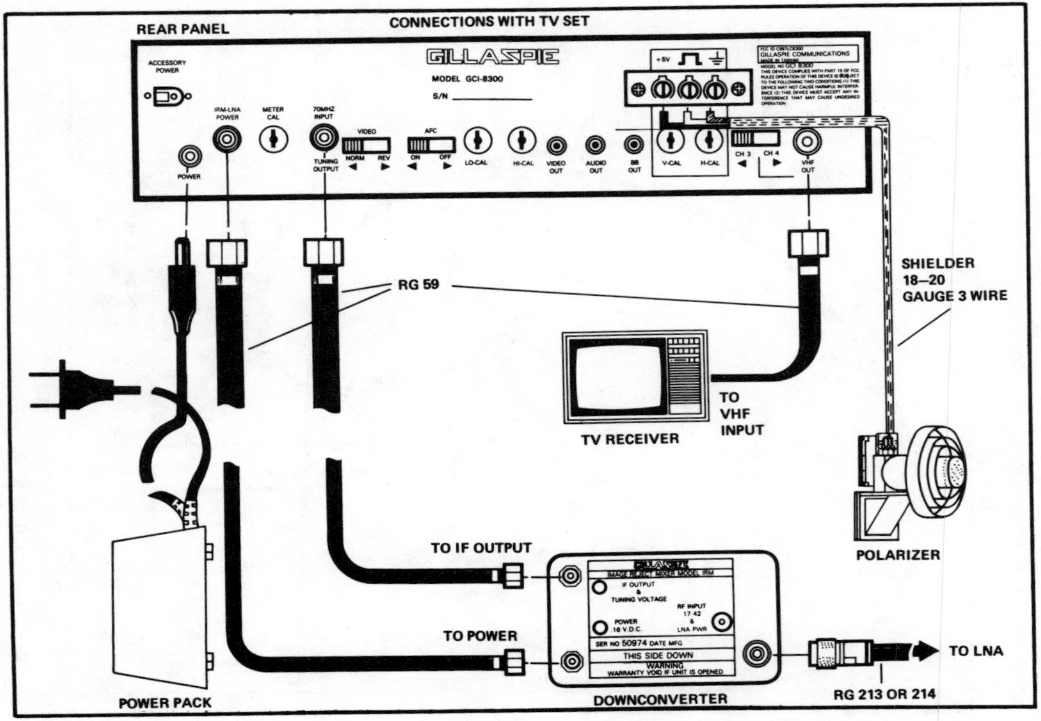

Fig. 7-3. Typical system configuration for the Gillaspie GCI 8300 receiver. Courtesy of Gillaspie Communications, Inc.

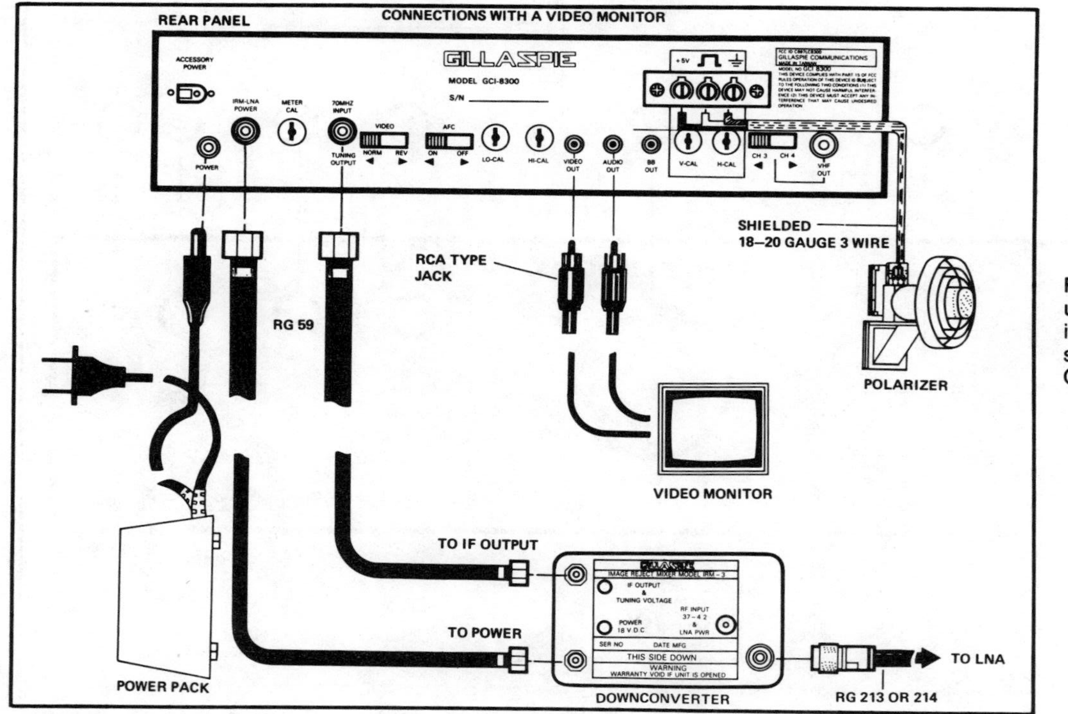

Fig. 7-4. Typical configuration using a composite video monitor instead of a standard television receiver. Courtesy of Gillaspie Communications, Inc.

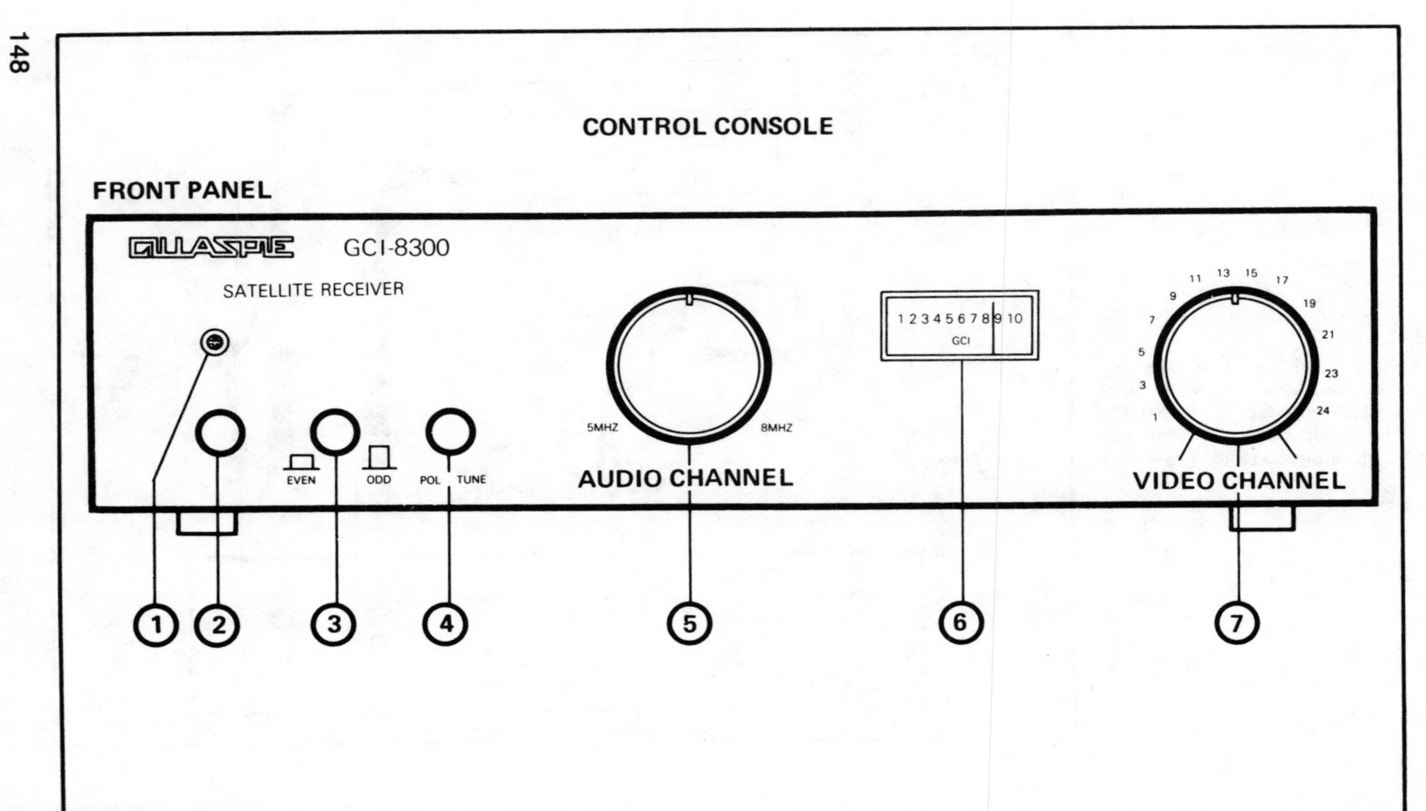
CONTROL CONSOLE
FRONT PANEL
GILLASPIE
GCI-8300
SATELLITE RECEIVER
EVEN
ODD
POL
TUNE
5MHZ
8MHZ
AUDIO CHANNEL
1 2 3 4 5 6 7 8 9 10
GCI
1
3
5
7
9
11
13
15
17
19
21
23
24
VIDEO CHANNEL
1
2
3
4
5
6
7

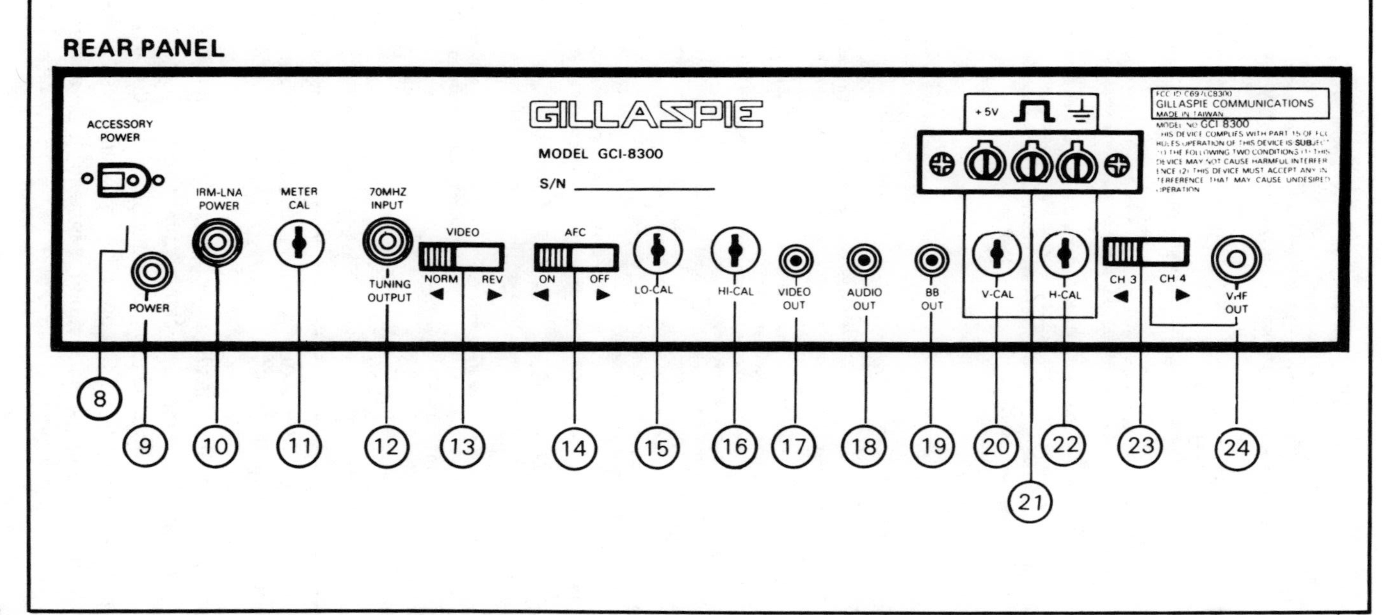

Fig. 7-5A. Front and rear panels of the GCI 8300 control console. Courtesy of Gillaspie Communications, Inc.

(1) "POWER LIGHT" - is on when "POWER SWITCH" is in "ON" position.
(2) "POWER SWITCH" - activates the Receiver.

Note: Power outlet to the Down Converter and LNA is active when unit is plugged in even when the Power Switch is in the "OFF" position. DO NOT plug unit into 115 V outlet until all external connections have been completed.

(3) "EVEN-ODD" Polarization position - when "IN", the polarizer position is set for even channels while "OUT", it is set for odd channels.

NOTE: Some satellite channels have the opposite polarity format.

(4) "POL-TUNE" - fine tune for polarizer
(5) "AUDIO CHANNEL" - selector knob is used to select the desired audio sub-carrier. Full counter clockwise rotation is approximately 5 MHz sub-carrier frequency. Television Audio is at the Selector knobs mid-range position (6.2 MHz).
(6) "SIGNAL STRENGTH METER/CARRIER LEVEL INDICATOR" - indicates the signal strength and is an invaluable tool for aligning the feed horn, antenna and tuning the receiver.
(7) "VIDEO CHANNEL" - for tuning the various satellite transponders. Counter clockwise rotation tunes the lower transponders and clockwise tunes the higher ones.

REAR PANEL CONTROL & CONNECTIONS

(8) "ACCESSORY POWER" outlet - provides 20 Vac for GCI-8400 stereo processor connection only.
(9) "POWER" connection - connect GCI 20 Vac 1.5 amp power pack wall outlet unit only.
(10) "IRM-LNA POWER" - connect RG-59/U coaxial cable to "Power +24 Vdc" connection on Image Reject Mixer Down Converter.
(11) "METER CALIBRATION" - adjusts the meter needle deflection for signal strength indications. Should be adjusted so strongest channel reads 5 to 7 on the signal strength meter (6). This control does not affect the receivers operating performance.
(12) "70 MHz INPUT" - connect RG-59/U coaxial cable from "IF OUTPUT & TUNING VOLTAGE" connection on Image Reject Mixer Down Converter.
(13) "VIDEO-NORM-REV" switch - should be in normal position for most uses. "REV" position inverts the video and sync for special broadcasts.
(14) "AFC ON/OFF" switch - should be "ON" for normal operation.
(15) "LO-CAL" video tune limit calibration — set "video channel" select to channel 1 and adjust "LO-CAL" to best picture quality for channel 1.
(16) "HI-CAL" video tune limit calibration - set "video channel" select to channel 24 and adjust "HI-CAL" to best picture quality for channel 24.
(17) "VIDEO OUT" (RCA-type phone jack) - provides 1 Vpp (negative sync) video output. Output impedance is 75 ohms, typical. Used when receiver is connected to the video input of video recorder or monitor.
(18) "AUDIO OUT" (RCA-type phone jack) - provides audio output to drive external.circuitry. Output is typically 2 Vpp, low output impedance. Used when receiver is connected to audio input of video recorder, monitor or hi-fi.
(19) "B B OUT" (RCA-type phone jack) - provides approximately 1 Vpp of raw baseband output (75 ohms). This is the video baseband signal before it is de-emphasized, low passed, or clamped.
(20) "V-CAL" for adjusting VERTICAL position of polarizer.
(21) Polarizer cable connections
(22) "H-CAL" for adjusting horizontal position of polarizer.
(23) "CH3/CH4" switch - modulation to channel 3 or 4 on T.V. set.
(24) "VHF OUT" connect RG-59/U coaxial cable to VHF antenna input on T.V. set.

Fig. 7-5B. Functions of the GCI 8300 controls and indicators. Courtesy of Gillaspie Communications, Inc.

The instructions for this alignment are well documented. Overall, it is a simple procedure that can be conducted by almost anyone.

Figure 7-2 shows the physical mounting of the down converter to the antenna, and Fig. 7-3 shows a typical overall system diagram. If you intend to use a video monitor instead of a standard television receiver, the pictorial diagram in Fig. 7-4 may be of help. Using a video monitor, the modulator output is bypassed, since the monitor accepts composite video and audio outputs. All of these figures are taken from the Gillaspie installation manual and are a good indication of the excellent documentation provided.

Figure 7-5 shows the front and rear panels of the control console with the various controls and indicator explanations. Figure 7-6 provides the specifications for the overall receiving system.

GILLASPIE MODEL 2001R SATELLITE RECEIVER

The Gillaspie 2001R is a quartz-synthesized stereo satellite receiver that can reproduce stereo audio. The quartz synthesization represents the ultimate in frequency stability, making this a truly state-of-the-art receiver design. The device offers 24 frequency-synthesized channels, microprocessor control of digital tuning, an automatic polarity switch, and a bright digital readout channel indicator. This unit offers tunable stereo audio, an easy-to-read signal strength indicator, and a full-function wireless remote controller. All keyed functions are reflected by status indicator lamps, enabling the user to see at a glance what modes are in use. The GCI

SECTION II – SPECIFICATIONS

Tuning Range: 3700 to 4200 MHz.

Local Oscillator Frequency: 3630 to 4130 MHz.

IF Frequency: 70 MHz

IF Output Level from Down Converter: −25dBm typical (with 50dB gain LNA) @ 75 ohms

Tuning Voltage: 4 to 15 VDC, typical

Noise Figure: 25db, typical

Image Rejection: −10db, minimum

Conversion Gain: 20 to 28db

Outputs:
- Audio – 2v pp into 10K ohm
- Video – 1v pp into 75 ohm
- Baseband Video – 1v pp, unterminated
- RF Built-in Modulator, TV Ch. 3 or Ch. 4, switchable

Video Polarity: Switchable, Normal or Inverted

Clamping: Keyed sync tip clamp cancels dispersal signal

Polarizor Control: Built-in Vertical/Horizontal Switch Tuning

Tunable Audio: 5.5 to 7.5 MHz

AFC: Switchable, ON/OFF

Signal Level Indicator: built-in illuminated meter (500 uA)

Fig. 7-6. Overall specifications for the GCI 8300. Courtesy of Gillaspie Communications, Inc.

2001R is a "pretty" unit. Its modern styling is designed to complement today's contemporary decor.

We can break this receiver down into three separate units that include the Image Reject Mixer, the Control Console, and the Wireless Remote Control. For all intents and purposes, we can say that the 2001R is a highly upgraded version of their 8300 model. The Image Reject Mixer offers the high stability of quartz-synthesization, stereo audio, and the wireless remote control. The Wireless Remote Control will be appreciated by many and is one of those conveniences that are often seen as options today.

The components of this receiver match those of the previous model as far as general operation is concerned. The only new component is the wireless, infra-red remote control. This is a hand-held unit from which all keyed functions on the GCI 2001R can be operated.

As with the previous model, the down converter is designed to attach directly to the LNA. However, if this is not satisfactory, it can be mounted elsewhere by means of a length of RG-213 or RG-214 coaxial cable. The connecting length should be no longer than fifteen feet. RG-259 coaxial cable is used to connect the output of the down converter to the control console. This run may be up to a maximum of 165 feet. If you must make a longer run than this (up to 800 feet), RG-6 coaxial cable is required.

Figure 7-7 shows the front and rear panels of the 2001R, along with a description of the functions of each control or indicator. The infra-red hand-held remote control unit can operate the system from a distance, and when used in conjunction with the GCI 2001C Antenna Controller, can serve as a complete system remote control. This includes the video and audio channels, as well as the antenna pointing system. This means that you never have to leave the comfort of your chair to change channels, fine tune a signal, or even redirect the antenna to another satellite. The Model 200 Remote Control is shown in Fig. 7-8. The eleven buttons on the lower half of the unit are used for antenna remote control, and the remaining buttons are generally used to control the receiver. The unit can activate all the satellite positions that are programmed into the control console's memory, as well as move the antenna east or west. By pressing the appropriate satellite name button and the appropriate satellite number, the antenna will move to the programmed position. Fine tuning can then be accomplished by activating the east/west buttons. Again, this applies only when the receiving system includes the Gillaspie Model 2001C Computerized Satellite Antenna Controller.

Obviously, the 2001R receiving system is not your run-of-the-mill receiver. It offers many innovations that address not only technical performance but comfort and ease of operation as well. Figure 7-9 provides the specifications for this system.

GILLASPIE MODEL 2001C ANTENNA CONTROLLER

Antenna controllers have been discussed in another chapter, but the discussion of the Model 2001C Computerized Satellite Antenna Controller is contained in this chapter because of the way it ties in with the 2001R receiver, and especially, the hand-held remote controller for both.

Shown in Fig. 7-10, the 2001C controller is a ruggedly engineered device that is designed to provide years of service. It is controlled by a microprocessor, which gives it the ability to program and recall up to 42 antenna positions. Antenna positioning is accomplished by pressing the East/West keys, and a special LOCK feature is provided to prevent unauthorized antenna movement. Factory-programmed codes are accessible from the remote keyboard or from the front panel keyboard to lock or unlock all functions. Also, factory-programmed east and west limits are provided to protect both the actuator and the antenna. Within the factory-programmed limits, users can program their own east/west limits to provide for customized installations. Preprogrammed antenna positions are easily recalled by pressing only two keys, as listed in published satellite TV program directories. The unit also contains an analog position indicator that is mounted on the front panel to monitor actuator performance while the antenna is being moved. A capacitive charge system is provided to preserve antenna position data during power failures for up to seven days. This unit is aesthetically and functionally compatible with the previously discussed 2001R satellite receiver.

The 2001C system consists of three separate units: the control console, the actuator motor drive, and the hand-held remote controller. The console is installed near the satellite receiver and consists of the controls to operate the system and the power to operate the motor drive. The actuator motor drive mounts on the antenna and, in response to the control console, provides movement to the antenna. The motor drive is a ballscrew type that is driven by 36 Vdc. It can provide an 18-inch stroke. The mechanical system

GCI 2001R FRONT PANEL

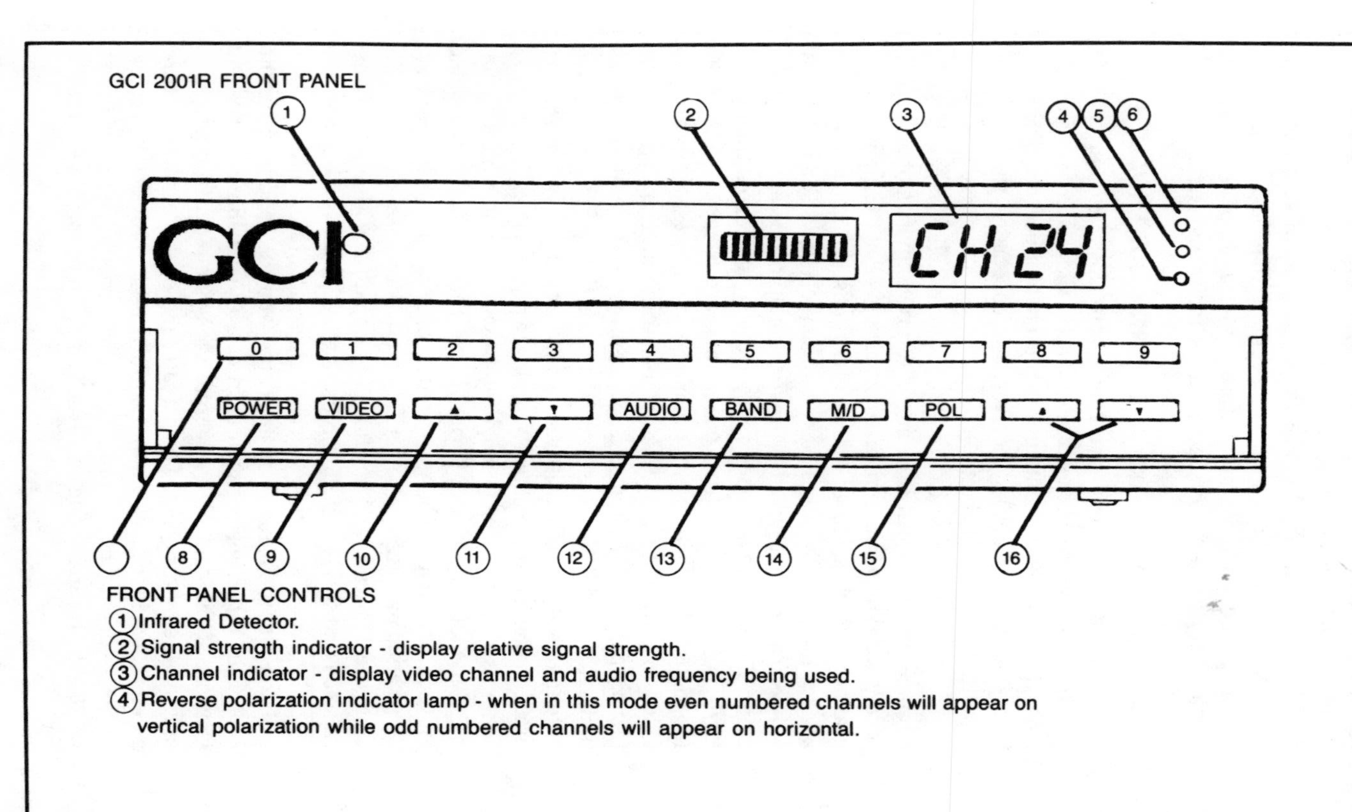

FRONT PANEL CONTROLS

(1) Infrared Detector.

(2) Signal strength indicator - display relative signal strength.

(3) Channel indicator - display video channel and audio frequency being used.

(4) Reverse polarization indicator lamp - when in this mode even numbered channels will appear on vertical polarization while odd numbered channels will appear on horizontal.

⑤ Remote ON indicator lamp - indicates that the infrared remote signal is being received from the GSI 200 hand held remote.

⑥ Normal polarization indicator lamp - when in this mode even numbered channels will appear on horizontal polarization while odd numbered channels are vertical.

⑦ Video channel select and audio frequency select buttons 0-9 when selecting video channels simply select the appropriate number 1 thru 24 and that channel will come up and the number will appear on the digital readout ②. If the GCI 2001R is in the audio mode it will be necessary to push the video button ⑧ prior to selecting a video channel.

⑧ Power - To turn power on and off. Remember external power is always on when the GCI 2001R is plugged into a wall outlet, even when the power switch is in the off position.

⑨ Video button - Used when placing the system in the video channel select mode, and to operate the system in auto scan.

⑩ Channel select up button - Press to select channels in a progressive manner.

⑪ Channel select down button - Press to select channels is a regressive manner.

⑫ Audio select button - Used when selecting right and left stereo frequency channels. Mono video channels are usually 6.9. This number must be put in when change to a mono video channel mode from a stereo mode.

⑬ Band - To select Narrow or Wide audio band width.

⑭ Discrete/Matrix (D/M) select button - To select discrete or matrix stereo audio format.

⑮ POL - Polarization mode select button - Used to place system in normal or reverse polarity.

⑯ ▲- ▼- Polarization skew adjustment - Used to adjust the polarity when antenna is moved from satellite to satellite in extreme locations. The skew adjustment ▲ or ▼ will fine tune the polarizer.

GCI 2001R REAR PANEL

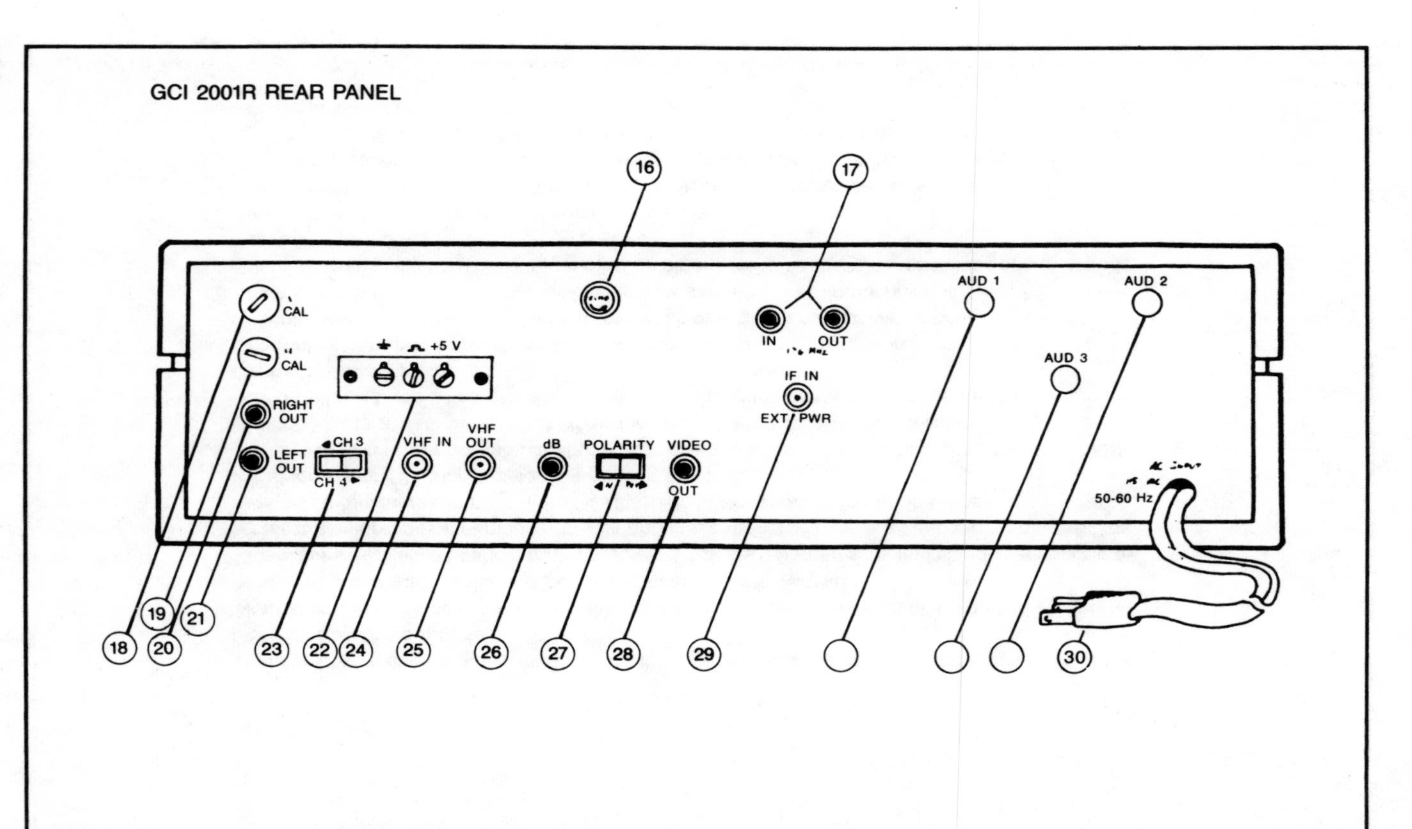

REAR PANEL CONTROLS

(17) 136 MHz IN and OUT - Used for optional interference filter connection.
(18) V-CAL - For adjusting vertical polarization
(19) H-CAL - For adjusting horizontal polarization
(20) Right out - Audio right channel for connecting to right channel of stereo unit.
(21) Left out - Audio left channel for connecting to left channel of stereo unit.
(22) Polarizer control cabler connector block.
(23) Channel 3 or 4 modulator output selector.
(24) VHF IN - Connect normal T.V. antenna.
(25) VHF OUT - Connect to VHF antenna input on standard T.V. set.
(26) Baseband out.
(27) Video polarity - Normal-reversed.
(28) Video out - Used when interfacing with a video monitor.
(29) IF output - External power out. Connected to the down converter with RG59u.
(30) AC input - connect to 115 Vac grounded wall outlet.

Fig. 7-7. Front and rear panels of the GCI 2001R Satellite receiver. Courtesy of Gillaspie Communications, Inc.

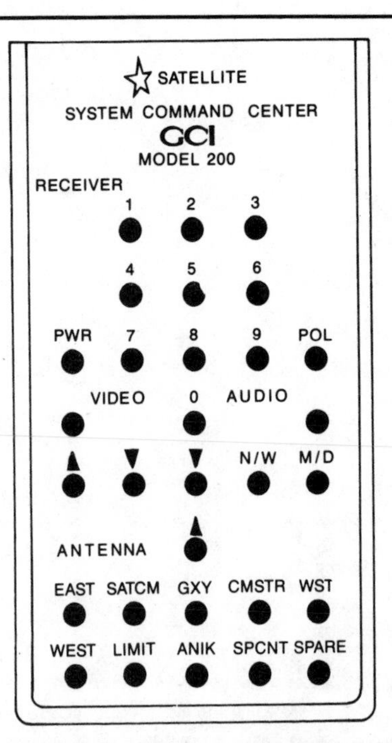

The system command center Model 200 can function all the GCI 2001R systems by remote control. When used in conjunction with the GCI 2001C Antenna Controller, the Model 200 functions as a complete system remote control, not only in video channel and audio frequency, but individual satellite select as well. The 19 buttons on the top half of the rough pad are used for the receiver functions and match the functions found on the 2001R. The Model 200 is battery powered by a standard 9 volt battery common to transistor radios. To replace the battery, slide the back cover of the 200 fof in the direction of the arrow. When the signal transmitter light becomes weak or non-functioning, it indicates that the battery needs replacing. The 200 remote should be pointed towards the GCI 2001R when functioning the remote.

Fig. 7-8. The Model 200 hand-held remote control. Courtesy of Gillaspie Communications, Inc.

is protected by a clutch that prevents damage to the actuator. The remote controller is a full-function infra-red remote. It was designed to be used with the 2001R Satellite Receiver and the 2001C Antenna Controller jointly and is powered by a 9-volt battery.

The GCI 2001C contains a special feature that prevents unauthorized antenna movement. The Lock feature disables the system from reading any satellite select input codes and prevents the antenna from changing satellites. The codes can only be activated from the control console, not from the remote. Figure 7-11 provides a breakdown on the mechanical control mounted at the antenna site.

Tuning Range: 3700 to 4200 MHz.
Local Oscillator Frequency: 3630 to 4130 MHz.
IF Frequency: 70 MHz.
IF Output Level from Downconverter: - 24 dBm typical (with 50 dB gain LNA) @ 75 ohms.
Tuning Voltage: 4 to 15 Vdc, typical.
Noise Figure: 25 dB.
Image Rejection: 10 dB.
Conversion Gain: 20 to 28 dB, typical.
Outputs:
Audio - 2 Vpp @ Lo Z.
Video - 1 Vpp @ 75 ohm Z.
Baseband Video - 1 Vpp, unterminated.
Rf Built-in Modulator, TV Ch. 3 or Ch. 4, switchable.
Video Polarity: Switchable, Normal or Inverted.
Clamping: Keyed sync tip clamp cancels dispersal signal.
Polarotor Control: Built-in, automatic, with vernier tuning.
A/B rf Switch: Built-in, for switching TV Antenna, CATV, VCR, etc., to television/monitor.
Tunable Audio: 5 to 8 MHz.
Signal Lever Indicator: Built-in 4" illuminated meter.
Remote Control: Infrared, wireless hand held unit.

Fig. 7-9. System specifications for the 2001R receiver. Courtesy of Gillaspie Communications, Inc.

GCI 9600 RECEIVER

For those TVRO enthusiasts who like excellent performance combined with sleek styling, the Gillaspie GCI 9600 Satellite Receiver may be just the ticket. Shown in Figure 7-12, the 9600 is a remote-control unit that offers excellent performance, economy, and a rich walnut cabinet. Video and audio quality are excellent, and the infrared remote control is simplified to just five buttons for all channel selections. A larger signal strength display on the front panel ensures ease of alignment. The 9600 has a scan tuning mode and is descrambler ready. Other features include sequential pushbutton digital tuning, video fine tuning, and continuous audio tuning. The hookups are compatible with electronic polarization controls, and the unit offers automatic polarity selection. The GCI 9600 contains a built-in VHF selectable channel 3 or 4 modulator. Figure 7-13 provides a complete table of specifications for this receiver.

CHANNEL MASTER

In late 1984, Channel Master released a pair of new receivers that should serve to widen the choices customers have in designing their own Earth station satellite system. Ever since Channel Master entered the TVRO Earth station field, their engineers have been working on designs that will reduce system cost while ex-

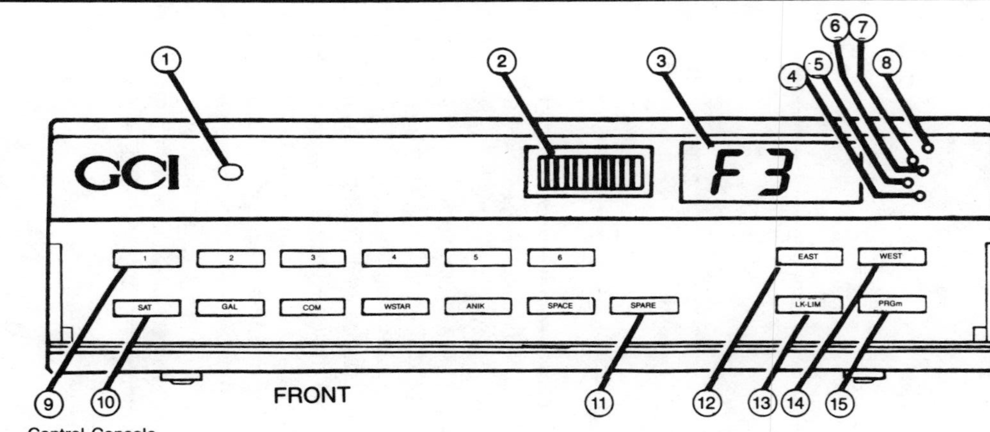

Control Console

1 Infrared Remote Signal Receiving Sensor.
2 Antenna Relative Position Indicator.
3 Digital Readout Indicators - displays antenna location and commands.
4 WEST Limit Indicator Lamp - indicates antenna is at its set west limit.
5 EAST Limit Indicator Lamp - indicates antenna is at its set east limit.
6 Program Mode Indicator Lamp - indicates system is in Program Mode.
7 System Lock Indicator Lamp - indicates system is in Lock Mode.
8 Antenna Model Indicator Lamp - indicates that antenna controller will receive instructions from remote (used in conjunction with 2001CR remote unit).
9 Satellite Number Input Keys - used in conjunction with satellite keys 9 when selecting or programming specific satellite.

10 Satellite Name Input Keys - used in conjunction with satellite numbers keys 8 when selecting or programming specific satellite. Example: For satellite SATCOM F3, select keys *SAT* and *3*; GALAXY 1, select keys *GAL* and *1*, etc.
11 Spare Satellite Select Key - used the same as satellite name input keys 9 for any existing or future satellite identification.
12 EAST Movement Key - for moving antenna in an eastward direction.
13 Limit - used when setting EAST or WEST limits.
14 WEST Movement Key - for moving antenna in a westward direction.

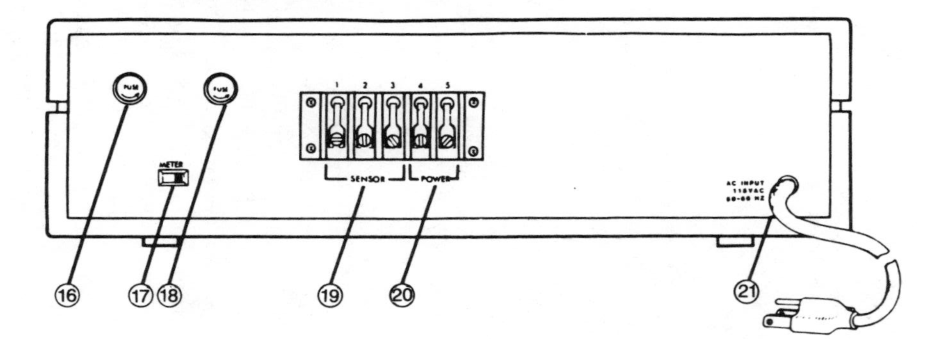

15 Program Mode Key - activated when system is to be programmed.
16 Dc Fuse Holder - replace with *5 amp* slow blow *only.*
17 Antenna Position Lamp Switch - to activate lamp.
18 Ac Fuse Holder - replace with *2 amp* fast blow *only.*

Fig. 7-10. The Gillaspie 2001C Antenna Controller. Courtesy of Gillaspie Communications, Inc.

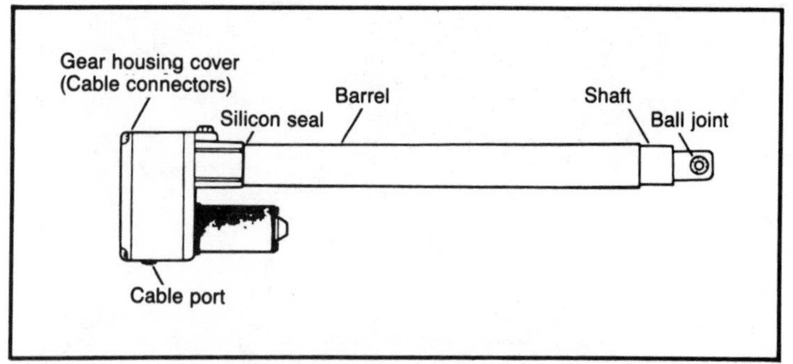

Fig. 7-11. The Gillaspie Mechanical Actuator. Courtesy of Gillaspie Communications, Inc.

panding capabilities and offering the widest choice of systems. Their new receivers are the product of this research and include the Model 6131 and 6136. Both receivers are fully-agile, 24-channel units that can provide excellent satellite broadcast viewing. Each offers different performance and convenience features, however.

The Model 6131 TVRO Earth Station Receiver from Channel Master is shown in Fig. 7-14. This may be the unit that will suit the majority of buyer's needs, as it offers an excellent combination of performance and cost. Selling for about $330, the Model 6131 contains an analog signal strength meter, audio tuning control, color-coded polarity selection, format choice, and the choice of channel 2 or 3 output.

At the opposite end of the TVRO Earth station receiver spectrum is the microprocessor-controlled Model 6136 shown in Fig. 7-15. This state-of-the-art receiver is designed as a remote-controllable centerpiece for an entire video entertainment system. The Model 6136 features programmable antenna control, stereo processor, digital audio and video tuning, automatic satellite dish positioning, block down conversion circuitry, hand-held infrared remote control, and parental lockout control functions.

Once the program and control keyboard on the sleek unit's top are used to preset dish position, polarity and audio format for each channel, the consumer can use the infrared remote control to actuate these functions as well as volume, fine tuning, mute, power on/off and time display.

A unique, multifunction LED can switch between displaying satellite location, transponder selection, audio frequency, signal strength and the time (with the built-in clock). Combining Model 6136 with its companion Model 6255 Satscan antenna controller

Fig. 7-12. The Gillaspie GCI 9600 Satellite receiver with remote control. Courtesy of Gillaspie Communications, Inc.

GCI9600 Satellite Receiver

Electrical Input Characteristics:

Frequency:	70 MHz
Signal Level Range:	−50 dBm to 0 dBm
Impedance:	75 Ohm
AC Power:	117V ac, 50/60 Hz, 40W

Electrical Output Characteristics
Video Output

Level:	1V p-p at 75 Ohms
Frequency Range:	60 Hz to 4.2 MHz
Polarity:	Normal/Inverted, Switch Selectable

Audio Output (Monaural)

Level:	1 V p-p at 10K Ohms
Frequency Range:	30 Hz to 15 kHz
Subcarrier Tuning Range:	5.5 to 7.5 MHz, continuously tunable

Unclamped Baseband Output:

Level:	250 mV p-p at 75 Ohm (min)
Frequency Range:	30 Hz to 8 MHz
De-Emphasis:	None

Modulated RF Output

Level:	0 dBm at 75 Ohm
Frequency:	vhf Ch 3 or 4, switch selectable
Polarizer Control:	+5V with pulse train for Servo Motor Control (1 or 2 m Second pulse for Horiz/Vert.)
External Power to LNA and Downconverter:	18-27 dc, 350 mA

Other Electrical Characteristics

Tuning Voltage to Downconverter:	+4V to +14 Vdc
IF Bandwidth:	22 MHz
Video Demodulation:	FM (CCIR Format 21.5 MHz p-p)
Demodulator Type:	Delay Line Discriminator
Video C/N Threshold:	7 db typical
Tuning Type:	Digitally controlled Microprocessor

Mechanical Characteristics
Input Connectors:

70 MHz RF:	Type F Female
Broadcast TV:	Type F Female
AC Power:	3 Prong Plug

Output Connectors

Modulated RF:	Type F Female
Video:	RCA Phono, Female
Audio:	RCA Phono, Female
Unclamped Baseband:	RCA Phono, Female
LNA/DC Power:	Type F Female
Polarizer:	3 Screw Terminal Strip (+5V dc, pulse, ground)
Dimensions:	16.5″W × 10.0″D × 3.75″H (41.9 cm × 25.4 cm × 9.5 cm)
Net Weight:	8 lb (3.64 kg)

Downconverter

Conversion Type: Single conversion
Input Frequency: 3.7-4.2 GHz
Conversion Gain: 20 dB nominal
Image Rejection: 16 dB nominal
Tuning Voltage: +4 to +14 Volt
IF Impedance: 75 ohms
Dc Power Input: 18 to 24 Volt
IF Connector: Type F Female
Dc Connector: Type F Female
Rf Connector: Type N Female

Fig. 7-13. The GCI 9600 Receiver specifications. Courtesy of Gillaspie Communications, Inc.

gives the consumer complete control over the system. The Model 6136 retails for $1,095, and the optional antenna controller can be had for slightly less than $500.

Channel Master's new offerings to their already broad satellite receiver line seem to be trend among receiver manufacturers today. When the first edition of this book was written, all receivers were generally in the $700 plus price range. Today, the price of receivers that are far better than these early offerings are typical-

Fig. 7-14. The Model 6131 TVRO Earth Station receiver. Courtesy of Channel Master.

Fig. 7-15. The Model 6136 TVRO Earth Station receiver. Courtesy of Channel Master.

ly less than half of the initial low-end price. However, manufacturers are generally offering microprocessor-controlled models that sell in the $1,000 plus range. Most of these are designed as central control units whereby all TVRO Earth station functions, including antenna positioning and tuning, are handled from one remote console mounted at the television receiver. This greatly expands the options available for the typical Earth station owner. A complete low-cost earth station today can sell for less than $1,500, whereas the price was at least twice this just a few years ago. However, the buyer who wants the best available can easily spend $5,000 or more for a truly elaborate setup. Of course, there are many options that will allow the TVRO Earth station that is custom-designed to fall anywhere within these price ranges.

ICM VIDEO

ICM Video is a division of International Crystal Manufacturing, Inc., and offers a line of sophisticated satellite television equipment. This equipment may not be suited to every personal Earth station owner, as it is designed more for cable TV, broadcast TV, apartment house, and hotel/motel installations. Their equipment is made to commercial specifications and is therefore more expensive. However, it also works under closer tolerances and will be ideal for the person who wants the best possible installation that can be had.

Figure 7-16 shows the SR-4600P Receiver which offers drift-free performance and the ability to return to the same channel after a loss of power. You'll notice that it is designed to be rack-mounted, as is most commercial satellite television equipment. Its electronic features make it suitable for unattended or remote operation.

The SR-4600P has a 30 MHz i-f bandwidth and a special video detector. These help to eliminate streaking on transitions and breakup. The video detector is a linear, highly efficient, drift-free quadrature detector. It yields a superior signal-to-noise ratio, extends the threshold, and requires no adjustment. The receiver uses what's known as a back porch video clamp for improved results. Most other receivers use diode clamping, which allows vertical sync distortion and some picture flicker. Vertical sync distortion is what sometimes causes the received picture to roll and distort horizontally. These problems are eliminated with the clamping technique used in this receiver.

The ICM SR-4600P has a dual conversion down converter,

Fig. 7-16. The SR-4600P receiver. Courtesy of ICM Video.

which means that no expensive isolators are needed in multiple receiver installations. The stabilized oscillators in this down converter give drift-free performance. The down converter can be plugged inside the receiver chassis for short cable runs from the antenna. Alternately, a separate waterproof down converter may be ordered for operation at the dish.

The receiver has screwdriver adjustments for video and audio

- Dual Conversion
- Separate or built-in downconverter
- Low Threshold quadrature video detector
- AGC with manual override
- AFC with manual override
- Frequency agile
- Tunable Audio
- Back porch clamped video
- Optional wired remote control
- Built-in polarizer controller
- Rack mountable

The SR-4600P is a stable, state of the art receiver and is ideal for cable TV, broadcast TV, apartment house, and hotel-motel installations. Its drift free performance, and ability to return to the same channel after a loss of power make it suitable for unattended and remote operations. Its 30 MHz IF bandwidth and special video detector eliminate streaking on transitions and eliminate breakup in highly saturated chroma.

The video detector is a linear, highly efficient, drift free, ICM exclusive, quadrature detector. It yields a superior signal to noise ratio, extends the threshold, and requires no adjustments.

The SR-4600P uses a back porch video clamp for improved results. Diode clamping, used in most other receivers, allows vertical sync distortion and some picture flicker. The vertical sync distortion sometimes causes picture rolling and flagwaving. These problems are eliminated with ICM's clamping technique.

The dual conversion downconverter means no expensive isolators are needed in multiple receiver installations. The downconverters' stabilized oscillators give drift free performance. ICM's DC-62 downconverter can be plugged inside the receiver chassis, or a separate weatherproof version, model DC-60, may be ordered for operation at the dish.

The SR-4600P has screwdriver adjustments for video and audio frequency and levels. Convenient LED indicators make exact center tuning easy. If the unit is not used for single channel operation, an optional wired remote control permits video and audio tuning plus polarity selection.

Additional features include: Single-Dual conversion switch (internal), clamped-unclamped video switch (internal), normal-invert switch (internal), operates Chaparral Polarotor I or M/A Com Omni Spectra polarizer, Skew control for Chaparral Polarotor I, 110VAC or 220VAC operation, 1¾" rack mounting, single board construction, front panel signal strength meter, optional channel 3/4 RF modulator, and optional stereo decoder.

Fig. 7-17. Rear panel of the SR-4600P receiver. Courtesy of ICM Video.

Specifications

Receiver:

Receive Frequencies	70 MHz
Input Levels	-60 dBm to -10 dBm
70 MHz Input Impedance	75 ohms
Threshold	8 db
IF Bandwidth	30 MHz
Maximum Detector Bandwidth	±20 MHz
AFC Range	±10 MHz
Video Bandwidth	4.2 MHz
Video Output Level	1V P-P, 75 ohms
Composite Baseband Output	.7V P-P, 75 ohms
Dispersion Removal	50 db
Meter Output	0-1 MA
Audio IF Bandwidth	280 KHz
Audio Frequency Response	30 HZ to 15 KHz
Tunable Audio Range	5-9 MHz
Audio Output Level	-40 to +10 db, balanced and unbalanced
Audio Output Impedance	600 ohms, balanced and unbalanced
Power Requirement	120/240 VAC, 50/60 Hz, 14 Watts typical
Dimensions	17" W x 1¾" H x 12¾" D
Shipping Weight	11 pounds
Accessories Included	Rack mount hardware

Downconverter — Model DC-60 and DC-62:

Receive Frequency	3.7 – 4.2 GHz, Frequency agile
Input Impedance	50 ohms
Minimum Input Level	-60 dBm
IF Frequencies (Dual Conversion)	830 MHz and 70 MHz
Frequency Drift from -25°to +75° C	±10 MHz
Gain	12 db
Tuning Voltage (via 70 MHz coax)	3 to 16 VDC
70 MHz Output Impedance	75 ohms
Power Required (Excluding LNA current)	24 to 32 VDC, 175 MA
Voltage Generated for LNA	15 VDC via "N" connector
Connectors Used (DC-60)	Input-N, 70 MHz out – F, Power in - 6 pin Cinch
Ambient Temperature Range	-25°C to +75°C
Dimensions (DC-60)	7" W x 4⅛" H x 3" D
Shipping Weight (DC-60 or DC-62)	4 pounds

Ordering Information:

Order downconverter separate from receiver, DC-60 dish mounted weatherproof version, or DC-62 chassis plug in version

Optional RF modulator, plug in, non cable/broadcast quality, order model RFC-68.

Optional remote control, wired with 20 ft. cable, audio and video tuning plus polarity selection, order model RC-20

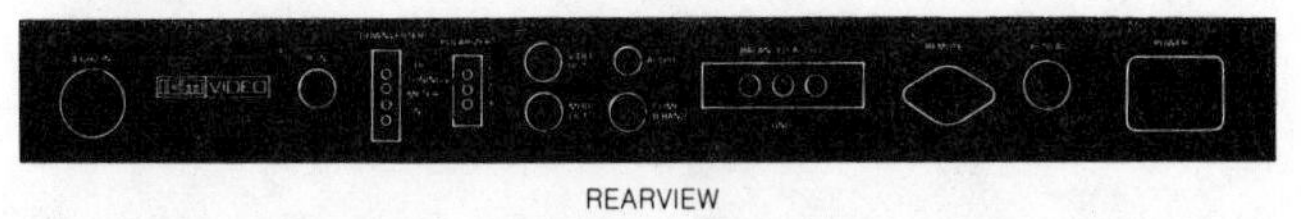

REARVIEW

Fig. 7-18. Specifications for the SR-4600P receiver. Courtesy of ICM Video.

frequency levels. LED indicators make exact center tuning easy. Of course, this discussion assumes a single-channel operation, as would be the case with a cable company or a hotel or motel. If the unit is not to be sued for single-channel operation, an optional wired remote control unit is available that permits video and audio tuning as well as polarity selection at the receiver.

Additional features include an internal single/dual conversion switch, clamped/unclamped video switch, normal/invert switch, and 110 or 220 Vac operation. Figure 7-17 shows the rear panel of this unit, along with its various components.

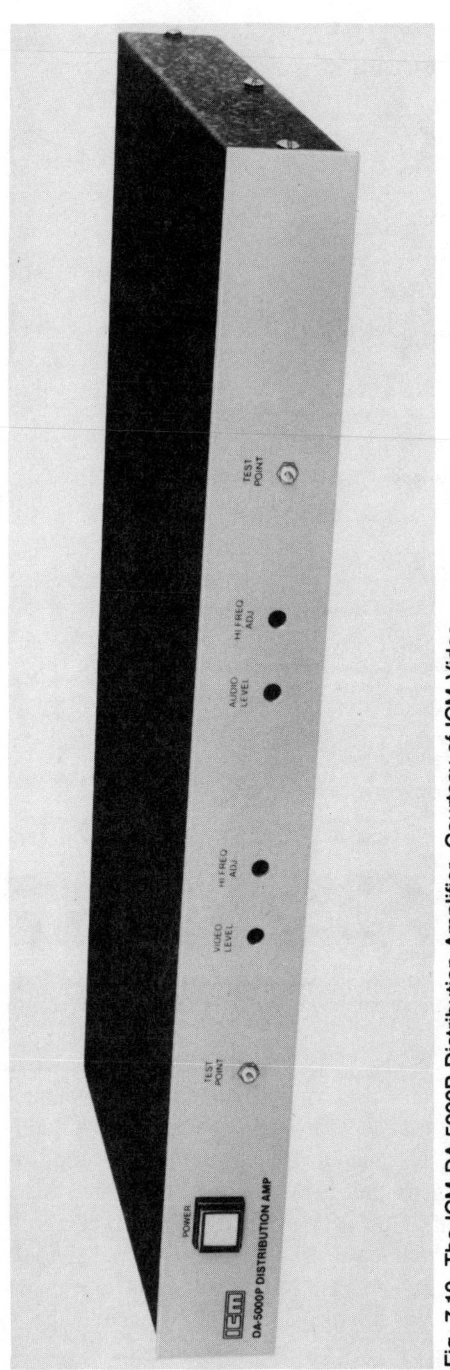

Fig. 7-19. The ICM DA-5000P Distribution Amplifier. Courtesy of ICM Video.

Again, this receiver is not intended for most home-owned TVRO Earth stations. Its applications are aimed more toward the commercial field. Figure 7-18 provides a list of specifications for this receiver and the down converters it is intended to be used with.

Assuming a multi-output operation, as would be the case with a motel or apartment building, a distribution amplifier is required. The ICM DA-5000P is shown in Fig. 7-19. This versatile distribution amplifier combines audio and video into one cabinet. Ten matched, isolated video outputs are provided from a single video input, along with ten matched, isolated audio outputs from a single audio input. The inputs are looping, or bridged, so that additional distribution amplifiers could be stacked to provide more than ten outputs. The front panel has a screwdriver adjustment on both video and audio channels for overall level and high frequency compensation. Test points on the front allow for oscilloscope monitoring of one of the video and audio outputs. Figure 7-20 shows the rear panel of this unit, and again, one must remember that distribution amplifiers are really aimed at commercial installations. However, since the readership may be comprised of some persons who will want to install satellite television service in their apartment houses or motels, a review of this equipment is included here. Figure 7-21 provides specifications on the DA-5000P Distribution Amplifier.

Another product from ICM Video may find usage in typical

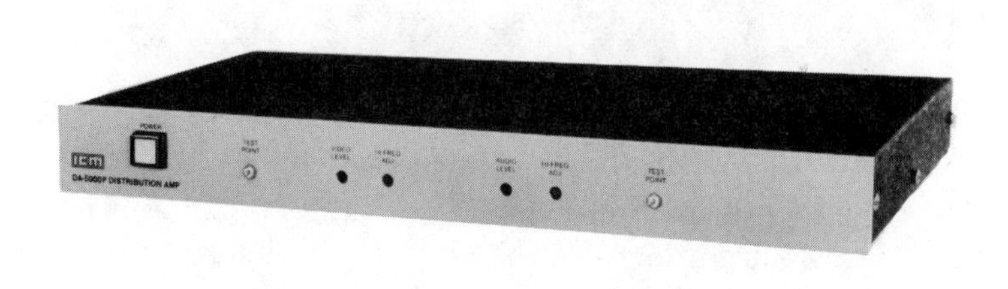

Fig. 7-20. Rear panel of the DA-5000P Distribution Amplifier. Courtesy of ICM Video.

Specifications

Video:

Bandwidth	10 MHz
Linearity	Better than 1%
Tilt	Less than 1%
S/N Ratio	56 db
Differential Phase	Less than ½°
Differential Gain	Less than 1%
Gain	Up to 6 db (front adjustment)
Hi Freq. Comp.	Up to 1500 ft. RG-59/U
Isolation Between Outputs	40 db
Connectors Used	BNC

Audio:

Bandwidth	50 KHz
T.H.D.	.05%
S/N Ratio	70 db
Gain	Up to 6 db (front adjustment)
Isolation Between Outputs	60 db
Input	High impedance, unbalanced
Outputs	600 ohm, unbalanced
Connectors Used	RCA phono

General:

Power Requirement	110VAC, 60 Hz (220VAC, 50 Hz capable)
Power Consumption	6 watts
Operating Temperature	10°C to 50°C
Dimensions	17" W x 12" D x 1.75" H
Shipping Weight	10 lb.
Accessories Included	Rack mount hardware

SPECIFICATIONS SUBJECT TO CHANGE WITHOUT NOTICE

REAR VIEW

Fig. 7-21. DA-5000P specifications. Courtesy of ICM Video.

TVRO Earth station installations. Shown in Fig. 7-22, this is the SA-50 Signal Purifier Module, which is a unique filter/amplifier device for the 3.7 to 4.2 GHz TVRO frequency spectrum. The SA-50 contains a five-pole bandpass filter that reduces or eliminates out of band interference from radar transmitters, image noise, etc. Its 3 dB bandwidth points are at 3.5 and 4.5 GHz. A screwdriver adjustment will set signal levels from +5 to −10 dB. A gain setting can be used to compensate for weak signals or to allow longer coaxial cable runs.

The SA-50 Signal Purifier is mounted at the LNA output and before any down conversion takes place. The device is extremely simple to hook up and operate. It has a female type-N input jack and a male output plug of the same type. This makes it easily insertable in the coaxial line. It receives its power through the coax and requires the same voltage as the LNA. It passes current through to the LNA. It can also be used as a dc block.

Because of the combination of gain/attenuation and filtering, SA-50 Signal Purifier can solve a wide variety of tough interference and/or weak signal problems. It is a device that can be used with any TVRO Earth station installation that encounters interference or weak signal problems.

Figure 7-23 shows the ICM Video DC-60 Down Converter. This is a dual conversion down converter that accepts an input of 3.7 to 4.2 GHz from the LNA. Its output is at the standard TVRO IF of 70 MHz. Dual conversion means no expensive isolators are necessary if multiple units are installed. The DC-60 has a built-in 3.7-4.2 GHz bandpass filter which reduces or eliminates out of band interference from microwave sources. Tuning voltage is fed up the 70 MHz coaxial cable line while 24 to 32 volts dc is fed via a separate cable for down converter and LNA power. This down converter requires 3 to 16 Vdc tuning voltage. Its input connector is a female type-N, while the 70 MHz output connector is a BNC type. The power connector is a 610 Cinch-Jones type. It will operate over a broad ambient temperature range of from -30° to +60° Centigrade.

Figure 7-24 shows the ICM Video TA-30 Tunable Audio. This

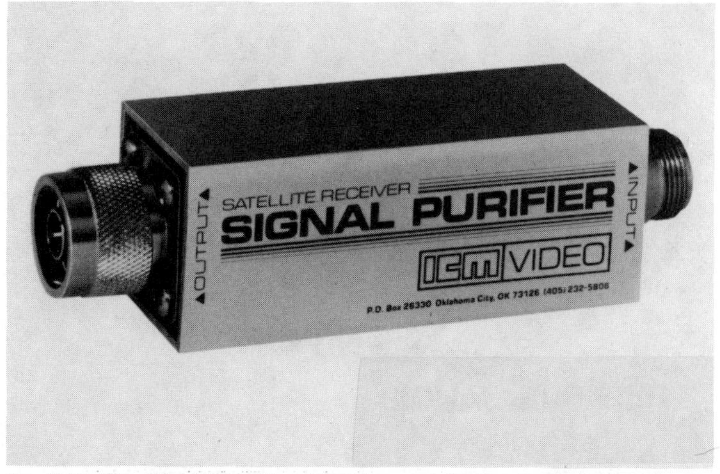

Fig. 7-22. The ICM SA-50 Signal Purifier module. Courtesy of ICM Video.

Fig. 7-23. The DC-60 down converter is housed in a weatherproof container. Courtesy of ICM Video.

can be added to any satellite receiver that is equipped with a subcarrier or baseband video output. The TA-30 allows simple tuning over the frequencies of 5.5 to 8.5 MHz. Its hookup is extremely simple and it contains RCA phono jacks for input and output. The

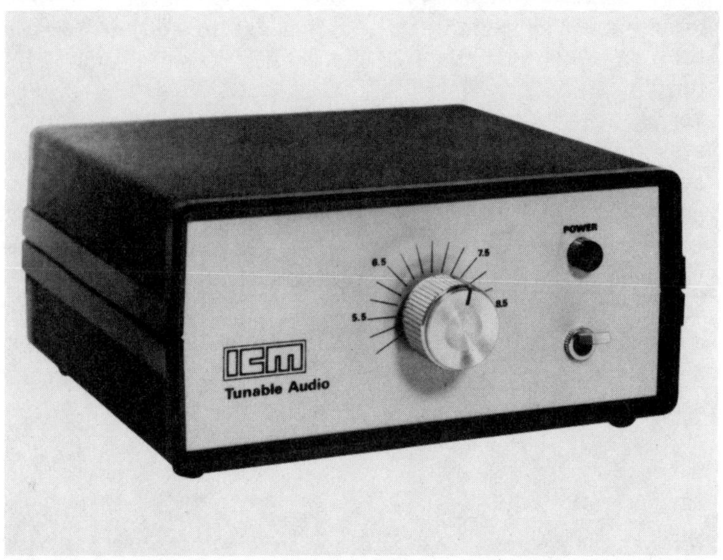

Fig. 7-24. The TA-30 Tunable Audio. Courtesy of ICM Video.

signal required to feed this device is usually labeled "subcarrier out", "baseband out", or "composite out" on most receivers. If a receiver does not have a jack that provides this signal, it can be easily modified to provide it. The TA-30 is designed to tune wideband audio channels only. Narrowband audio services will be unsatisfactory, as they will sound weak and noisy.

AVCOM

Avcom is a satellite TV company located in Richmond, Virginia. This company specializes in receivers and down converters. They also offer a number of complete installation packages that start at approximately $1,700. Avcom has been in the TVRO Earth station business since its beginnings and they are constantly offering new products, along with a host of accessory equipment.

Avcom COM-2 Series

The COM-2 series from Avcom consists of the COM-2A and COM-2B, which are low-cost TVRO receivers. Both models have the same specifications. The COM-2A, however, features a comprehensive remote control console, whereas the COM-2B is a single unit that mounts at the receiver. Each of these receivers is ideally suited for low-cost installations where maximum reliability and high performance are important. The COM-2B has operator controls on the front panel, extended threshold performance, a sophisticated tunable audio system with wide and narrow i-f filters, a video inversion switch, and a remote down converter. The tunable audio system allows reception of special audio feeds. Many options are available for the COM-2B, including the COM-2A remote control panel.

Other options for both models include a protective down converter housing, crystal-controlled modulator, and a sensitive signal strength meter on the optional remote control. Each unit features a continuous transponder tuning knob combined with a carefully designed automatic frequency control, which makes transponder selection extremely easy. Since there are no detents on this knob, no realignment is necessary due to seasonal changes. In other words, the tuning knob is continuously fine-tuned as opposed to having built-in stops that are internally tuned. Figure 7-25 provides specifications for this series of receivers. Again, both the 2A and 2B receivers are identical, save for the remote control on the 2A.

Downconverter, RDC-11 Remote (included with COM-2A)	
Input Frequency	3.7 to 4.2 GHz
Input Impedance	50 ohms
Input Level	-55 to -25 dBm
Bandwidth	500 MHz
Flatness	± .8dB/100 MHz
i-f Output Frequency	70 MHz
i-f Output Bandwidth	36 MHz
i-f Output Impedance	75 ohms
Image Rejection	23 dB nominal
Supply Voltage	+23 V.D.C.
Tuning Voltage	-7 to -12 V.D.C. nominal (via IF coax)
Receiver, COM-2A	
i-f Input Frequency	70 MHz
i-f Input Impedance	75 ohms
i-f Input Level	-50 to -20 dBm
i-f Bandwidth (standard) (other bandwidths available)	28 MHz
Threshold Level	Better than 8 dB C/N
Video Specifications	
De-emphasis	CCIR 405-1, 525 lines
Frequency Response	to 4.2 MHz
Dispersion Removal	43 dB nominal
Output Level	1 volt p-p
Output Impedance	75 ohms
Output Polarity (Polarity is internally reversible)	Standard, Negative Sync
Tunable Audio Specifications (Standard)	
Subcarrier Frequency	4.5 to 8.0 MHz
Bandwidth, Dual	280 KHz wide, 120 KHz narrow
De-emphasis	75 Microseconds
Frequency Response	100 Hz to 15 KHz
Output Impedance	600 ohm balanced or unbalanced
Output Level	1 Volt, adjustable

Fig. 7-25. Specificiations for the Avcom COM-2 Series receivers. Courtesy of Avcom of Virginia, Inc.

Avcom COM-3 Series

The next step up the receiver line is the Avcom COM-3 and COM-3R receivers. These are high-performance, state-of-the-art receivers that offer excellent satellite reception with small aperture antennas. Many features have been incorporated into these receivers to provide ease of operation and at the same time yield powerful and versatile Earth station receiver capabilities.

Special threshold extension circuitry is used to enable the receiver to produce outstanding quality video even in the fringes of a satellite footprint. For international satellite reception, optional threshold peaking and selectable dual i-f filters, optimized for this purpose, provide excellent results.

The COM-3 model is an all in one unit, while the COM-3R is remote control equipped. Each of these receivers offers a detented channel selector to allow fast, positive transponder selection. A high-performance, tunable audio demodulator with selectable i-f filters is a standard feature. The down converter can either be installed in the receiver itself or ordered as a remote control unit at no extra charge. Scan-tune is provided to automatically sweep the satellite frequencies to enable rapid, initial antenna orientation and to locate new satellite activity. Dual video outputs and a video polari-

RF Input	
Input Level	–55 to –25dBm
Frequency	3.7 to 4.2 GHz
Impedance	50 Ohms
Return Loss	≈ 15 dB
Noise Figure	≈ 12 dB
Image Rejection	25 dB nominal
Connector Type	Type N
IF	
IF Frequency	70 MHz
IF Bandwidth (standard) (other bandwidths available)	30 MHz
IF Input Level	–50 to –20 dBm
Threshold Level	Better than 8 dB C/N
Dynamic Range of Threshold Extension	> 20 dB
Video	
Impedance	75 Ohms
Return Loss	> 25 dB
Polarity	Switch Selectable, Black to White, Normal or Inverted
Clamping	Fully active, > 40 dB
Output Level	1 v p-p, adjustable
De-emphasis	Standard NTSC
Audio Demodulator	
Tuning Range	4.2 to 8.1 MHz
Bandwidth, Dual	280 KHz wide, 120 KHz narrow
De-emphasis	75 Microseconds
Frequency Response	30 Hz to 15 KHz
Output Impedance	600 Ohm balanced or unbalanced
Output Level	1 volt, adjustable
Mechanical	
Height	5″
Width	17″
Depth	17″
Weight	16.5 lbs.
Power Requirement	
100 to 130 VAC 60 Hz	40 Watts

Fig. 7-26. Table of specifications for the Avcom COM-3 Series receivers. Courtesy of Avcom of Virginia, Inc.

ty inverting switch is standard. Remote control of video inversion is optional. Either unit includes an internal dc power block, excellent threshold sensitivity, 24-position switch-selectable tuning, and an LNA power supply. Figure 7-26 shows a list of specifications for this series of receivers.

Avcom COM-65T

The COM-65T satellite receiver and Avcom's BDC-60 block down converter comprise a high-performance, semi-agile dual conversion satellite receiving system designed specifically for commercial applications. Modular circuit packaging results in a compact and highly reliable receiver in a 3-inch standard rack mount configuration. Highly stable oscillators eliminate frequency drift and allow operation over wide temperature ranges. The 270-770 MHz block down conversion frequency enables complex and versatile systems to be configured using low-cost cable and components. Special threshold extension circuitry offers superior video quality with small aperture antennas.

The COM-65T receiver uses Avcom's unique "Group Card" concept, which configures the COM-65T to receive eight groups, each group covering three of 24 possible transponders. Each of the eight groups is selected by a card that plugs into the front panel. The cards can be changed in seconds and a complete set is furnished. There are several important benefits derived from the use of group cards. Simplification of the receiver's design allows fewer parts to be used, particularly some of the electrically fragile ones. The lowered parts count reduces cost and increases receiver reliability dramatically. Image rejection can also be improved over other competitive equipment. The receiver is more stable and maintenance is greatly simplified.

Standard features of the COM-65T include tunable audio with wide and narrow audio i-f filters, a complete set of group cards, normal or inverted video switch, and precise signal strength metering for antenna optimization. The COM-65T may be ordered with optional threshold peaking and dual i-f filters for receiving international transmissions. Other options include remote video polarity switch, clamper disable switch, and 600 ohm balanced audio output. Figure 7-27 shows the specifications for the COM-65T, while Fig. 7-28 provides specifications for the BCD-60 block down converter. The specifications on the 65T are quite similar to those of the 66T. Figure 7-29 shows a typical COM-60 series system configuration.

RF Input	
Input Level	–70 dBm to –25 dBm
Frequency	270 to 770 MHz
Impedance	75 Ohms
Return Loss	> 15 dB (270-770) MHz)
Noise Figure	≈ 15 dB with attenuator
Image Rejection	> 45 dB
Connector Type	Type F
IF	
Manual Gain Control	24 dB variation
Frequency	70 MHz
Bandwidth	30 MHz
Dynamic Range of Threshold Extention	> 20 dB
Video	
Impedance	75 Ohms
Return Loss	> 25 dB
Polarity	Switch Selectable, Black to White, Normal or Inverted
Clamping	Fully active, > 40 dB
Output Level	1 v p-p, adjustabie
De-emphasis	Standard NTSC
Audio (Dual 6.8 and 6.2 MHz standard) COM-65(T) incorporates a tunable audio demodulator	6.8 and 6.2 MHz standard. Others optional. Tunable demodulator in COM-65T
Frequency Response	30 Hz to 15 KHz ± 1 dB
De-emphasis	75 Microsecond
Output Level	.7 volt p-p
Harmonic Distortion	< 1.8%
Operating Temperature	5° to 40°C
Mechanical	
Height	3.5″
Width	19″
Depth	14″
Weight	12 lbs.
Mounting	
Power Requirements	Standard 19″ Rack Mount
100 to 130 VAC 60 Hz	45 Watts

Fig. 7-27. Specifications for the Avcom COM-65T receiver. Courtesy of Avcom of Virginia, Inc.

Avcom COM-20T

The COM-20T satellite receiver, when coupled with the RDC-20/RDC-21 remote down converters, forms an economical and high-performance satellite receiver system for cable, SMATV, radio stations, and other dedicated use applications. State of the art microwave oscillator technology allows the innovative down converters used with the COM-20T to maintain high stability over temperature extremes. Tunable audio and rack mounting result in a versatile receiver for diverse applications. Many options are available, including internal down converter, clamper disable, and remote polarity video control. Figure 7-30 shows a typical system interconnection using two down converters. Figure 7-31 provides a list of specifications for the Avcom COM-20T satellite video receiver.

As of this writing, two new products have just been released by Avcom. First, there is the COM-23T satellite receiver for international reception. This new receiver is intended for dedicated channel reception applications in conjunction with the RDC-20/RDC-21 remote down converters. This system results in an economical, high-performance satellite receiver for commercial applications. The COM-23T receiver is highly stable and features dual i-f controls and threshold peaking. This should be more than enough to meet the demands that specialized international satellite reception requires. Many options are available, and by the time you read these words, a lot more information should be available on this new product.

Also new from Avcom is the SPM-3 stereo processor. This is a high-performance, economical stereo demodulation system that will decode the stereo signals from satellite-delivered programming. The SPM-3 stereo processor has a selectable audio i-f bandwidth for those special narrow band subcarriers, an instant monaural stereo switch, tunable audio demodulators covering all subcarrier frequencies, and standard outputs for easy connection to any ste-

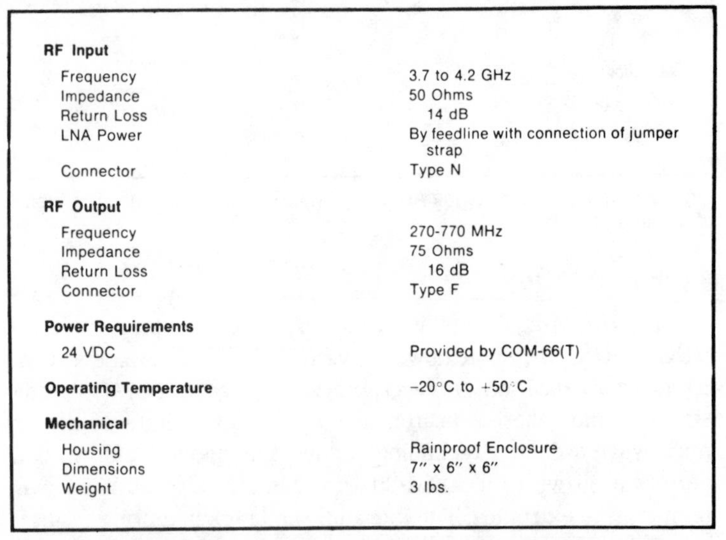

RF Input	
Frequency	3.7 to 4.2 GHz
Impedance	50 Ohms
Return Loss	14 dB
LNA Power	By feedline with connection of jumper strap
Connector	Type N
RF Output	
Frequency	270-770 MHz
Impedance	75 Ohms
Return Loss	16 dB
Connector	Type F
Power Requirements	
24 VDC	Provided by COM-66(T)
Operating Temperature	−20°C to +50°C
Mechanical	
Housing	Rainproof Enclosure
Dimensions	7″ x 6″ x 6″
Weight	3 lbs.

Fig. 7-28. Avcom BCD-60 block down converter specifications. Courtesy of Avcom of Virginia, Inc.

reo amplifier. Also featured on the SPM-3 stereo processor are LED indicators for "power on" reception of MPX stereo signals and selection of L+R and L−R tunable demodulators.

UNIDEN UST 1000 SATELLITE RECEIVER

A brand-new offering to the satellite receiver market is the Uniden UST 1000. Featuring audio and video fine tuning, the receiver is attractively styled, offers polarity reversal and channel scanning features. Both of these are served by appropriate indicators. There is also a skew control on the front panel and a built-in modulator. The UST 1000 is built to be durable, is reasonably priced, and is extremely easy to operate. Figure 7-32 provides a list of specifications for this receiver and its matching UST 500 Down converter.

M/A-COM RECEIVERS

M/A-COM has long been involved in the TVRO business, and they have recently announced two new receivers. They are the T-1 and H-1. Each of these receivers has microprocessor controls, that allow one-step tuning. It is usually not necessary to repeat the process every time the receiver is turned on. Another feature is microprocessor memory retention that allows the programmed events to remain intact even during a power failure. Some of the features of these two receivers are parental supervision channel lockout, a single infrared remote control for the new M/A-COM antenna positioner, acceptance of both mechanical and electronic polarization controls, and programmable audio. On the H-1 series, up to 72 audio memories are available. Dynamic noise reduction is used on the H-1 receiver, while Dolby is used on the T-1. The T-1 comes in a very slim, space-age case with a light metallic finish. The H-1 stands a bit higher and is housed in a matte black case. Figure 7-33 provides a table of specifications for the H-1 receiver, and Fig. 7-34 contains the specifications for the T-1 receiver. Both of these units are available from M/A-COM Cable Home Group, Hickory, North Carolina.

LOW NOISE AMPLIFIERS

It's very difficult to write a product description of the various makes of low noise amplifiers (LNAs) on today's market. For all intents and purposes, one general description fits all makes. This

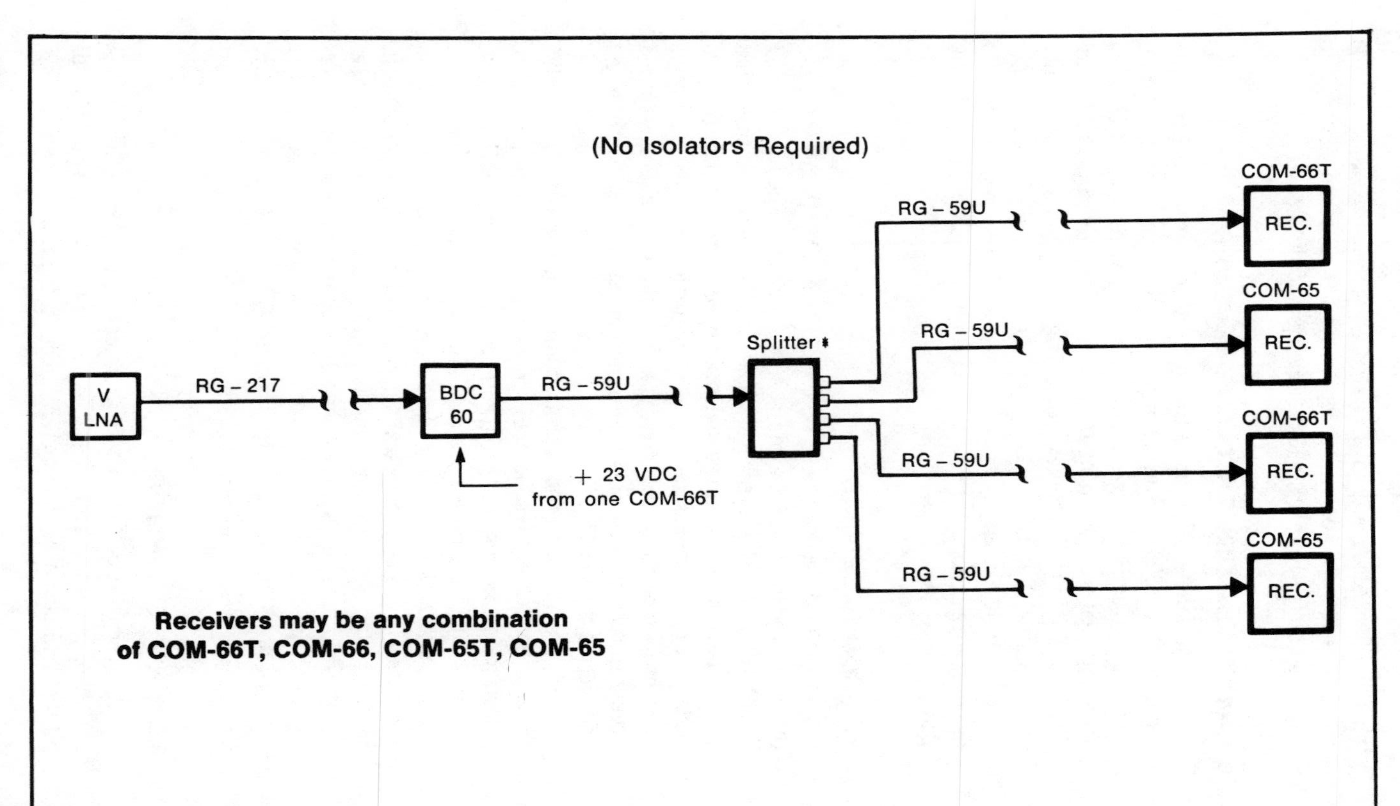
(No Isolators Required)
V
LNA
RG – 217
BDC
60
RG – 59U
+ 23 VDC
from one COM-66T
Splitter *
RG – 59U
RG – 59U
RG – 59U
RG – 59U
COM-66T
REC.
COM-65
REC.
COM-66T
REC.
COM-65
REC.
Receivers may be any combination
of COM-66T, COM-66, COM-65T, COM-65

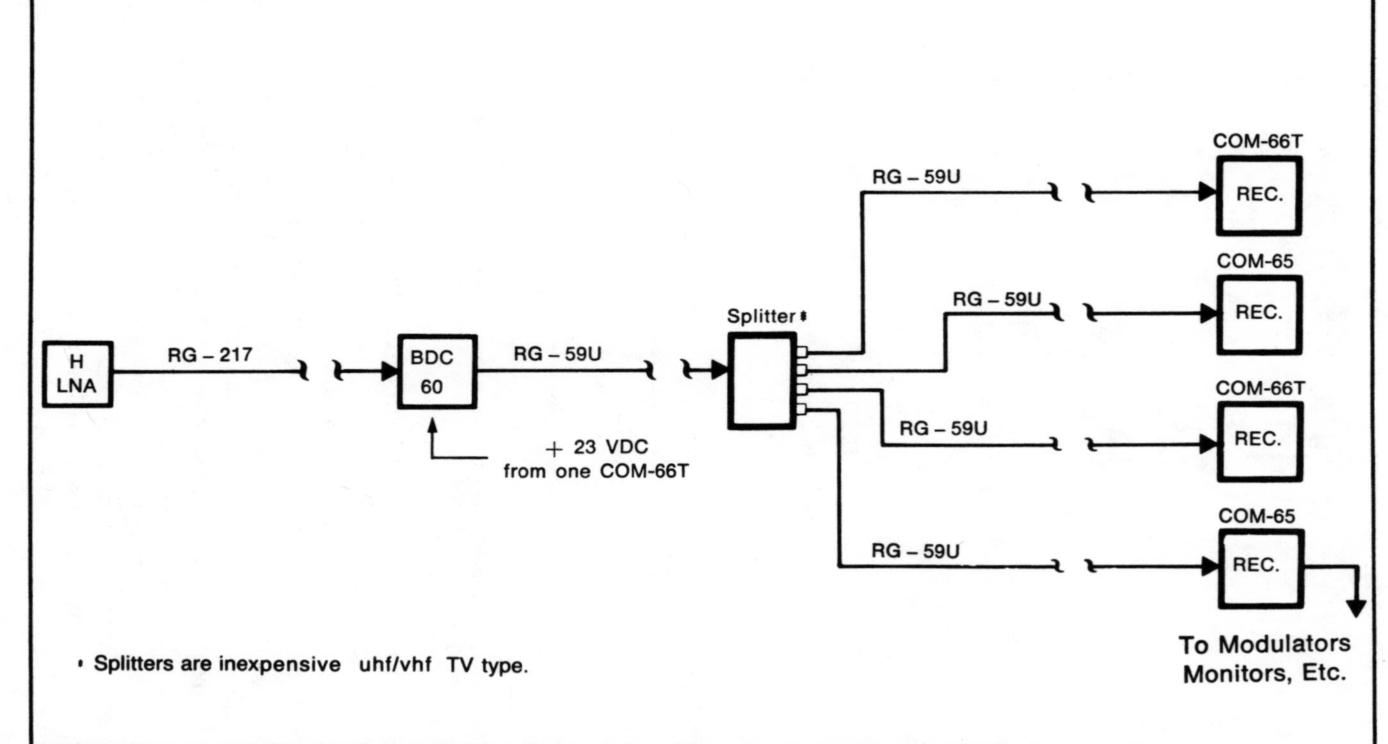

Fig. 7-29. Typical COM-60 Series system configuration. Courtesy of Avcom of Virginia, Inc.

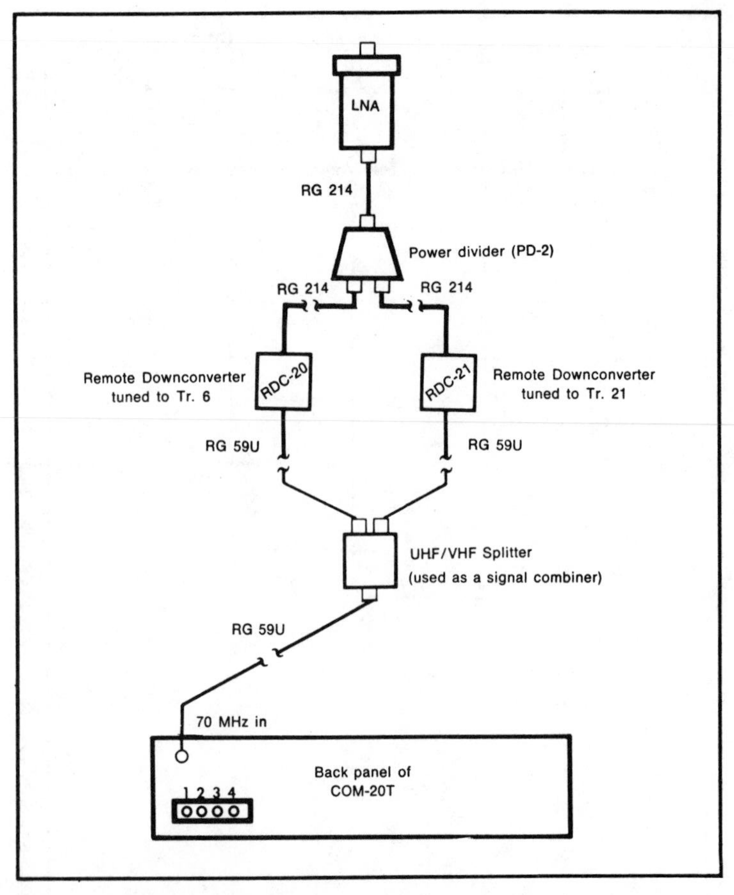

Fig. 7-30. Typical system interconnections using two down converters. Courtesy of Avcom of Virginia, Inc.

is not to say, however, that all LNAs are alike. The real description of the LNA is garnered from the technical specifications, which address gain and noise figures. Most LNAs made today offer typical gains of about 50 dB minimum. Noise figures (measured in degrees Kelvin) typically range from a low of 60° to a high of 120°. The 100° LNA is considered to be a standard in the TVRO field.

To overcome the boredom of describing umpteen different LNAs from as many different companies, only one is included in this section. You should be aware, however, that while there may be many different brands of LNAs on today's TVRO market, most are made by a handful of manufacturers and distributed to a larger number of companies who label these generic LNAs with their name. Specifications are the main guide to choosing an LNA, although one must be concerned with mechanical details such as weatherproofing, connectors, etc.

Shown in Fig. 7-35, the GCI 3742 Low Noise Amplifier from Gillaspie features rugged mechanical construction required for this type of device. A minimum gain of 50 dB is guaranteed, and the unit is available in four different noise figure options. These are 85°, 90°, 100°, and 120°. A built-in filter and input isolator is used to reject outside interference and ensure stable operation. Voltage regulator and reverse bias protection assures electrical integrity. The housing of the GCI 3742 is cast aluminum and weatherproof. It is sold with a full one year guarantee, providing that the protective seals are not broken. These seals are broken whenever the unit

Downconverter, RDC-20 (High Stability)	
Input Frequency	3.7 to 4.2 GHz
Input Impedance	50 ohms
Input Level	-55 to -25 dBm
Flatness	± .8dB/100 MHz
IF Output Frequency	70 MHz
IF Output Bandwidth	36 MHz
IF Output Impedance	75 ohms
Image Rejection	23 dB nominal
Supply Voltage	+23 V.D.C.
Tuning Voltage	Mechanically tuned high stability microwave oscillator
IF Input Frequency	70 MHz
IF Input Impedance	75 ohms
IF Input Level	-50 to -20 dBm
IF Bandwidth (standard) (other bandwidths available)	28 MHz
Threshold Level	Better than 8 dB C/N
Video Specifications	
De-emphasis	CCIR 405-1, 525 lines
Frequency Response	to 4.2 MHz
Dispersion Removal	43 dB nominal
Output Level	1 volt p-p
Output Impedance	75 ohms
Output Polarity (Polarity is internally reversible)	Standard, Negative Sync
Tunable Audio Specifications (Standard)	
Subcarrier Frequency	4.5 to 8.0 MHz
Bandwidth, Dual	280 KHz wide, 120 KHz narrow
De-emphasis	75 Microseconds
Frequency Response	100 Hz to 15 KHz
Output Impedance	600 ohm balanced or unbalanced
Output Level	1 Volt, adjustable

Fig. 7-31. Specifications for the Avcom COM-20T receiver. Courtesy of Avcom of Virginia, Inc.

IF SPECIFICATIONS	
IF Frequency:	70 MHz
IF Impedance:	75 ohms
IF Bandwidth:	25 MHz @ - 3 dB
Threshold:	Less than 8 dB C/N
VIDEO SPECIFICATIONS	
De-Emphasis:	CCIR 405-1,525 lines
Frequency Response:	20 Hz to 4.2 MHz
Output Level:	1 volt p-p adjustable
Output Impedance:	75 ohms
Output Polarity:	Negative sync, (switchable)
Composite Output:	20 Hz to 8 MHz
AUDIO SPECIFICATIONS	
Subcarrier Frequency:	5 to 8 MHz selectable; 6.2 and 6.8 MHz preset
Frequency Response:	50 Hz to 15 kHz
Output Impedance:	600 ohm unbalanced
Output Level:	0 dB nominal
Harmonic Distortion:	Less than 2% T.H.D.
Rf Re-Modulator	
Channels:	3 or 4 switchable
Frequency Stability:	Quartz controlled
POLARIZATION CONTROL	
Interface:	Unirotor™ and SPDT relay
Adjustment:	Continuously controllable from panel "SKEW" knob
Polarity:	SATCOM or WESTAR polarity switchable

CONTROLS	
Front Panel:	Detent channel selector; channel indicator; (in channel selector knob); video fine tuning; variable audio tuning only; antenna polarity mode switch; tuning meter; channel scan mode switch; automatic polarity selector; polarity indicator; SKEW control (Unirotor); power on/off switch
Rear Panel:	Vertical adjustment; horizontal adjustment; switchable channels 3 or 4; video polarity switch; SKEW angle limit adjustment
Bottom Panel:	Tuning meter sensitivity control; cable compensation control
UST 500 WEATHER-SEALED DOWN CONVERTER	
Input Frequency Range:	3.7 ~ 4.2 GHz
Output Frequency:	70 MHz
Power Supply from Receiver	+15 volts dc 150 mA
Input Connector:	N connector, 50 ohms
Output Connector:	F connector, 75 ohms
Image Rejection:	20 dB minimum

One Year Limited Warranty

Fig. 7-32. Specifications for the Uniden UST 1000 Receiver and UST 500 down converter. Courtesy of Uniden.

cover is removed. Generally speaking, LNAs are not serviceable by anyone other than factory personnel. Therefore, there should never be a reason to go tampering inside. Each GCI 3742 is fully burned in, which means the units have been pre-tested in standard operation before being sold. Figure 7-36 provides the most important electrical and mechanical specifications of the GCI 3742.

AM I SMART ENOUGH TO INSTALL MY OWN EARTH STATION?

The heading of this section is a question many persons ask themselves when contemplating the installation of their own TVRO

GENERAL	
RF Input	
Frequency	3.7 to 4.2 GHz
Noise Temperature	120°K max.
Level	—88 dBm ±5 dB
Connection	CPR229G Flange
Image Rejection	greater than 40 dB
IF Input	
Frequency	950 to 1450 MHz
Impedance	75 ohms
FM Static Threshold	8 dB C/N min.
Second IF Section	
Frequency Output	403 MHz
Bandwidth	Over 27 MHz
Noise Temperature	15 dB Nominal
Gain Control	AGC
Tuning Control	AFC
Video	
Bandwidth	10 Hz to 4.2 MHz
Polarity	non-inverted
Dispersion Clamp	Greater than 40 dB
Impedance	75 ohm
Output Level	1V p-p
Connectors	Type F
Baseband Output	1V p-p, 75 ohms
Audio	
Subcarrier	5.0 to 8.5 MHz
Modes	Monaural, Direct and Matrix Stereo
De-Emphasis	75 microseconds
Output Level	2.5V p-p
Impedance	600 ohms, unbalanced
Connectors	RCA
PRIMARY POWER	
Source	114V AC (60 MHz)
Fuse	1 Ampere
ODU POWER	
Voltage	+15 to +21.2V DC
Current	150 mA (max.)
Fuse	0.5 Ampere
Mechanical	
Size	17⅓″ (W) x 13⅓″ (D) x 3⅓″ (H)
Weight	Less than 14 lbs.

Fig. 7-33. M/A-COM H-1 receiver specifications. Courtesy of M/A-COM.

GENERAL	
RF Input	
Frequency	3.7 to 4.2 GHz
Noise Temperature	120°K max.
Level	−88 dBm ±5 dB
Connection	CPR229G Flange
Image Rejection	greater than 40 dB
IF Input	
Frequency	950 to 1450 MHz
Impedance	75 ohms
FM Static Threshold	8 dB C/N min.
Second IF Section	
Frequency Output	140 MHz, upperside heterodyned
Bandwidth	Over 27 MHz
Noise Temperature	15 dB Nominal
Gain Control	AGC
Tuning Control	AFC
Video	
Bandwidth	10 Hz to 4.2 MHz
Polarity	non-inverted
Dispersion Clamp	Greater than 40 dB
Impedance	75 ohm
Output Level	1V p-p
Connectors	RCA
Baseband Output	1V p-p, 75 ohms
Audio	
Subcarrier	5.0 to 8.5 MHz
Modes	Monaural, Direct and Matrix Stereo
De-Emphasis	75 microseconds
Output Level	2.5V p-p
Impedance	600 ohms, unbalanced
Connectors	RCA
PRIMARY POWER	
Source	115V AC ±10% (60 Hz)
Fuse	1 Ampere
ODU POWER	
Voltage	+15 to +21.2V DC
Current	150 mA (max.)
Fuse	0.6 Ampere
Mechanical	
Size	16½″ (W) x 11¾″ (D) x 2″ (H)
Weight	Less than 12 lbs.

Fig. 7-34. Specifications for the T-1 receiver. Courtesy of M/A-COM.

Fig. 7-35. The GCI 3742 Low Noise Amplifier is available in a number of different noise figures. Courtesy of Gillaspie Communications, Inc.

Earth station. Since the price of packaged Earth stations has fallen dramatically in the last several years, fewer persons are installing their own and are opting to have this done by a dealer. In many instances, this is preferable, assuming that you have purchased a packaged system from a dealer who normally does installations.

GCI-3742 Low Noise Amplifier

Electrical Specifications:

Input frequency:	3.7 - 4.2 GHz
Noise figure: 3742-120	120°K (Meets minimum
3742-100	100°K specifications at
3742-90	90°K every in band
3742-85	85°K frequency)
Noise figure variation:	0.01 dB/°C
Gain:	50 dB minimum
Gain Flatness:	±1 dB
Gain Slope:	0.015 dB/MHz
Input VSWR:	1.25 typical
Output VSWR:	1.5 typical
Supply Voltage:	15 to 24 volt dc
Supply Current:	120 mA via rf cable:

Mechanical Characteristics

Rf Input:	CPR 229 waveguide flange
Rf Output/dc Input:	Type "N" female

Environmental Specifications

Vibration:	G = 5; Frequency = 50 ± 2 Hz; t = 5 minimum; Direction X, Y, Z
Shock:	G = 1; Direction X, Y, Z
Operating Temperature:	−30 to +60° C
Humidity:	95% RH

Fig. 7-36. Electrical and mechanical specifications of the GCI 3742. Courtesy of Gillaspie Communications, Inc.

Here, the price of the installation is built into the overall price of the system, and this can sometimes mean a savings.

However, it has been my experience that by carefully selecting the individual components of an Earth station and installing them yourself, an even greater savings can be realized. Also, you have the advantage of looking at the specifications of an antenna made by one manufacturer, an LNA made by another, and a receiver made by a third manufacturer. Maybe the manufacturer who makes the best receiver doesn't offer the best LNA, or maybe the manufacturer of the best antenna supplies a mediocre receiver. By mixing and matching, you may be able to come up with the best system for your own particular needs. Of course, you should realize that rarely does a single manufacturer make all of the components of the Earth station. Packaged Earth stations often consist of LNAs made by another company. Quite frequently, the company's name does not even appear on the LNA, but is replaced by the name of the company offering the package.

In any event, a bit of technical knowledge is required these days in order to properly decipher the various specifications that accompany each Earth station component. This should place no great strain on anyone, as a couple of hours of study should be all that's required to properly rate any piece of Earth station equipment. We've already learned that a 120 ° LNA is not as sensitive as a 100 ° unit. This same principle carries over to the receiver and antenna in sensitivity and gain figures, respectively.

The question "Am I smart enough to complete my own installation?" often arises after a particular combination has been chosen. In my opinion, if you're familiar with some of the simpler terminology surrounding Earth stations and can wield a screwdriver, possibly a soldering iron, and a few other hand tools, you are probably capable of making your own installation.

Unfortunately, you often don't know exactly what you're getting into until the equipment arrives and you leaf through the installation manuals. Ideally, you will have already asked for (and received) a copy of the installation manual (usually available for a modest fee) before actually deciding whether or not to go ahead with the installation.

In order to give beginners an idea of what a portion of an installation would be like, the following section of this chapter includes a reprinting of the installation manual for installing the Model 9600 Satellite Receiver from Gillaspie Communications, Inc., of Sunnyvale, California. Gillaspie has been in the business for a long time and offers a wide range of products for the satellite television enthusiast. They are especially noted for their clear, concise instructions, which may be partially attributed to the fact that at one time, Gillaspie offered a number of kits for home construction that addressed the needs of the TVRO Earth station owner.

In reviewing their installation manual, I found it to be indicative procedurally and generally to other satellite receivers. If you can understand the instructions provided in the following pages, you should have no problem installing your own TVRO Earth station, regardless of the particular components you choose. I have opted to reprint instructions for a receiver installation rather than those for an antenna. The assembling of an antenna is almost entirely mechanical, whereas the installation of a receiver requires some mechanics but more general electronics proficiency. In many cases, the installation of a satellite receiver will be no more difficult than the installation and alignment of a fairly sophisticated stereo system in your home.

Installation

Caution: Do not plug the control console into a live 110 Vac power source until instructed to do so. External power to down converter and LNA is present whenever the unit is plugged in, regardless of whether the power is on or off.

1. Polarization Device

Make secure all electrical connections to the polarizer. Follow the instructions supplied by the manufacturer. **Do not** mount the polarization device to the antenna at this time.

2. Down Converter

A. The down converter (see Fig. 7-37A and 7-38) is installed behind the antenna with mounting bracket supplied. Be sure to orient the connector side of the down converter so it faces **downward.** This is necessary to prevent moisture from collecting inside the unit.

B. Connect a length of RG-213 or 214 coaxial cable, up to 15 ft. long, fitted with Type N connectors, to the RF INPUT on the down converter. The other end of the cable will attach to the LNA RF Output connector.

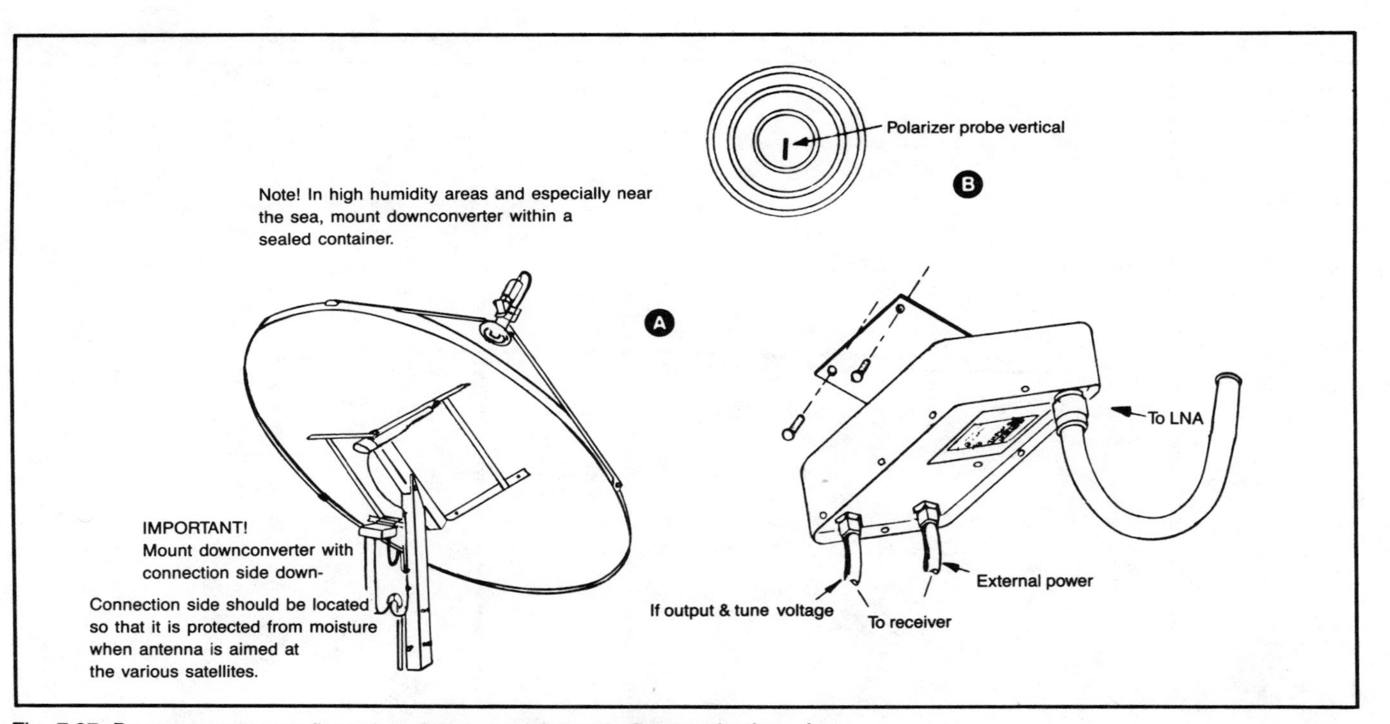

Fig. 7-37. Down converter configuration. Courtesy of Gillaspie Communications, Inc.

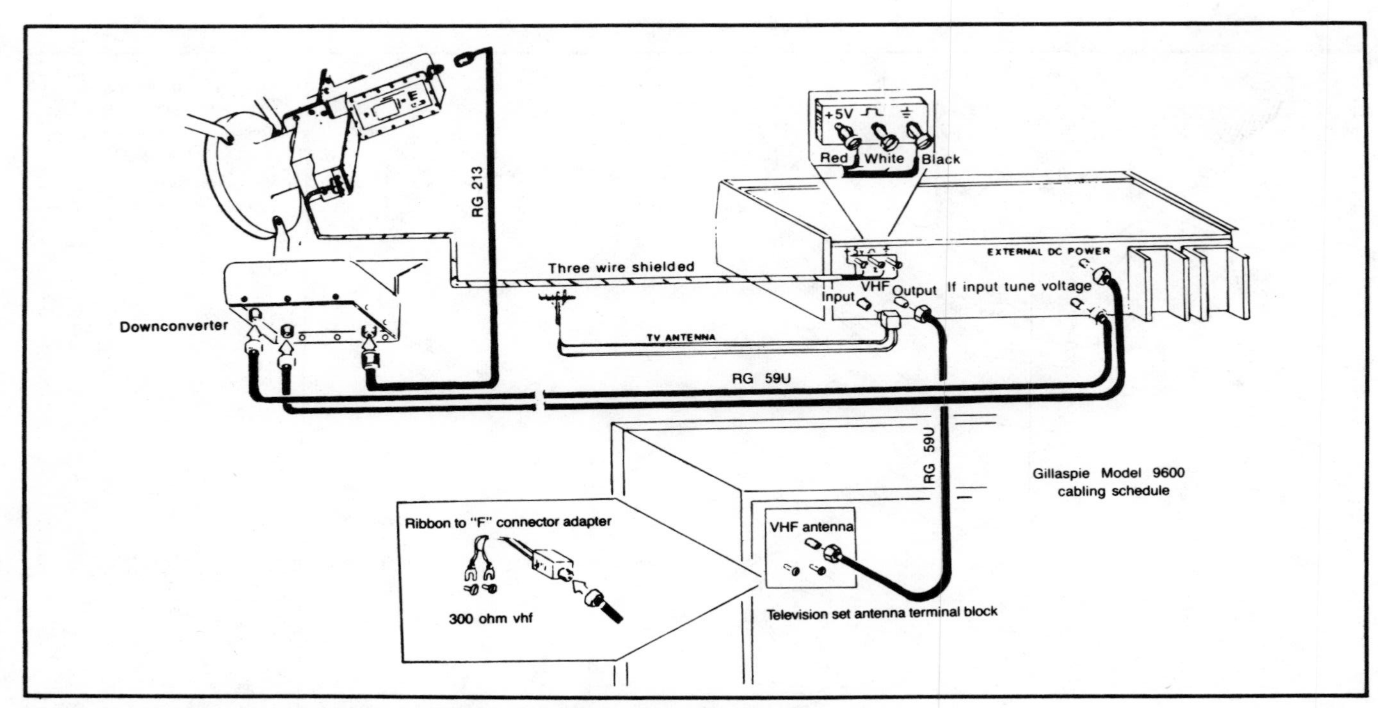

Fig. 7-38. Complete system arrangement. Courtesy of Gillaspie Communications, Inc.

C. Connect a length of RG-59/U coaxial cable or its equivalent, up to 165 ft. long fitted with Type F connectors between the down converter IF OUTPUT and the IF INPUT on the Model 9600 Control Console.

D. Connect a length of RG-59/U coaxial cable or its equivalent, up to 165 ft. long fitted with Type F connector between the down converter POWER and the EXTERNAL DC POWER on the Model 9600 Control Console.

Where longer lengths of cable are required (up to 800 ft.) to interconnect the 9600 Control Console and the down converter, use type RG-6/U coaxial cable (i.e. Beldon 9248, 8228, or equivalent).

Note: Identify one length of RG-59/u cable with tape on each end to ensure proper connection after cable has been installed.

3. Control Console

A. Connecting Television/Monitor (See Fig. 7-38 and 7-39): The Model 9600 can be used with a standard television set (black & white or color) or an Audio/Video Monitor.

To facilitate the use of a television set, connect any convenient length of RG-59/U coaxial cable up to 50 ft. and fitted with suitable connectors between the vhf OUTPUT connector on the rear panel of the Control Console and the vhf INPUT connector on the back of the television set.

If an Audio/Video monitor is to be used, connect suitable cables between the Audio Output-Video Output on the rear panel of the Control Console and the corresponding inputs on the monitor. Consult the Owner's Manual on the monitor for proper cable type to be used.

B. Modulator Output Channel (See Fig. 7-40): The switch enabling selection of either TV Channel 3 or TV Channel 4 output from the built-in modulator in Model 9600 is located on the bottom of the control console. Place this switch to the channel of your choice in accordance with the directions shown in Figure 7-40.

C. Connecting and Adjusting Polarizer: The Model 9600 is designed for use with a three wire polarization device. On the REAR PANEL of the Control Console are terminals marked Feed Rotator. + 5 V ⎍ ⏚. Following the instructions accompanying the Polariza-

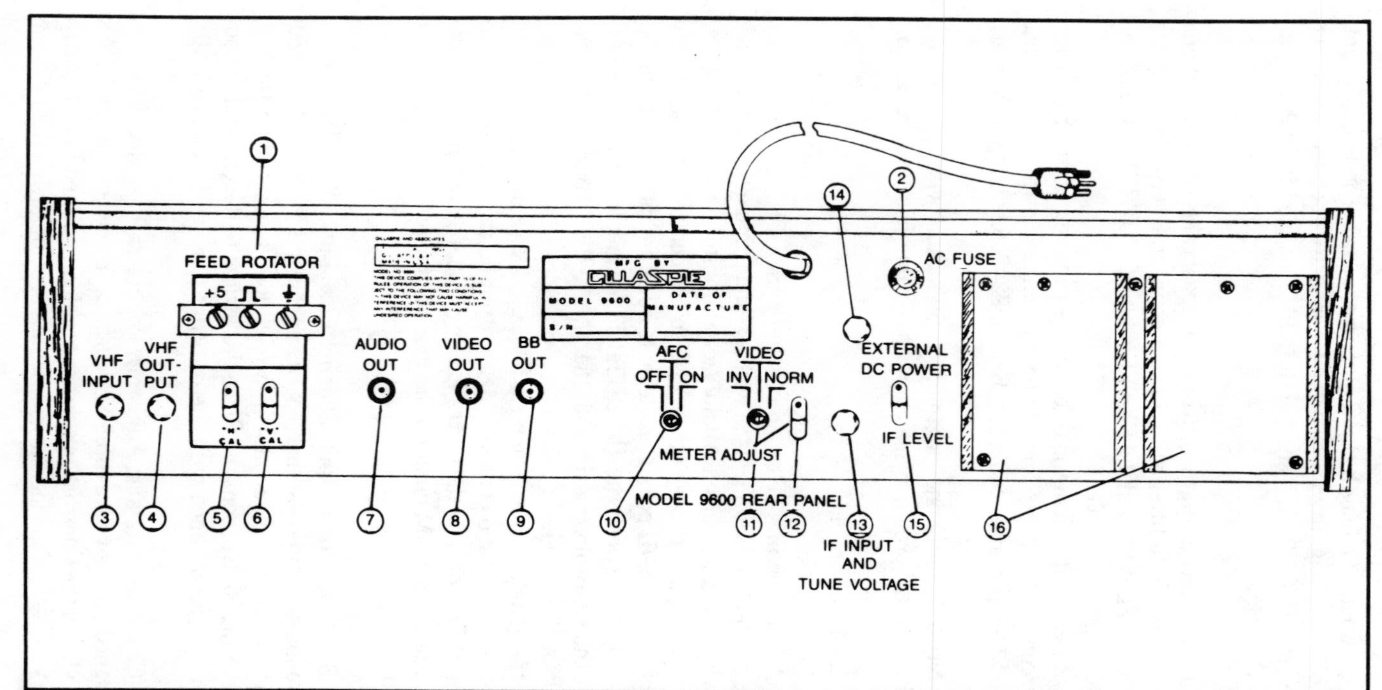

Fig. 7-39. Rear panel of GCI 9600 Receiver. Courtesy of Gillaspie Communications, Inc.

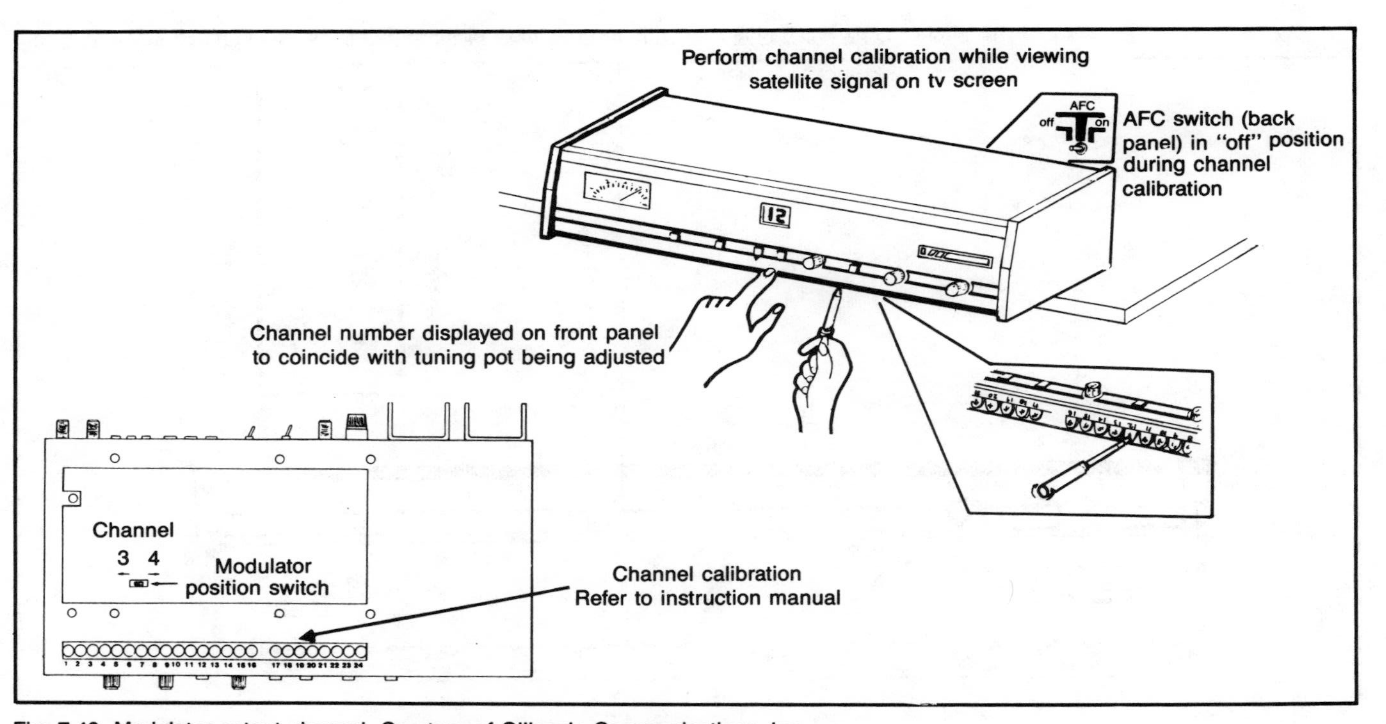

Fig. 7-40. Modulator output channel. Courtesy of Gillaspie Communications, Inc.

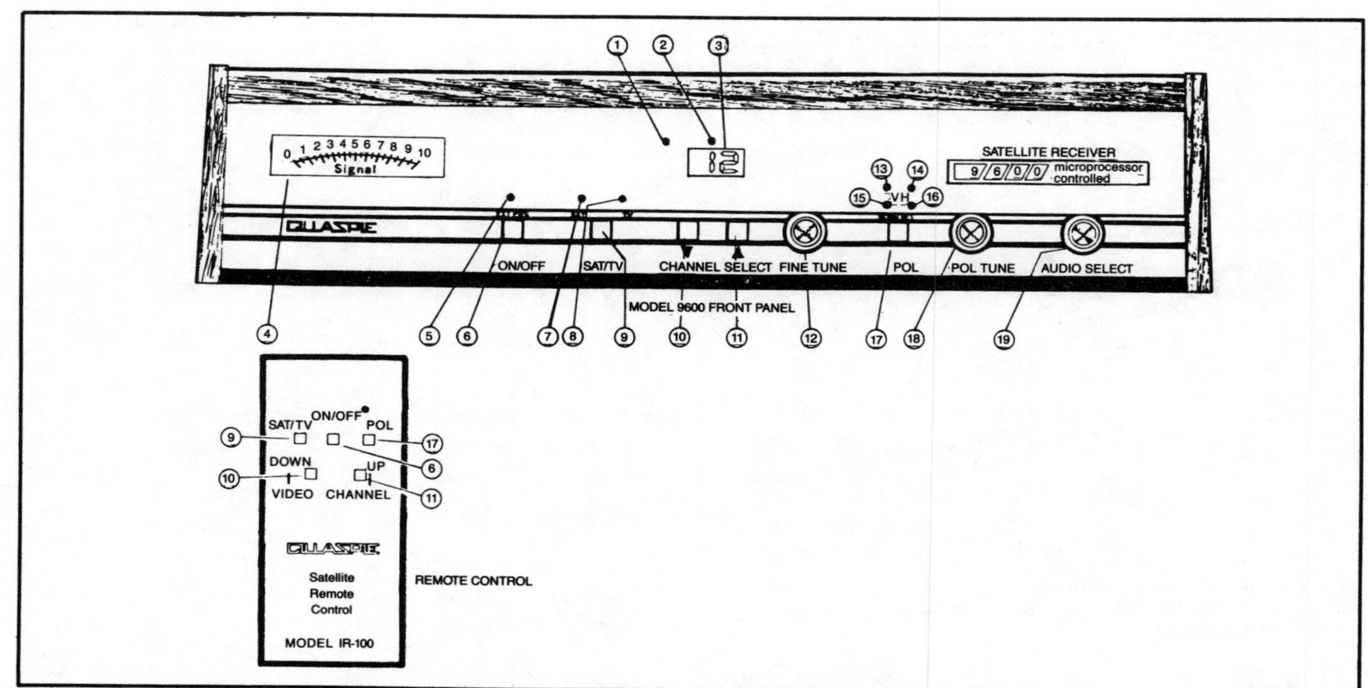

Fig. 7-41. Front panel of GCI 9600 and Satellite Remote Control. Courtesy of Gillaspie Communications, Inc.

tion device to be used, connect it to the appropriate terminals.

Feed Rotator Connections

		"FEED ROTATOR" TERMINALS ON THE REAR PANEL OF THE CONTROL CONSOLE.
RED	1	+5v
WHITE	2	⎍
BLACK	3	⏚

The distance between the antenna site and the point where the Model 9600 Control Console is installed will determine the wire size necessary to pass the proper operating current to the Feed Rotator device. The maximum resistance allowable for a given length of cable for this purpose is 1.3 ohms dc. The following table will assist you with the selection of the proper wire size for the three wire cable required:

#22 gauge wire—1.6 ohms/100 ft.—max. run. 80 ft.
#20 gauge wire—1.0 ohms/100 ft.—max. run. 130 ft.
#18 gauge wire—64 ohms/100 ft.—max. run. 200 ft.

Temporarily connect the control console to a 110 Vac power source. Set the Pol. Tune control to mid-range. Observe the status lamps for the polarization device to assure that they indicate V and Normal (Fig. 7-4, #13 and #15). Disconnect the Control Console from power source.

The Feed Rotator and LNA can now be mounted to the antenna. Observe the position of the probe inside the throat of the polarization device. Mount the device so as to cause the probe to appear perpendicular to the ground, i.e., the dish pad. See Fig. 7-37B. This procedure should be optimized for a satellite at the center of the orbital arc or the satellite which the system will most often view.

Once the connections to and mechanical alignment of the polarization device have been accomplished, the following electrical adjustments can be made to optimize the position of the probe in the feed rotator.

Connect the Model 9600 Control Console to a 110 Vac power

source. The external power status lamp on the front panel will glow. Allow 10 to 15 minutes initial warm up time to elapse prior to performing any further adjustments. This assures proper long term stability and channel calibration. Remember, power to the down converter is always present as long as the control console is connected to a power source.

Turn the Model 9600 receiver on by pressing the On/Off switch on the front panel of the Control Console.

Prior to performing the following steps, be sure the Pol. Tune control on the Front Panel of the control console is adjusted to its mid-range (12 o'clock) position.

Vertical—Select a vertically polarized signal and adjust V-CAL control on the rear panel of the Control Console for greatest signal strength indication on front panel tuning meter.

Horizontal—Upon completion of V-CAL adjustment, switch to a horizontally polarized signal. Status lamps should read H and Normal. Adjust H-CAL control on rear panel of Control Console for greatest signal strength indication on front panel tuning meter. This completes the connections to and adjustments of the polarization device.

D. Channel Calibration (Fig. 7-40): The Model 9600 Receiver is completely aligned and calibrated at the factory. However, due to variations in cable length between the control console and the down converter at the installation, a minor touch-up of the calibration controls may be necessary.

Depending upon the environmental temperature the 9600 may require 30 to 45 minutes of operation to become temperature stable. Proper channel calibration can be accomplished only after the 9600 has reached temperature stability.

Before proceeding with the calibration procedure, pre-set the following controls as indicated: (Rear Panel) AFC Switch OFF; Video Switch NORM; (Front Panel) Channel Select, 1; Fine Tune, midrange; POL., V-NORMAL; POL Tune, mid-range; SAT/TV., SAT.

The Channel Calibration adjustments are located on the bottom of the Control Console near the front panel. They are identified 1 through 24. As shown in Fig. 7-40, select channel one by operating the channel select control on the front panel.

Using a small insulated chisel point screw driver, adjust 1 control to cause channel one to appear on the screen of the TV/Monitor.

Select Channel 3 with the Channel Select control on the front panel and adjust channel calibration control 3 for best video quality. Repeat this step for remaining vertically polarized channels, 5 through 23.

Select channel 2 with the Channel select control. The automatic polarizer control circuitry will switch the Feed-Horn polarizer to Horizontal polarization. Adjust the Channel Calibration Control marked 2 for best video quality on the TV/Monitor. Repeat this step for remaining Horizontally polarized channels, 4 through 24.

Cycle the system through each channel. Minor adjustment of the fine tune control may be required. This completes the Channel Calibration procedure on the Model 9600.

In Case of Difficulty

1. If your newly installed system does not perform correctly, check the following items to ensure nothing has been overlooked or miswired.

A. Check RG-213 between down converter and LNA for continuity and short circuits.

B. Check RG-59 cables between the down converter and the control console for continuity and short circuits. Be sure each cable terminates at the proper connector.

C. Observe the external power indicator lamp on the control console. It should be lit whenever the receiver is plugged into a live 110 Vac power source.

D. If a television set is used as a monitor, be sure the channel selected matches vhf output channel of the Model 9600 i.e., CH3 or CH4. Check the RG-59 cable between the control console vhf Output and the television's vhf Input for continuity and short circuits. Observe the position of the SAT/TV switch to ensure it is in the SAT mode.

Remember, the Model 9600 Satellite Television Receiver contains no customer controls other than those specified in this manual. When difficulty is experienced with your satellite television receiver system, refer any service to competent service personnel.

SUMMARY

Once you have chosen your receiving components, their installation and alignment is usually a fairly simple matter. In most installations, the LNA is simply bolted to the back of the feedhorn and the down converter is attached to the output of the LNA. The coaxial cable is run from the down converter to the receiver, and that's just about it. In installations where the antenna is located very close to the satellite receiver, the down converter can be mounted inside the receiver chassis. Some say that the receiving electronics are the most critical components of a TVRO Earth station, but this is not necessarily true. When dealing with microwave frequencies, every portion of the receiving station becomes critical. An excellent receiver will in no way make up for a poor LNA or antenna. Likewise, a poor antenna will simply not be overcome by using a great LNA. The entire system must be chosen with good specification ratings in mind.

In some cases, a small antenna can be made more workable by using an LNA with a lower noise figure rating, but one must remember that the antenna is the device that actually intercepts the signal from the satellite and the receiving electronics can only act upon what the antenna passes. Choose your Earth station components carefully and match them quality-wise. It would be ridiculous to select the cheapest antenna you can find and match it with the best, most expensive receiving electronic available. In some areas, it would be totally unnecessary to pay the extra expense of obtaining a 70 ° LNA when a 100 ° unit is adequate. In other areas it would be ridiculous to purchase a 120 ° LNA that would be only marginal, especially when a 100 ° LNA would not cost significantly more. Don't take shortcuts. An extra $100 spent on the initial system may save you many hundreds of dollars and hours of reinstallation time in the future.

Also remember that the interconnections between the LNA and down coverter, the down converter and the receiver, and possibly the receiver and the modulator are all quite critical. If you will be wiring your own cables (installing your own connectors), make certain you know how this is done. If not, save yourself a lot of headache and buy prewired cables. A faulty connector installation can cause high losses and can also allow moisture to enter the cables, making your system useless after a short period of time.

There are many, many good receivers available today that don't cost a fortune. Don't buy the first thing you see, but take a general look at all that's available within your price range and make your decisions based upon specifications and the conveniences that you need. Don't buy something just because it looks stylish or fancy.

Remember, the specifications tell it all. This technical material is what should be the main attraction for any prospective TVRO Earth station owner.

Chapter 8

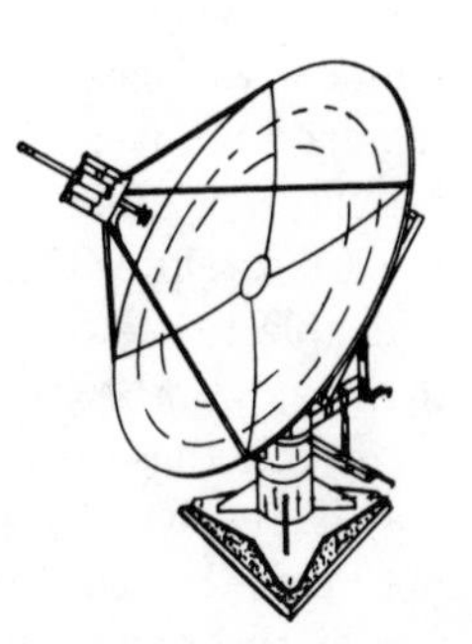

Satellite Location Techniques

ONCE YOU HAVE ARRIVED AT A DESIGN FOR YOUR EARTH station, it is absolutely essential to decide which satellite or satellites you desire to receive signals from. This would have been a simple process a decade or so ago before the communications satellites began appearing en mass over the Earth. Today, there are many, many satellites which broadcast television signals back to Earth.

All TV satellites are put into special orbits over the Earth which cause them to hover motionless in the sky. They are located some 22,000 miles above the surface of the Earth and rotate in sync with our home planet. As far as we on the Earth are concerned, the satellites are motionless because they rotate with the Earth. These are called geostationary or geosynchronous satellites due to their orbits, which are stationary or synchronous with the Earth's rotation. All of the entertainment satellites we are interested in are located high above the equator, but each one is stationed at a different longitude. Various satellites are positioned in very exacting orbits in order for their transmissions to cover different parts of the Earth.

In order to receive TV signals from these satellites, it is necessary to use a dish-shaped antenna which is aimed directly at the satellite whose signals are to be received. If the aiming is even a hair off, the received signal will be very poor or nonexistent. This assumes that all Earth station equipment is properly set up and functioning as it was designed to do.

There are many geostationary satellites, and the ones we're interested in broadcast special programming. Home Box Office, Showtime, Superstation, and The Movie Channel are some examples of what can be received in the entertainment line from these special satellites. Other services include cable and pay TV programs, uncut feature-length movies, sports, educational and religious programs, and news. There are also geostationary communications satellites which carry long distance and overseas telephone conversations. Other orbiting bodies include navigation satellites, weather satellites, and even military spy satellites. From the TV satellites alone, there are presently more than 75 channels available to viewers in North and South America and almost as many as in other parts of the world.

While these entertainment satellites broadcast signals which may be received by viewers in North and South America, this does not mean that all viewers can receive them. Your particular location on the globe will determine which satellites can be effectively "hit" by the antenna. The terrain immediately surrounding the Earth station location will also be a determining factor and applies especially to the quality of reception. Obviously, if your receiving antenna is located on one side of a mountain and the orbiting satellite is on the other side, it would be necessary for the transmitted signal from space to travel through the mountain in order to be received. This is not possible.

As is the case with most antennas, the greatest reception efficiency will be had from systems which mount the parabolic dish antenna in a high, clear location. This should also provide the capability of receiving several satellites by simply changing the position of the dish antenna. In the above example, the top of the mountain would have been the most ideal location for the antenna, but losses between the antenna and the receiver would be quite prohibitive, since a sizeable travel distance would be required.

When using simple antennas such as dipoles and verticals for shortwave communications work, it is usually a simple matter to mount the elements fairly high in the air, thus avoiding nearby obstructions. This is not the case with Earth station antennas, which can be described as massive in size as compared with the shortwave antenna. To mount one of the microwave dishes high in the sky would require a massive tower which could easily cost ten times as much as all of your Earth station equipment. Also, signals received from an orbiting satellite are line-of-sight. This means that the signal travels directly from the satellite to the Earth without any appreciable bending or skip. Shortwave signals often bounce from the Earth to the ionosphere and then back again. They may do this many times in a single transmission; thus, it is possible to communicate with receiving stations which are completely blocked from the transmitting antenna by the curvature of the Earth.

Fortunately for all persons interested in their own personal Earth stations, there are many satellites in equatorial orbits. In most situations, at least one and probably many of these will transmit signals which can be received at your location, regardless of where it might be. It is necessary to find out what is available in your area before you can accurately determine your station location.

This is easily accomplished and for a minimal fee by obtaining a *computer printout* of the satellites which can be received from your area. Each printout will be different for various station locations across the United States. You will that some satellites are completely *hidden* from your location, while others will serve as prime reception targets for your antenna.

How do you obtain a computer printout? In many instances, the manufacturers of complete Earth stations will supply one for your particular location when you express an interest in purchasing some of their products. This is not usually done free of charge, although some companies will refund the price of the computer search if you actually make a purchase. For example, Heathkit offers a full site survey packet which has full information on their Earth station and will help determine if your site is acceptable for installation. The price is currently $30.00 and you receive a computer printout which shows satellite "look angles" for your address, a compass for determining the directions of different satellites, and an inclinometer to help determine the degree of elevation between the Earth's surface and the satellite to assure a clear line-of-sight path for reception. Upon receiving this kit, you can make a good determination as to which satellites may be received in your area and, using the inclinometer, you can also find out if there are any major local obstructions such as mountains, towers, etc. in the signal path. Should you decide to order Heathkit's Earth station, the $30.00 Site Survey kit price is deducted from the equipment price. This example of obtaining a site survey is typical of the many reputable manufacturers of satellite Earth stations.

A less expensive route to go when you simply desire a computer printout for your area is offered by Satellite Computer Service, 1808 Pomona Drive, Las Cruces, New Mexico, 88001. This company offers an independent computer analysis of your location,

GROUND STATION AT: ROBERT J. TRAISTER

SAT LON	AZ	EL	LOSS IN DB AT 1 GHZ	4 GHZ	8 GHZ	12 GHZ
160 W	***	**	***	***	***	***
159 W	***	**	***	***	***	***
158 W	***	**	***	***	***	***
157 W	***	**	***	***	***	***
156 W	262	1	186	199	207	211
155 W	262	2	186	199	207	211
154 W	261	2	186	199	206	211
153 W	260	3	186	199	206	211
152 W	260	4	186	199	206	211
151 W	259	5	186	199	206	211
150 W	258	5	186	199	206	211
149 W	258	6	186	199	206	211
148 W	257	7	186	199	206	211
147 W	256	8	186	199	206	211
146 W	256	8	186	199	206	211
145 W	255	9	186	199	206	211
144 W	254	10	186	199	206	211
143 W	254	11	186	199	206	211
142 W	253	12	186	199	206	211
141 W	252	12	186	199	206	211
140 W	251	13	186	199	206	211
139 W	251	14	186	199	206	211
138 W	250	15	186	199	206	211
137 W	249	15	186	199	206	211
136 W	248	16	186	199	206	211
135 W	248	17	186	199	206	211
134 W	247	18	186	199	206	211
133 W	246	18	186	199	206	211
132 W	245	19	186	199	206	210
131 W	245	20	186	198	206	210
130 W	244	21	186	198	206	210
129 W	243	21	185	198	206	210
128 W	242	22	185	198	206	210
127 W	241	23	185	198	206	210
126 W	240	24	185	198	205	210
125 W	239	24	185	198	205	210
124 W	239	25	185	198	205	210
123 W	238	26	185	198	205	210
122 W	237	26	185	198	205	210
121 W	236	27	185	198	205	210
120 W	235	28	185	198	205	210

Fig. 8-1A. A computer printout will provide you with information regarding satellite reception in your area. Courtesy of Satellite Computer Service, (continued through page 211).

SAT LON	AZ	EL	LOSS IN DB AT 1 GHZ	4 GHZ	8 GHZ	12 GHZ
119 W	234	28	185	198	205	210
118 W	233	29	185	198	205	210
117 W	232	30	185	198	205	210
116 W	231	30	185	198	205	210
115 W	230	31	185	198	205	209
114 W	229	32	185	198	205	209
113 W	228	32	185	198	205	209
112 W	227	33	185	198	205	209
111 W	226	34	185	198	205	209
110 W	225	34	185	198	205	209
109 W	224	35	185	198	205	209
108 W	222	35	185	198	205	209
107 W	221	36	185	198	205	209
106 W	220	37	185	198	205	209
105 W	219	37	185	198	204	209
104 W	218	38	185	198	204	209
103 W	216	38	185	198	204	209
102 W	215	39	185	197	204	209
101 W	214	39	185	197	204	209
100 W	213	40	185	197	204	209
99 W	211	40	185	197	204	209
98 W	210	40	185	197	204	209
97 W	208	41	185	197	204	208
96 W	207	41	185	197	204	208
95 W	206	42	185	197	204	208
94 W	204	42	185	197	204	208
93 W	203	42	185	197	204	208
92 W	201	43	185	197	204	208
91 W	200	43	185	197	204	208
90 W	198	43	185	197	204	208
89 W	197	44	185	197	204	208
88 W	195	44	185	197	204	208
87 W	194	44	185	197	204	208
86 W	192	44	185	197	204	208
85 W	191	44	185	197	204	208
84 W	189	45	185	197	204	208
83 W	188	45	185	197	204	208
82 W	186	45	185	197	204	208
81 W	184	45	185	197	204	208
80 W	183	45	185	197	204	208
79 W	181	45	185	197	204	208

*** MEANS THAT A GEOSTATIONARY SATELLITE AT THE GIVEN POSITION IS NOT VISIBLE FROM THIS GROUND STATION.
SATELLITES AT LONGITUDES NOT LISTED ARE NOT VISIBLE FROM THIS GROUND STATION.

FRONT ROYAL

SAT			LOSS IN DB AT			
			1	4	8	12
LON	AZ	EL	GHZ	GHZ	GHZ	GHZ
78 W	180	45	185	197	204	208
77 W	178	45	185	197	204	208
76 W	177	45	185	197	204	208
75 W	175	45	185	197	204	208
74 W	173	45	185	197	204	208
73 W	172	45	185	197	204	208
72 W	170	44	185	197	204	208
71 W	169	44	185	197	204	208
70 W	167	44	185	197	204	208
69 W	166	44	185	197	204	208
68 W	164	44	185	197	204	208
67 W	163	43	185	197	204	208
66 W	161	43	185	197	204	208
65 W	160	43	185	197	204	208
64 W	158	43	185	197	204	208
63 W	157	42	185	197	204	208
62 W	155	42	185	197	204	208
61 W	154	42	185	197	204	208
60 W	152	41	185	197	204	208
59 W	151	41	185	197	204	208
58 W	150	40	185	197	204	209
57 W	148	40	185	197	204	209
56 W	147	39	185	197	204	209
55 W	146	39	185	197	204	209
54 W	144	38	185	198	204	209
53 W	143	38	185	198	204	209
52 W	142	37	185	198	204	209
51 W	141	37	185	198	204	209
50 W	140	36	185	198	205	209
49 W	138	36	185	198	205	209
48 W	137	35	185	198	205	209
47 W	136	35	185	198	205	209
46 W	135	34	185	198	205	209
45 W	134	33	185	198	205	209
44 W	133	33	185	198	205	209
43 W	132	32	185	198	205	209
42 W	131	32	185	198	205	209
41 W	130	31	185	198	205	209
40 W	129	30	185	198	205	210
39 W	128	30	185	198	205	210
38 W	127	29	185	198	205	210

SAT LON		AZ	EL	LOSS IN DB AT 1 GHZ	4 GHZ	8 GHZ	12 GHZ
37	W	126	28	185	198	205	210
36	W	125	28	185	198	205	210
35	W	124	27	185	198	205	210
34	W	123	26	185	198	205	210
33	W	122	25	185	198	205	210
32	W	121	25	185	198	205	210
31	W	120	24	185	198	205	210
30	W	119	23	185	198	205	210
29	W	118	23	185	198	206	210
28	W	118	22	185	198	206	210
27	W	117	21	186	198	206	210
26	W	116	20	186	198	206	210
25	W	115	20	186	198	206	210
24	W	114	19	186	199	206	210
23	W	114	18	186	199	206	211
22	W	113	17	186	199	206	211
21	W	112	17	186	199	206	211
20	W	111	16	186	199	206	211
19	W	111	15	186	199	206	211
18	W	110	14	186	199	206	211
17	W	109	14	186	199	206	211
16	W	108	13	186	199	206	211
15	W	108	12	186	199	206	211
14	W	107	11	186	199	206	211
13	W	106	11	186	199	206	211
12	W	105	10	186	199	206	211
11	W	105	9	186	199	206	211
10	W	104	8	186	199	206	211
9	W	103	7	186	199	206	211
8	W	103	7	186	199	206	211
7	W	102	6	186	199	206	211
6	W	101	5	186	199	206	211
5	W	101	4	186	199	206	211
4	W	100	4	186	199	206	211
3	W	99	3	186	199	206	211
2	W	99	2	186	199	207	211
1	W	98	1	186	199	207	211
0	E	97	0	186	199	207	211
1	E	***	**	***	***	***	***
2	E	***	**	***	***	***	***
3	E	***	**	***	***	***	***

anywhere in the world. Their computer will tell you which satellites can be received at your location, where to point your antenna, and how strong the signals are likely to be. The cost for this service is $19.95 (add $2.00 extra if you want it sent by air mail). I wrote to this firm and was able to obtain a computer printout for my area. This printout is shown in Fig. 8-1A. In looking over the specifications provided, they at first appear to be rather complex. A closer examination shows that the information is very basic and easily understood by even the most non-technical individual. Satellite longitude, elevation, azimuth, and loss in decibels at various frequencies are provided in a clear, understandable manner. The computer even indicates which satellite longitudinal locations cannot be received. The computer printout is used with a list of geostationary satellite longitudes and frequencies, which is also provided by Satellite Computer Service. See Fig. 8-1B. this allows the user to determine from the list and computer printout combination which satellites are available to his or her particular area.

It can be seen from the printout that only a very few (seven) satellite locations would not be generally usable for the author's location in the Southeast. Satellite Computer Service determines your geographic coordinates from your address and gives them to the computer. It then uses these numbers to solve a series of complex mathematical equations involving spherical trigonometry and logarithms. It is worth mentioning here that this particular mathematical analysis does not just cover TV satellites, but also every possible geostationary satellite location and does so in one-degree increments of longitude.

The first step in the computer's process of evaluating your location is to determine which satellite longitudes are not "visible" from your location. These are then eliminated from the printout. Obviously, the word visible is used loosely here to describe those satellites which are below the horizon from a location, plus making it impossible to receive transmissions from them.

The satellite longitudes which are visible from your location are then listed on the computer printout, along with some very important numbers. These numbers will tell you where to look in the sky to "see" each satellite from your location. They also give an indication of how strong or weak the signals from each particular satellite are likely to be at your location.

To better understand how to use the printout to determine which satellites you will be able to receive transmissions from, refer to the previously mentioned List of Geostationary Satellite

Longitudes and Frequencies (Fig. 8-1B) which Satellite Computer Service includes as part of their package. On the list, identify a satellite that you are interested in. For example,let's use my printout and assume that we wish to receive SATCOM I, which is a television satellite owned by RCA and offers several popular movie channels. By referring to the listing of longitudes and frequencies, we

SAT LON	NAME	OWNER	USE	FREQ GHz
90 E	Statsionar-6	USSr	TV/Tel.	4
99 E	Statsionar-T	USSR	TV	0.714
99 E	Ekran-1	USSR	TV	0.714
99 E	Ekran-2	USSR	TV	0.714
110 E	BSE	Japan	TV	12
125 E	STW-1	P. R. China	Exp./Mil.	4
130 E	KIKU-2	Japan	Exp.	0.136, 1.7, 11
135 E	CSE	Japan	Exp.	4, 20
135 E	ECS-2	Japan	TV/Tel.	4, 32
140 E	GMS-1	Japan	Weather	0.137, 1.7
140 E	GMS-2	Japan	Weather	0.137, 1.7
140 E	Loutch-4	USSR	Mil.	11
140 E	Statsionar-7	USSR	TV/Tel.	4
140 E	Volna-6	USSR	Commo	1.5
145 E	ECS	Japan	Exp.	4, 32
172 E	FLTSATCOM-4	USA	Mil.	2.2, 7
174 E	INTELSAT-IV-F8	INTELSAT	TV/Tel.	4
175 E	DSCS-2-F8	USA	Mil.	7
175 E	DSCS-2-F10	USA	Mil.	7
175 E	DSCS-2-F12	USA	Mil.	7
175 E	DSCS-2-F13	USA	Mil.	7
176.5 E	Marisat	COMSAT	Maritime	0.25, 1.5, 4
179 E	INTELSAT-IV-F4	INTELSAT	TV/Tel.	4

Abbreviations

Commo.	VHF-UHF aircraft and mobile communications, Telemetry, digital data, etc.
Exp.	Experimental, misc. research.
TV/Tel.	Television, Telephone, miscellaneous communications.
Mil.	Military communications.
Maritime	Shipboard navigation and communication.
Notes:	Not all satellites listed are active at this time. Some are on standby status for various operational reasons. This list is believed to be accurate at the time of publication. Nevertheless, the user should be aware that satellites are occasionally relocated, and new ones are placed into operation. Information about such changes is published in a number of aviation and space journals.

Portions of this material were derived, with permission, from the Comsat Technical Review, Volume 10, Number 1, "Geosynchronous Satellite Log for 1980," by W. L. Morgan.

Fig. 8-1B. List of geostationary satellite longitudes and frequencies. Courtesy of Satellite Computer Service, (continued through page 215).

SAT LON	NAME	OWNER	USE	FREQ GHz
0 W	Meteosat 1	ESA	Weather	0.136, 1.7
0 W	Meteosat 2	ESA	Weather	0.136, 1.7
1 W	INTELSAT-IV-E	INTELSAT	TV/Tel	4, 11
2.6 W	INTELSAT-IV-A	INTELSAT	TV/Tel	4, 11
4 W	INTELSAT-IV-F2	INTELSAT	TV/Tel	4, 11
11.6 W	Symphonie-2	France / W. Ger.	Exp.	4
13 W	DSCS-2-F7	USA	Mil.	7
13.5	Statsionar-4	USSR	TV/Tel	4
14 W	Volna-2	USSR	Commo	1.5
14 W	Loutch-1	USSR	TV/Tel	11
14.2 W	Ghorizout-2	India	TV/Tel	4, 11
15W	Marisat	COMSAT	Maritime	0.25
15 W	SIRIO-1	Italy	Exp./Met.	0.136
18 W	NATO-3A	NATO	Mil.	2.2, 7
18.5 W	INTELSAT-IV-F1	INTELSAT	TV/Tel.	4
21.5 W	INTELSAT-IV-F3	INTELSAT	TV/Tel.	4
23 W	FLTSATCOM-3	USA	Mil.	0.25, 7
24.6 W	INTELSAT-IV-A-F1	INTELSAT	TV/Tel.	4
25 W	Volna-1	USSR	Commo	0.28, 1.5
25 W	Stasionar-8	USSR	TV/Tel.	4
25 W	Loutch-P1	USSR	TV/Tel.	11
25 W	Gals-1	USSR	Mil.	7
27.5 W	INTELSAT-IV-A-F2	INTELSAT	TV/Tel.	4
34.5 W	INTELSAT-IV-A-F4	INTELSAT	TV/Tel.	4
41 W	TDRS	USA-NASA	Commo	2.2, 13
44 W	LES-9	USA	Exp.	7, 32
50 W	NATO-3C	NATO	Mil.	2.2, 7
70 W	FLTSATCOM	USA	Mil.	0.25, 7
70 W	ATS-5	USA	Exp.	0.136
75 W	Hughes-1	USA	TV/Tel.	4
75 W	SMS-2	USA	Weather	0.136, 1.7
85 W	LES-6	USA	Research	—
86.9 W	Comstar-D3	COMSAT	TV/Tel	4, 20, 30
90 W	GOES-1	USA	Weather	0, 136, 1.7
90.9 W	Westar-3	WU-USA	TV/Tel	4
95 W	Comstar-2	COMSAT	TV/Tel	4, 18, 28
99 W	TDRS	WU-USA	Commo	4, 12, 13
99 W	Westar 1	WU-USA	TV/Tel	4
100 W	LES-9	USA	Exp.	7, 30
100 W	FLTSATCOM-1	USA	Mil.	0.25, 7
103.9 W	ANIK-A1	Canada	TV/Tel.	4
105.2 W	ATS-3	USA	Exp.	0.136, 4
106 W	SBS-B	USA	Tel / Commo	12
107 W	GOES-2	USA	Weather	0.136, 1.7
108.9 W	ANIK-B1	Canada	TV/Tel.	4, 12
110 W	LES-8	USA	Exp.	7, 30
112.5	ANIK-C1	Canada	TV/Tel.	12
113 W	SMS-1	USA	Weather	0.136. 1.7
113.9 W	ANIK-A3	Canada	TV/Tel.	4
114 W	ANIK-A2	Canada	TV/Tel.	4, 12
116 W	ANIK-C2	Canada	TV/Tel.	12
116 W	CTS	USA/Canada	Exp.	12
118.9 W	SATCOM-2-F1	RCA-USA	TV/Tel	4
119 W	SBS-B	USA	Tel/Commo	12
122 W	SBS-A	USA	Tel / Commo	12

SAT LON	NAME	OWNER	USE	FREQ GHz
123.5 W	Westar 2	WU-USA	TV/Tel	4
127.8	Comstar-D1	COMSAT	TV/Tel	4, 18, 28
130 W	DSCS-2-F9	USA	Mil.	2, 7
132 W	SATCOM III	RCA-USA	TV/Tel.	4
134.9	SATCOM-1-F1	RCA-USA	TV/Tel.	4
135 W	GOES-3	USA	Weather	0.136, 1.7
135 W	GOES-4	USA	Weather	0.136, 1.7
135 W	SMS-2	USA	Weather	0.136, 1.7
135 W	DSCS-2-F11	USA	Mil.	2, 7
135 W	DSCS-2-F14	USA	Mil.	2, 7
135 W	NATO-3B-F2	NATO	Mil.	2.3, 7
140 W	ATS-6	USA	Exp.	1.5, 4, 20, 30
149 W	ATS-1	USA	Exp.	0.136
170 W	Volna-7	USSR	Commo	0.25, 1.5
170 W	Statsionar-10	USSR	TV/Tel.	4
170 W	Loutch-P4	USSR	Mil.	11
170 W	Gals-4	USSR	Mil.	7
171 W	TDRS	WU-USA	Commo	2.2, 13
0 E	Nordsat	Nordic Nations	TV/Tel.	12
10 E	OTS-2	ESA	Exp.	0.138, 11
35 E	Raduga-3	USSR	TV/Tel.	4
35 E	Statsionar-2A	USSR	TV/Tel.	4
35 E	Raduga-4	USSR	TV/Tel.	4
40 E	Marecs-A	ESA	Maritime	1, 5, 4
45 E	Loutch-P2	USSR	TV/Tel.	11
45 E	Statsionar-9	USSR	TV/Tel.	4
45 E	Gals-2	USSR	Mil.	7
45 E	Volna-3	USSR	Commo	0.25, 1.5
49 E	Symphonie-1	France/W. Ger.	Exp.	4
53 E	Ekran-3	USSR	TV/Tel.	0.714
53 E	Ekran-4	USSR	TV/Tel.	0.714
53 E	Gorizout-3	USSR	Mil.	4, 7
53 E	Loutch-2	USSR	TV/Mil.	11
54 E	DSCS-2-F4	USA	Mil.	8
56.5 E	INTELSAT-III-F3	INTELSAT	TV/Tel.	4
60 E	INTELSAT-IV-A-F6	INTELSAT	TV/Tel.	4
60.2 E	INTELSAT-IV-A-F5	INTELSAT	TV/Tel.	4
61.4 E	INTELSAT-IV-F1	INTELSAT	TV/Tel.	4
63 E	INTELSAT-IV-A-F3	INTELSAT	TV/Tel.	4
70 E	GOMS	USSR	Weather	0.136, 1.7
70 E	STW-2	P. R. China	Exp./Mil.	4
71 E	Insat	India	Exp.	2.5, 4
73 E	Marisat	COMSAT	Maritime	0.25
74 E	Insat-1A	India	TV/Tel.	2.5, 4
75 E	FLTSATCOM	USA	Mil.	2.2, 7
77 E	Palapa-2	Indonesia	TV/Tel.	4
80 E	Statsionar-1B	USSR	TV/Tel.	4
80 E	Raduga-1	USSR	TV/Tel.	4
80 E	Raduga-2	USSR	TV/Tel.	4
83 E	Palapa-A1	Indonesia	TV/Tel.	4
85 E	Gals-3	USSR	TV/Tel.	7
85 E	Loutch-P3	USSR	TV/Tel.	11
85 E	Raduga-2	USSR	TV/Tel.	4
85 E	Volna-5	USSR	Commo	0.25, 1.5
90 E	Loutch-3	USSR	TV/Tel.	11

can determine that this particular satellite's longitude is 134.9 W. Make a note of this figure, making sure to pay careful attention to whether the longitude is east (E) or west (W). If the indicated longitude contains a decimal part, as in the case of our example, round it off to the nearest whole number. This would make the longitudinal location of the SATCOM I satellite 135W. Also, make a note of the satellite frequency, which in our case is 4 GHz.

The next step is to go to the computer printout (Fig. 8-1A) and look up the same satellite longitude, paying careful attention to whether the longitude is east or west. Referring to Fig. 8-1, it can be seen that 135W does appear on the printout, which gives us a preliminary indication that we should be able to receive signals from the SATCOM I satellite. However, if a particular longitude did not appear on the printout, this means that we would not be able to "see" that satellite and, therefore, would not be able to receive signals from it.

Now that we know that the longitudinal position of the SATCOM I satellite makes it possible for us to receive its signals, some additional checks are in order. Notice on the printout that the next two columns of numbers after the satellite longitude are labelled AZ and EL. These are the azimuth and elevation numbers which tell the user exactly which direction to look in order to "see" the satellite. Before explaining how to use these numbers, however, it is important to discuss some information that should be kept in mind during this process.

When using the term "looking" with regard to "seeing" a satellite, this refers to the direction that the antenna would be aimed in order to point at a particular satellite. Of course, you can't really see the satellite with your eyes, since most are located more than 22,000 miles away. Nevertheless, in order to make sure that you can receive signals from the satellite, you will have to stand in the location that you plan to mount your antenna (in the backyard, on the roof, etc.) and actually look in the direction that the computer printout provides.

The reason for doing this is that your satellite TV antenna must have an absolutely clear shot at the satellite. There can't be obstructions of any kind in the way. Hills, buildings, trees, houses, or even a telephone pole will completely block out the signal. Therefore, even if the computer printout shows that you may be able to receive signals from a particular satellite, you must make sure that there are no local obstructions that would prevent you from doing so. Usually, such problems can be overcome by relocating your antenna site. In the case of nearby hills or large buildings, however, you may not be able to move far enough to be able to use a particular satellite.

Also, a good rule of thumb to follow is that if the listed elevation (EL) figure is 10 or less, the signal will most likely be of poor quality even if the antenna has a clear view of the satellite. This is due to the fact that the antenna will be pointing so close to the surface of the Earth that it will pick up a great deal of man-made interference. Unless you can afford to spend an additional amount of money on some quite expensive equipment, it is wise to avoid those signals from a satellite which have an elevation figure of less than 10.

Another point to keep in mind is that some satellites have highly directional transmitting antennas which concentrate their signals toward very specific areas. In cases such as this, you may find that you are within "view" of the satellite, but you might not be within its strongest area of coverage. Some of the ANIK satellites, for example, beam their signals primarily toward the northern provinces of Canada. In this example, a viewer in Canada would probably receive much stronger signals than would a viewer in Mexico, even though the distance from the satellite is farther and the elevation angle is lower.

With this information in mind, let's return to the process of determining whether or not we will be able to receive a good signal from the SATCOM I satellite at my location. Both azimuth (AZ) and elevation (EL) represent angles, or directions, which are measured in degrees. The AZ figure provides the compass direction, from 0 to 360 degrees, measured clockwise from true north. Refer to Fig 8-2 for a simplified diagram of compass directions. In this view, we are looking down on a man who is facing north (0 degrees AZ). The AZ figures are drawn around him. It can be seen that South is an AZ of 180 degrees, North-East an AZ of 45 degrees, and so forth. The azimuth figure for the SATCOM I satellite, which is at a longitude of 140 W, is 251 degrees. In other words, we must first face true north (0 degrees azimuth) and then, using a compass, determine the location of 251 degrees azimuth and face in that direction.

Keep in mind that the azimuth figures are based on true north, which in most locations is somewhat different from magnetic north. If you don't know how to locate true north directly, you can use an inexpensive dimestore compass to locate magnetic north, but it will be necessary to convert the printout AZ figure to a magnetic

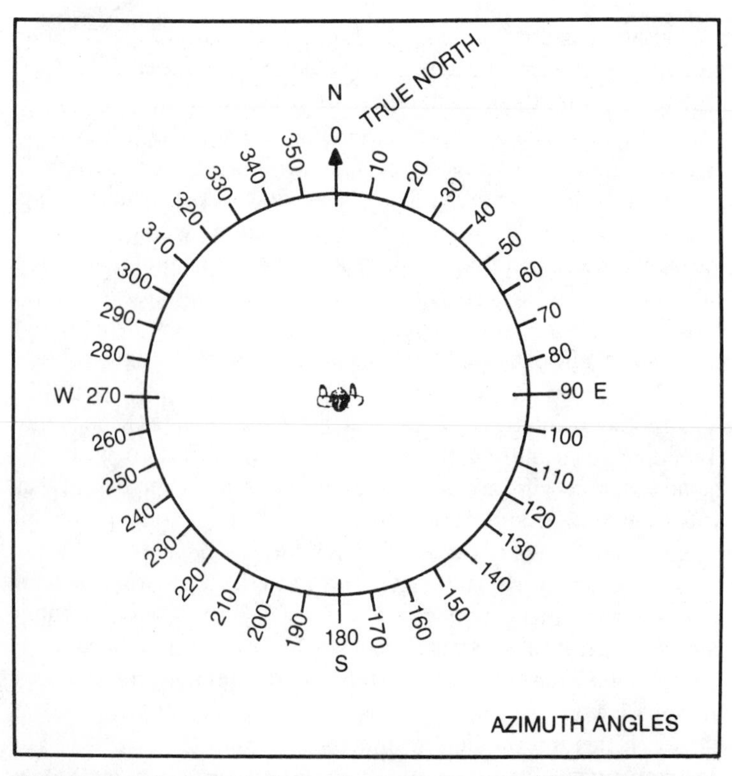

Fig. 8-2. Azimuth angles, indicating true north. Courtesy of Satellite Computer Service.

azimuth. This is done by referring to Fig. 8-3 and marking your approximate position on the map. In the case of my location, which is in Virginia, it would be between the dotted lines indicating 4 degrees and 8 degrees. These dotted lines indicate the amount of magnetic variation in the different parts of the country. Now, follow the dotted line which is closest to my location, which is the 8 degree line, out to the end. This 8 degree figure will be used to estimate the amount of magnetic variation for my area. It is also important to make note that this 8 degree variation is westerly (W), as indicated on the map. In our example, since the variation is westerly, the 8 degree figure is added to the computer printout AZ figure of 251, or 251 + 8 = 259. Therefore, 259 would be the magnetic AZ figure to use. However, if the variation were easterly, the indicated amount would be subtracted from the computer printout AZ figure.

To illustrate further, suppose you live in Denver, Colorado, and the satellite that you're interested in has a computer printout AZ figure of 163. Referring to Fig. 8-3, it can be seen that the magnetic variation at Denver is about 13 east (E). Thus, this amount is subtracted from the printout AZ, or 163-13 = 150. Therefore, 150 would be the magnetic AZ figure to use.

In yet another example, suppose you live in Erie, Pennsylvania and the satellite that you're interested in has a computer printout AZ figure of 191. Again referring to Fig. 8-3, the magnetic variation at Erie is 7 west (W). Therefore, this amount is added to the printout AZ, or 191 + 7 = 198. The figure of 198 AZ would be the magnetic AZ figure to use.

If you happen to live outside the United States, it will be necessary to obtain the magnetic variation figures for your particular area from another source. This information is generally available from either a public library or possibly a nearby airport.

Now that we have determined which direction it will be necessary to face, the next step is to find out the elevation at which to look. Elevation(EL) angles are measured from the horizontal (0 degrees) to the vertical (90 degrees). In other words, if you look straight ahead toward the horizon, this is an elevation of 0. If you tilt your head back until you are looking straight up, the elevation is 90. Although the computer printout does provide the exact elevation angle, most persons will find it a bit difficult to estimate this angle by sight alone. However, this can be determined using a few

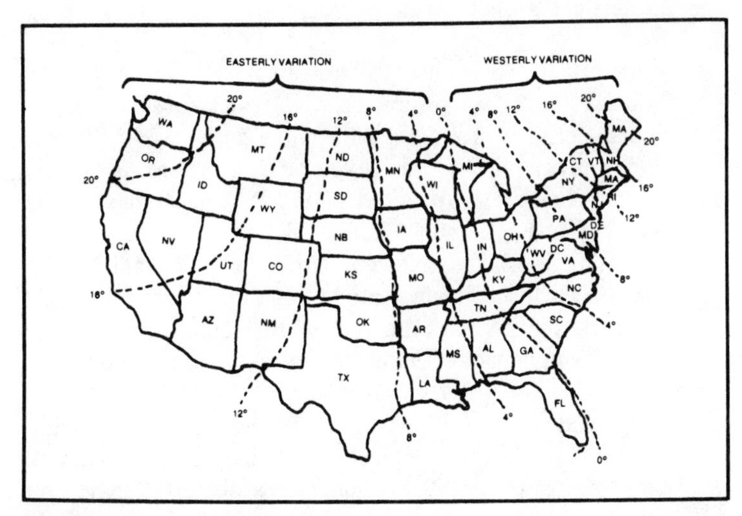

Fig. 8-3. The dotted lines on this map indicate magnetic variation. Courtesy of Satellite Computer Service.

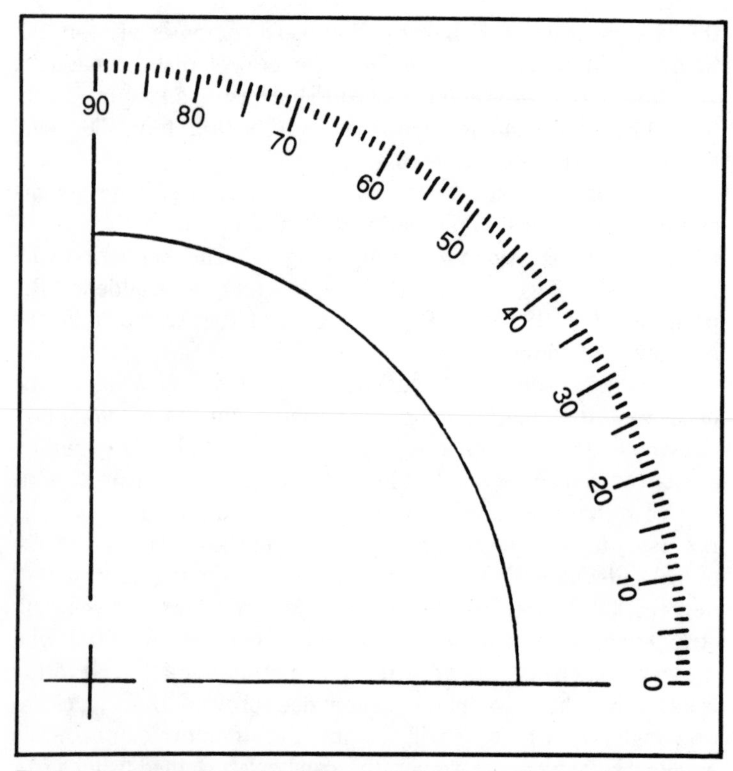

Fig. 8-4. A protractor is used to make a cardboard angle of your location's elevation. Courtesy of Satellite Computer Service.

simple materials with a much greater degree of accuracy.

Shown in Fig. 8-4 is a protractor, which can be used to measure elevation angles. The protractor shown here can be traced and then cut out, or you can purchase an inexpensive plastic protractor from an office supply store. Now, referring to Fig. 8-5, use the protractor to mark off an angle equal to the satellite elevation angle, which in our example is 13, on a piece of cardboard. A triangle is then cut from the cardboard as indicated by the mark made using the protractor. Be sure to mark the letters "EL" on the corner of the triangle that you measured with the protractor so that you won't forget which corner of the triangle to refer to.

Now, stand at the location which has been selected for mounting of the antenna. Face the direction of the satellite's azimuth, which in our example is 259 and hold the cardboard triangle even with your eyes, keeping the bottom side horizontal (parallel to the ground). Sight up the slope of the triangle as shown in Fig. 8-5.

You should now be looking directly toward the satellite, although you already know that it won't be possible to actually see it.

What we are looking for at this point is any type of obstruction that might block the satellite's signals. If, when sighting along the slope of the triangle, your view is blocked by obstructions of any kind, such as trees, hills, houses, etc., then you will not be able to receive signals from the satellite. If, however, all you see in this direction in the sky, you should be able to receive this particular satellite's transmission. In the case of my location when performing this procedure there was a mountain range in the vicinity, but it did appear to be approximately 4 or 5 degrees below our proposed antenna aiming point. If it was found that there were some obstructions that would prevent reception of a particular satellite, it would have been necessary to either change the location of the antenna or possibly provide some means of elevating it at the original proposed site.

This completes the procedure for determining whether or not you will be able to receive signals from a particular satellite. Of course, this process will have to be repeated for each additional satellite that you may be interested in. This will involve preparing a new cardboard elevation triangle each time.

Once you have selected a satellite and determined that you can receive signals from it at your particular location, the final step will be to once again refer to the computer printout in order to check

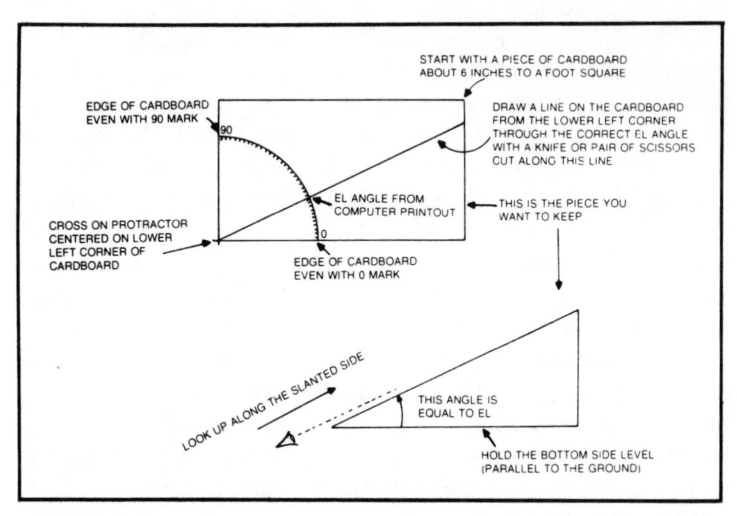

Fig. 8-5. Using the protractor, the proper elevation is marked off. Courtesy of Satellite Computer Service.

the strength or weakness of the signals. Remember that when originally checking the computer printout, we made a note of the satellite frequency. Again referring to the printout, notice that there are four columns of numbers after the figures for SAT LON, AZ, and EL. These numbers are labeled "LOSS IN DB AT 1 GHz, 4 GHz, 8 GHz, and 12 GHz."

Go to the column which most closely corresponds to the frequency noted earlier, which in our example is 4 GHz. However, if your frequency was 5 GHz, for example, you would go to the column labeled 4 GHz also. In that column (taking care to stay on the same line as the SAT LON that was used earlier to find the azimuth and elevation), you will find a number which indicates how strong or weak the signal will be for you location. The lower the dB LOSS number, the stronger the TV signal will be. The higher this number is, the weaker the TV signal will be. The lowest dB LOSS number that will be encountered on the printout will be about 183. This indicates the best (strongest) signals. The highest dB LOSS number will be about 215, which indicates the worst (weakest) signals. The weaker signals will require a larger antenna and a more expensive converter to receive them with any degree of quality. Since the dB LOSS number for my location is 186, this is a good indication that the signals from the SATCOM I satellite can be received quite strongly at the proposed location for my earth station.

It should be kept in mind when requesting a computer printout for a proposed Earth station site that a single printout is calculated for only one specific location. Any additional proposed sites will require a separate printout. Also, if the requested printout does not include a particular geostationary satellite's longitude (SAT LON), this means that you cannot receive transmissions from that satellite at your proposed location. If a satellite's longitude does appear on the printout but its elevation figure is less than 10, you might be able to receive signals from it, but they will probably be of poor quality. Take care when setting up your TVRO antenna. Be sure that it is aimed directly at the satellite you wish to receive transmissions from using the azimuth and elevation figures as a guide and making sure that there are no obstructions between the antenna and the satellite toward which it is pointing. Following the guidelines provided here and installing your Earth station properly, you should be able to receive quality transmissions from your antenna arrangement.

It is necessary to reemphasize the fact that site selection will play an important part in the reception of satellite signals. Just because you have a *window* to an entertainment satellite in your area does not necessarily mean that you can receive it from any site in the same vicinity. As was previously mentioned, local obstructions, especially those lying close by the Earth station, can effectively close that window and make reception impossible. In these situations, moving the Earth station antenna to a slightly different location may open the window fully. Chapter 10 deals more fully with the selection of Earth station sites.

It is mandatory that all persons wishing to install their own satellite Earth station obtain the data and perform the calculations necessary to determine which satellites beam signals that are receivable at a particular site. If a satellite that you desire to receive is not available in your area, there is nothing you can do to correct this situation short of moving to an area which has an open window. This will undoubtedly be many miles (or several hundred) from your home location. Even the best ground station equipment available will do nothing to open a window which does not exist.

Fortunately, there are many satellites available to all areas of North and South America, and the enterprising Earth station builder will certainly find one or more which broadcast the programs desired.

Satellite location data should be obtained before you decide to actually purchase the components to build the Earth station. This is especially true if you desire to receive signals from one particular satellite. If there is no window to this satellite in your area, then you may decide to forego the project entirely. This chapter has overviewed the methods used to obtain computer printouts and to properly determine the direction, elevation and azimuth for the antenna installation. Most manufacturers of satellite ground station equipment are in a position to supply this information (usually for a fee) or to advise you as to where this information can be obtained.

Chapter 9

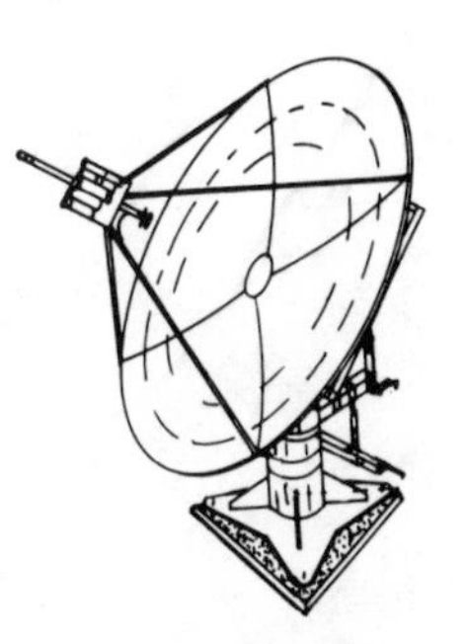

A Commercial Cable Television System

PREVIOUS CHAPTERS IN THIS BOOK HAVE DESCRIBED PERsonal Earth stations and their application to home use. Many areas in the United States offer cable television systems which supply service to customers who access their facilities. Many potential buyers of personal Earth stations will wonder about the differences in their systems when compared with a commercial installation and what advantages are to be gained by the latter.

In order to present a good idea of what is involved in a commercial installation, I contacted Telaco, Inc., a cable TV company which serves the Front Royal, Virginia area. Telaco, Inc. is owned by Mid Atlantic and has been in business for several decades. Only recently has satellite TV been added to the cable, which formerly offered only vhf and uhf channels. The vhf/uhf antenna site was located several miles from Front Royal, Virginia. While this was an excellent site for this type of reception, it was totally unsuitable for an Earth station due to mountains which would not allow a clear window on the satellite which supplies Home Box Office, ESPN, The Super Station, and other programs which were to be offered.

An in-town location was obtained for the building of the Earth station. Fortunately, this was only a couple of hundred feet away from my offices. The manager of Telaco,Inc. was quite cooperative in explaining the operation of his company's system and even gave a personal tour of the facility, pointing out each piece of equipment and phase of operation in the order in which they are used.

The first difference noted between commercial installations and personal Earth stations was the fact that the parabolic dish antenna is much larger (see Fig.9-1). Most dish antennas used for personal TVRO Earth stations are ten to twelve feet in diameter. The

Fig. 9-1. The Telaco, Inc. commercial Earth station's dish measures a full five meters or sixteen feet wide. Courtesy of Telaco, Inc.

Telaco, Inc. dish is a full five meters (sixteen feet) wide and presented a massive appearance upon close examination. Unlike most personal Earth stations, this antenna was not readily adjustable in regard to elevation and azimuth. A more or less permanent mounting technique was used, which bolts everything into place. Elevation and azimuth can be adjusted, but not in a short period of time. This requires unbolting several connections and/or manually "horsing" the antenna base into the desired position. Of course, easy manual adjustment is not necessary in commercial installations, because the antenna, once adjusted, is left in the fixed position permanently. The manager pointed out that it would soon be necessary to make a major readjustment in the antenna's position because of a new satellite which will be used by the same programmer. This satellite will take over the handling of the same programs presently offered by Satcom I. Often, satellite users will switch as the orbit of one satellite in use begins to decay. Commercial Earth stations require excellent windows. These deteriorate directly in relationship to time in orbit. When the programmer leaves this satellite, other services will begin leasing it. These latter uses are not as critical as commercial television reception, so the satellite will still be used for a long time to come.

The new satellite that Telaco, Inc. will use lies at a slightly different point on the horizon. When it comes time to switch, the antenna will simply be readjusted and will function as before, but from a new satellite.

Another difference can be readily discerned by looking at the feed horn/LNA assembly at the focal point. Whereas most personal Earth stations have a single LNA here, the commercial station has two. Satellite signals are transmitted either as horizontally polarized waves or those which are vertically polarized. Most personal Earth stations use a single LNA and feed horn but incorporate a standard antenna rotor which will allow the assembly to be turned 90°. If the system is set up to receive horizontally polarized signals, the rotor, when activated, will perform the 90° shift, which converts the system to vertical polarization. Since an LNA can easily cost $800 or more, it is far more cost efficient to use an inexpensive rotor system rather than two low noise amplifiers. You might wonder why the commercial facility doesn't do the same thing. There's a good reason. It just won't work. At your personal station when you must switch from vertical to horizontal polarization, you do it from a remote control located atop your TV. But remember, you and your family are the only users of this personal system and on-

Fig. 9-2. The Telaco low noise amplifier (LNA) /feed horn assembly at the antenna focal point.

ly one channel can be selected at a time. In a commercial facility, there are thousands of users. Some of them will be tuned to one channel offered by the cable company, while others will simultaneously be receiving another available channel. Suppose the first channel is transmitted by vertical polarization, while the other is horizontally polarized. If a single LNA is in the vertically polarized position, then those persons wishing to receive the horizontal channel will be out of luck.

Unlike personal Earth stations, cable companies offer several satellite channels simultaneously. This requires a different receiver for each channel and the capability (at the antenna) for simultaneous reception of both horizontally and vertically polarized transmissions.

Since two separate feed horn/LNA combinations are used at the antenna site, there are two transmission lines back to the bank of receivers. The line from the horizontal LNA is attached to receivers set up for satellite transponders which transmit horizontally polarized signals. The same is true of the other line, which feeds signals to receivers set up on vertically polarized channels. Figure 9-2 shows the LNA/feed horn assembly at the antenna focal point.

At this point, we can say that one of the major differences in cable systems as opposed to personal Earth stations is the fact that

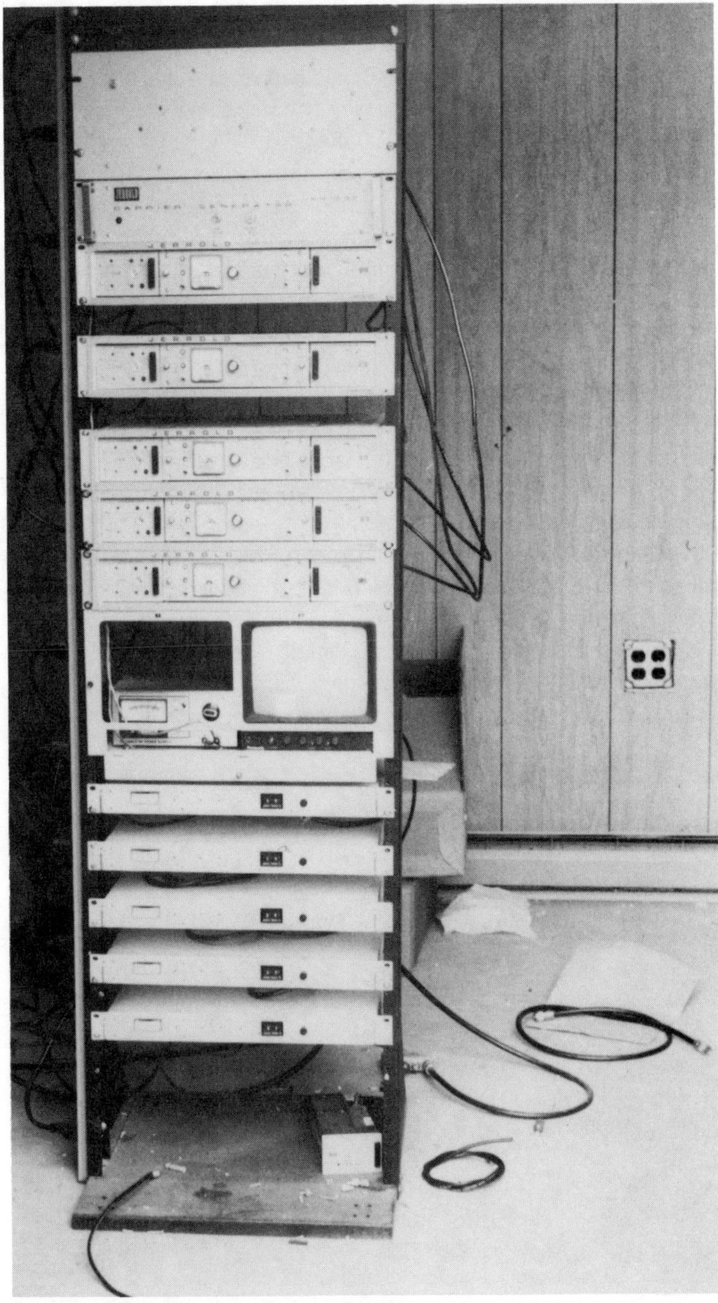

Fig. 9-3. The instrumentation rack contains receivers, modulators, monitors, and amplifiers.

several simultaneous channels are offered. This makes the entire system far more complex, as it is necessary to add (in addition to the two LNAs) more receivers and modulators, one of each for every channel. this is one of the reasons why cable television Earth stations usually cost well over $100,000.

Figure 9-3 shows the instrumentation rack which is enclosed in a small building at the Earth station site. Contained here are the receivers, modulators, monitors, and some amplifiers for the Earth station. Additionally, the combining network which brings in the vhf/uhf channels from the other antenna site is found at this location. This latter network performs basically the same function as was described for combining signals in an earlier chapter on multiple access Earth stations.

The cable used to bring the signals the short distance from the antenna to the electronic apparatus is a special low-loss coaxial type. This is shown in Fig. 9-4. Connections are made to the back of the panel shown earlier.

Before discussing the bank of receivers and modulators, it is important to make another distinction here between equipment intended for general public uses and that which is intended for commercial purposes. The latter is usually built to much closer tolerances and is generally more rugged and dependable than the former. Commercial receivers and modulators are designed for

Fig. 9-4. The cable used in this commercial installation is a special low-loss coaxial type.

untended operation. Your personal Earth station equipment is not. Also, one must consider the duty factor. Few personal Earth stations will be operated 24 hours a day, 365 days a year. Most cable companies, on the other hand, offer full time operation day in and day out. Obviously, their equipment must be designed to withstand this amount of usage. While circuits of equipment intended for the two purposes (personal and commercial) may seem theoretically comparable, the actual choice of components used will differ greatly. Commercial electronic equipment usually is designed with larger power supplies and with commercial quality components which will provide a longer and more dependable service life. These are often called commercial grade or military grade components.

Obviously, the choice of select and often specialized circuits and components has to cost more. The cable company is paying extra for this equipment because it needs more services from its devices than you would as a personal Earth station owner. Commercial receivers can cost more than an entire personal Earth station facility. And remember, the cable company may need six or more, depending upon the number of channels it is to offer its customers. A different modulator will be needed for each channel as well. The modulators used by Telaco, Inc. cost in excess of $3,000 each.

The receiver and modulator act in the same manner as those incorporated in personal Earth station installations. The video and audio output from the detected microwave signals are superimposed on the carrier generated by the modulator. Another difference makes itself known at this point. before discussing this, a review of the portion of the cable company Earth station presented thus far reflects a lot of equipment, installation time, and cost. And remember, we still haven't gotten the signal to the home television receiver yet.

The difference which occurs at this point when comparing cable Earth stations with those intended for personal use includes the massive distribution system which must be run to every customer's home. However, the modulator at the antenna site does not usually have an output at a frequency which can be received directly by the standard television set. Usually, the output frequency of the modulator lies above the vhf television band and below the uhf band. Why is this done? First of all, if the output were on one of the vhf channels, this might mean the dropping of one presently received station for every satellite channel offered. Today, cable companies often present a range of programs on all channels 2 through 13.

Secondly, cable companies almost always offer satellite television reception as an option to customers who are already on the cable and are presently receiving the vhf channels. customers have the option of paying an additional fee each month for the satellite channels; but if they don't want them, they can continue to receive vhf channels and pay no more.

In order for the cable company to offer satellite reception as an option, they purchase modulators which produce frequencies that fall outside of the normal television bands. These signals are on the line which attaches to your television receiver, but you can't detect them without a converter. The cable company installs a converter at each home which accepts the modulator output and changes it to a frequency within the television band. This is usually somewhere in the uhf channel area. Here is another major expense for the cable company, because if 6,000 customers want satellite reception, 6,000 converters are necessary, one at each set.

Since their inception, cable television companies have had problems with some disreputable persons stealing their signals. In other words, a few persons who did not subscribe to the company found ways of intercepting their output. this was most often done by tapping into one of the coaxial lines.

Most of the stealing occurred in homes which subscribed to the service and who then attached more television sets to the single incoming line. Cable companies usually charge a minimal fee for each additional television in a home and must use splitters which are company-installed to provide for the additional outputs.

Like anything else, if there is a way to get something for nothing, even though it may be against the law, some persons will jump to take advantage of the situation. The converters which the cable companies use to provide satellite channel reception in each home are available to the general public through special electronic outlets. Most are reputable firms, but a few specialize in offering equipment specifically for the individual who wants to beat the system. These companies often advertise in questionable manners. To prevent this situation from becoming widespread, many cable companies will install frequency traps between the main signal line and any home which is taking cable service. When the subscriber signs up for satellite reception, the technician who installs the converter also removes the frequency trap. This latter device effectively blocks the satellite signals from the homes which do not subscribe to the optional service. It is unfortunate that these companies find it necessary to go to all this trouble, but statistics would prove that it's absolutely essential. Less than one-half of one percent of the subscribers would even consider doing anything like stealing from a cable company, but the company has no way of knowing exactly who comprises the dishonest minority. All customers pay for their potential dishonesty by mutually footing the bill for all those frequency traps, their cost being passed along to the subscribers through increased service rates.

A converter unit is installed by the cable company at each satellite subscriber's set. This is a type of modulator which accepts the input signal from the cable and then converts it to a channel within the uhf television band.

Another problem is often encountered by cable companies in that some subscribers still have older television models which do not offer uhf tuning. Here a special converter may be used which allows one vhf channel to be used for reception of all the satellite channels. This converter has its own channel knob which selects all of the satellite programs being offered by the company. The output of this unit may be on any vhf channel 2 through 13 and there is a cutout switch so that the vhf channel may also be used to receive the vhf station which is replaced by the satellite signals during that mode of operation.

In addition to all of the equipment necessary to operate a successful cable TV company, the redundancy factor must be considered. Most commercial equipment has a bit of redundancy built into it, but most companies still keep a spare receiver, modulator, LNA, and other vital equipment on hand in case of emergency. Even the finest equipment can fail unexpectedly and cable customers get mad when they can't receive what they're paying for. Redundancy is another necessary expense of cable company operation.

While there are many differences in cable company operation, let's now examine the similarities between it and a personal TVRO Earth station. Both use dish antennas, LNAs, receivers and modulators. Both receive signals in the same manner and from the same satellites. Both ultimately offer outputs which are designed to be received by standard television sets. The only major differences include back up devices, redundancy equipment, special tuning arrangements, cost, and size.

Cable companies must obtain a stronger signal than is necessary for practical personal Earth station uses. A lot of equipment is attached to the output of the antenna and this tends to lessen the level of the output to each receiver. Also, the cable company must provide constant and dependable reception during all weather conditions. For example, a large cable company's parabolic dish antenna is far more firmly seated on its mounting site than would be a typical personal Earth station antenna. Both will probably have similar wind survival ratings, but the commercial installation will be far less affected as to reception by moderately high winds. The personal Earth station antenna will probably induce a bit of flutter and fading when winds gust to over 50 miles per hour. This won't occur in the system discussed in this chapter. Also, during periods of heavy rain, high temperatures, snow and ice, the increased signal strength provided by the large commercial dish will allow the cable company to provide a signal to users which appears no different than those received during optimum weather conditions. The commercial antenna actually provides more gain than is needed most of the time. But the backup is needed in order to assure that the same quality of signal is available even under the most adverse conditions. This will not be true of most personal Earth stations. Ninety percent of the time, the received image and audio on a television receiver operated from a personal Earth station will be as good as that which is provided by the cable company. But you will notice some difference when adverse weather conditions are in effect. Sure, you could purchase a larger antenna and LNAs with better noise figures, but you're talking about doubling or tripling the price of your personal Earth station. This is simply not cost effective for the small advantage these extras provide.

It can be seen that while the principles involved in satellite television reception are basically the same in cable company installations as well as in personal TVRO Earth stations, the demands placed upon the commercial station are far more numerous. Both fields are closely allied and the TVRO experimenter is urged to be on the lookout for commercial surplus equipment as it becomes available. In some instances, it is more profitable for a cable company to simply replace malfunctioning equipment which is out of warranty than to have it repaired. These units are then relegated to a dusty shelf in a back room and can often be purchased for next to nothing. A visit to your local cable company may result in making acquaintances that will aid you in your future personal TVRO Earth station effort.

I would like to thank Telaco, Inc. and especially its manager, Richard Burke, for the assistance offered in researching this chapter. I think the reader will find that most cable company personal are more than willing to give advice and lend experience in addressing problems which can develop with personal Earth stations. Actually, the two fields are a bit competitive, but there is little chance that in the near future, personal Earth stations will cause any significant financial losses to the cable companies. Satellite television reception is still a new field. Most persons in this area of endeavor are pleased by the experimentation which is being conducted by major companies, manufacturers, and individuals. This is something we all have in common.

Chapter 10

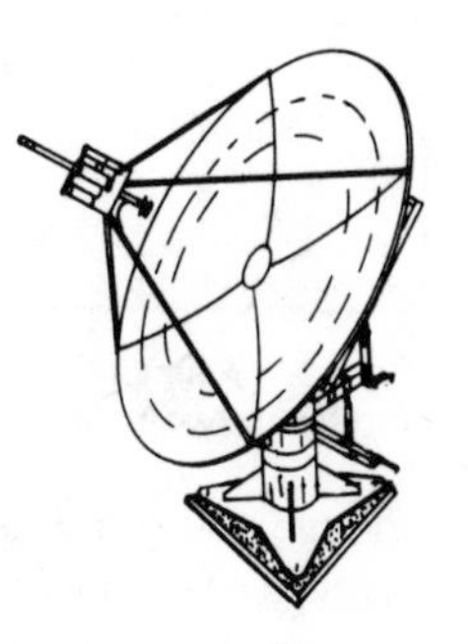

Specific Site Selection

As IS ALWAYS THE CASE WHEN SETTING UP ANY ANTENNA system, the physical site is of utmost importance. Previous chapters in this book have outlined methods of choosing a general site that has windows open to the satellites you wish to receive. This selection was done with aid of a computer printout for a general geographic area. This lets you know whether or not it is possible to receive certain satellites, but it does not mean necessarily that adequate reception will be maintained at a specific site within this general area. The computer printout does not take into account smaller mountains, metal towers, large trees, and other obstructions which can severely interfere with the satellite transmission path. This must be taken into account when you choose a specific mounting site on your property because one location may be poor for certain satellites, while another a few feet away may be ideal.

First of all you will want to obtain the computer printout. Then with a compass and inclinometer, determine the exact position in the sky toward which your antenna must be aimed for reception of a particular satellite. Building a simple sighting device may help you. This can be constructed from a small piece of lumber or dowel rod, which is fitted with two nails. This is shown in Fig. 10-1. By sighting along the nail heads, as shown in Fig. 10-2, you will be in a good position to see whether or not you have any major obstructions.

If you live in an urban area, buildings and metal towers often

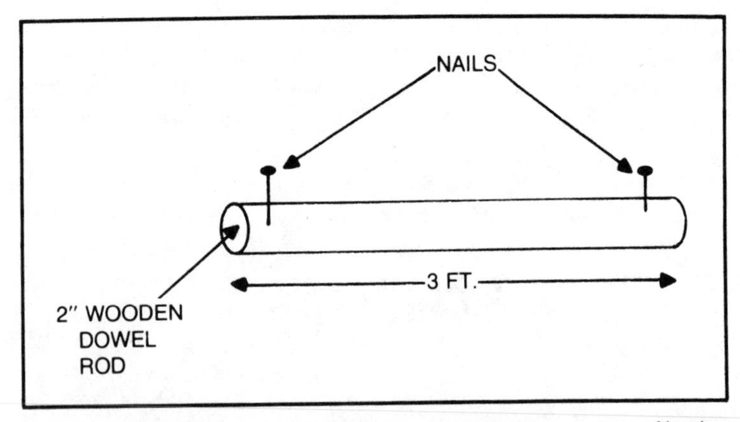

Fig. 10-1. A simple sighting device can be constructed from a piece of lumber fitted with two nails.

abound and can present major interference problems. But moving the dish site a few feet in one direction or another will often remedy this problem, at least for a particular satellite. Fortunately, many urban apartment buildings are constructed with flat roofs which can make ideal mounting sites for parabolic dishes. When located atop a high building, many terrestrial obstructions are cleared and many more clear windows to various satellites should be available.

The antenna's view through the window can be thought of as being the size of the dish diameter. For this reason, you will want

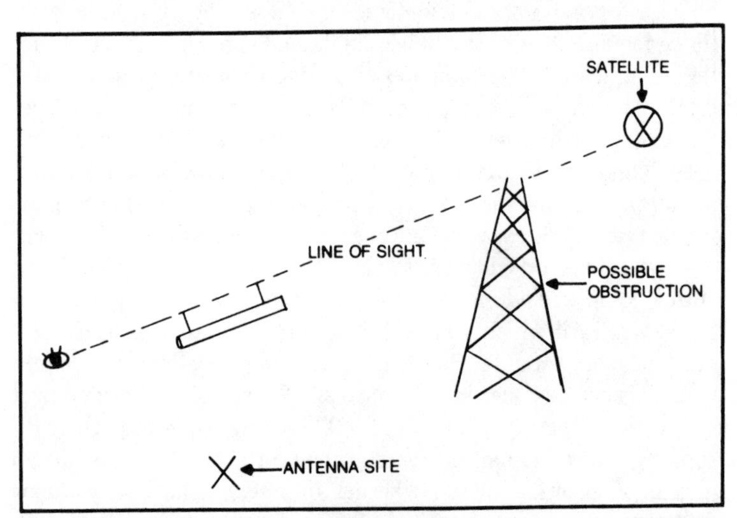

Fig. 10-2. By sighting along the nail heads you can determine if there are any obstructions between your site and the satellite.

to make certain when sighting on a satellite with your homemade wooden device that the window is open to all portions of the antenna. The best way to do this is to place the sighting device at the exact center of the mounting site. Once you have obtained a clear window, move the site one-half the diameter of the antenna to the right and left of a line which is perpendicular to the satellite signal path. For a ten-foot dish antenna, you would move the site five feet to the left of center and then five feet to the right of the same point, taking sightings at each additional location. This is shown in Fig. 10-3. Be sure to keep the site at the same elevation and azimuth for each reading. This will assure a completely open window for the antenna to intercept the transmissions of a particular orbiter. Repeat these steps for each additional satellite you wish to receive. It may take many small repositionings of the antenna's center mounting site before the best location is determined.

Small trees and bushes are normally not considered to be major signal obstructions, although in marginal areas these should be avoided whenever possible. Any object or structure which is made of metal, especially those that are very large, can play havoc with reception, so these are to be avoided in all systems regardless of relative signal strength at your particular geographic location. It may be necessary to locate the antenna at a higher physical site to overcome these, although in some instances a site which is lower in elevation may be able to look under these potential sources of interference. Think of the sighting routine as firing a gun at a target. In other words, you want to hit the satellite with the fired projectile. If there are no major objects in the way, your bullet will hit its mark. Any obstructions lessen the chances of a successful hit.

Hills and mountains are often encountered in rural areas and

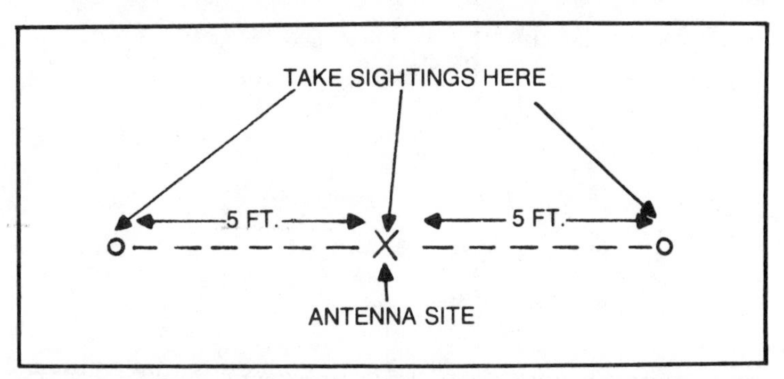

Fig. 10-3. To ensure that all portions of the antenna have a clear window, move the site five feet to the right and left of the center of the dish antenna.

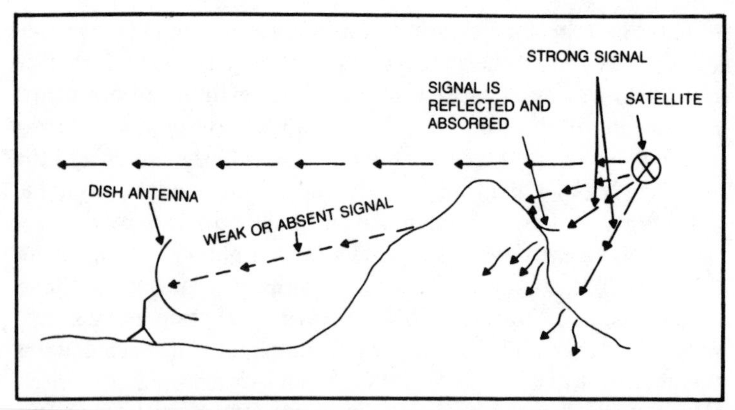

Fig. 10-4. You may have reception problems if your antenna site is located at the base of a mountain or hill.

present major interference problems. This is especially true if your antenna is mounted at the base of one of these, as shown in Fig. 10-4. The only solution is changing sites if satellites from behind the mountain are to be received. You might move the dish to the top of the mountain or to a point farther away from the base, as is shown in Fig. 10-5. Make sure there is plenty of clearance between the mountain ridge and the antenna sighting line to assure maximum signal reception. When site relocation is not possible, you will simply have to be content or receive the signals of those satellites which lie out of the shadow of the hill or mountain.

Once you have obtained a good potential site location, you will

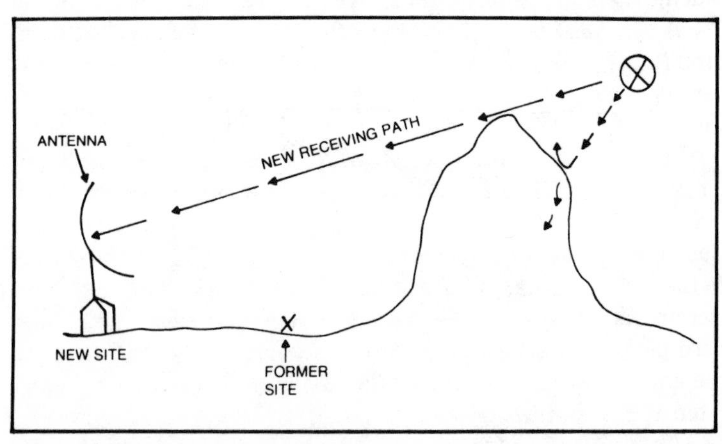

Fig. 10-5. It may be necessary to relocate your antenna site to a point farther away from the base of a mountain.

want to examine some of the other physical characteristics that may prove to be very important in the future. There are very few demands made upon a TVRO Earth station site when compared with some other types of antennas used for operation at a lower frequency. Some types of antennas must be mounted over soil which exhibits good electrical conductivity. This is not even a consideration with TVRO installations. The parabolic dish does not depend upon soil conductivity to operate properly.

Putting aside the unimportant electrical characteristics, the physical condition of the site will involve the firmness of the soil, the presence of large underground rocks, accessibility, etc. When installing most dish antennas, it is necessary to bore three or more holes several feet into the earth. These will later be filled with concrete which will serve as anchors for the antenna mount. Obviously, if you have very rocky soil conditions, it's going to be difficult to drill these holes on your own. Hiring a commercial crew to come in and do it for you can be rather expensive.

TVRO Earth station enthusiasts should be very careful when the dish antenna must be mounted in swampy areas. The high water content of the soil can cause the concrete footers to shift during changing weather conditions. If the base shifts even slightly, the antenna will be out of alignment with the satellite and reception will be greatly reduced or possibly terminated completely. Certainly, the antenna can be readjusted, but this may have to be done quite often as soil conditions change. Swampy areas may be successfully used, but it is often necessary to sink the concrete footers much deeper. This adds to the site preparation expense.

Along these same lines, you will want to check land contours around the proposed site location. If you choose a low spot which serves as a natural drainage and collection point for runoff water, the same problem mentioned above can exist.

How secure is your proposed site? Is is off the beaten path? Or is it near an area which is quite accessible to passersby, who may want to vandalize an attractive setup like a TVRO Earth station? If an easily accessible site is chosen, it would be wise to go to the extra expense of constructing a fence or other type of barrier around the installation to prevent unauthorized access. Standard practice is to place a few "Danger! High Voltage" signs on the antenna and Earth station. This may be of some help, but is often an exercise in futility if a physical barricade is not used as well.

Another site selection item which must be considered is that of trees and other tall objects which are located nearby. While these may not interfere with the satellite transmission window, in the spring a leafy growth can expand the size of a tree by a fair amount. Appropriate trimming during the growing season will provide an easy correction of this problem. Another major factor to be considered about the Earth site surroundings is whether or not large trees could fall on the antenna and equipment should a heavy wind bring them down. One sizeable tree tearing through your Earth station can instantaneously reduce the entire system to rubble. An Earth station is an expensive undertaking and must be protected at all times from all dangers.

In growing residential neighborhoods and in commercial districts, it might be a good idea to run a check at your local city hall to see if any building permits have been issued for nearby structures which could become potential sources of interference. That empty lot next door may present no problem now, but when a house or other building is erected there, it could totally ruin your perfect window. Once an Earth station is installed, it can be thought of as semi-permanent. While the location can be moved, this will require the drilling of more holes for establishment of the new base and, of course, the labor involved in the dismantling and reassembling of the station. While you're at City Hall, also check on the zoning regulations to see if any local ordinances prohibit the erection of an Earth station antenna. In some areas, permits may be required. In others, you may have to appear before a Board of Zoning Appeals in order to obtain a variance to a restrictive ordinance. These are not hard to get, but you have to stress the fact that there is no high voltage danger, nor any chance of the structure damaging property should it fall from its mount. Most antenna ordinances involve height limitations which most Earth-mounted TVRO antennas do not even come close to exceeding.

The problem mentioned here and some possible solutions are generalized in that they could apply to any part of the country. By regionalizing this discussion, a few basic facts can be presented. Since the satellites we are interested in are in geostationary orbits over the Equator, those persons living closer to the Equator will generally have fewer problems with earth-mounted objects, since the antenna elevation angle will be increased. In other words, the antenna will be pointed higher in the sky. Antenna sites in the northern part of this hemisphere will have to be aimed lower on the horizon. This can bring distant mountains and other natural objects into the path of the satellite transmission window. This is illustrated in Fig. 13-6. Of course, when an antenna is pointed higher

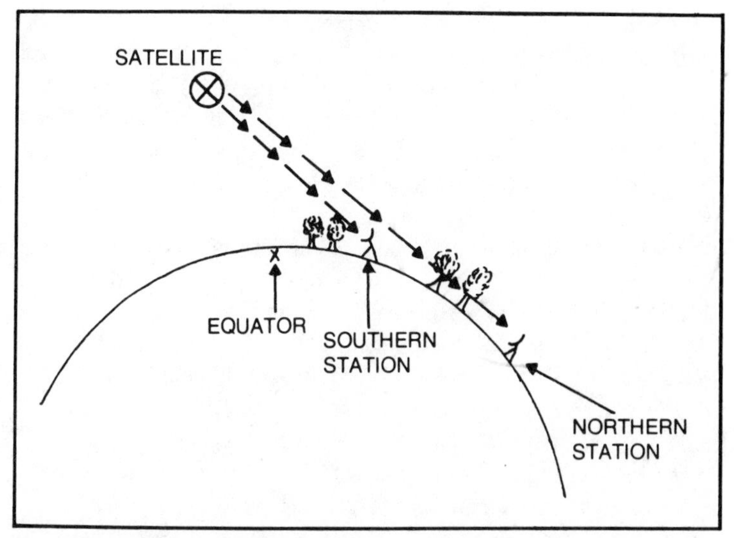

Fig. 10-6. Antenna sites in the northern part of this hemisphere will be aimed at a lower point on the horizon.

in the sky, this can result in other problems. This does not apply so much to parabolic TVRO antennas as it does to spherical designs discussed in earlier chapters. The spherical dish is not adjusted: rather, the feed horn is moved for reception of different satellites. If the spherical dish is aimed toward a high portion of the sky, this will mean that the feed horn will have to be mounted a higher distance from the ground than when the dish is lowered to a point closer to the horizon. This is shown in Fig. 10-7.

In rural locations, site access is sometimes a problem period,

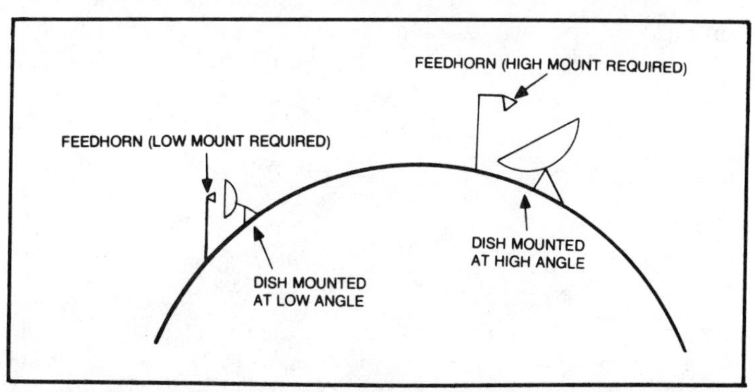

Fig. 10-7. When a spherical dish is pointed higher toward the sky, it will be necessary for the feed horn to be mounted higher from the ground.

Remember, you will have to truck the heavy components to the site, and while the finished Earth station consumes very little area, several times this amount of area will be needed as a work site to put the whole thing together. A small porch roof may make an excellent mounting site for an Earth station, but it will be difficult and potentially dangerous to assemble one there, since all of the parts will have to be spread out and assembled piece by piece. Additionally, two or three men will be required to assemble the antenna panel sections. Small space problems can be overcome during the assembly, but the extra preparations and precautions must be planned in advance. Remember also that you will probably have to travel to the site every so often for preventive maintenance and to adjust the antenna to receive a different satellite. During heavy snowstorms, dish antennas must be cleaned periodically to avoid excessive ice and snow buildup which can cause structural damage as well as interfere with reception. If you find it necessary to locate the antenna some distance from the house and use line amplifiers to boost the signal strength caused by the increased transmission line lengths, make certain you can get to the site in all types of weather. Even if you have motorized antenna positioning controls which are activated at the viewing location, regular on-site inspections are still necessary.

Another consideration is electric power. If the antenna is mounted a short distance from the home, this can probably be handled by means of a waterproof extension cord. But if larger distances are involved, you may wish to run an underground service to the antenna site. This is especially true when motorized base drives are to be used. In some areas, a separate electrical installation may require a permit and/or an electrical inspection. This is to make certain that your installation conforms to all safety regulations.

Wind problems have been discussed many times in this book and it bears repeating again under this site selection discussion. Most commercially manufactured Earth station antennas are designed to withstand 100 mile per hour winds when mounted on approved bases and in compliance with the manufacturers' instructions. This does not necessarily mean, however, that the antenna will efficiently receive signals under these conditions. As structurally sound as most Earth stations are, high winds can cause the dish antenna to vibrate and play havoc with the received signal. If the site you have selected is regularly exposed to high gusting wind velocities, you may want to consider building a wind break to one side of the dish directly in the path of prevailing winds. A local contractor may be able to help you with the design and construction. In no instances should the wind break be allowed to interfere with the sighting path of the antenna. Figure 10-8 shows how a protective structure might be installed in relationship to the Earth station. This structure is not absolutely essential but will provide more reliable reception characteristics from your Earth station during high wind conditions.

The problems and corrective measures discussed in this chapter assume that the TVRO Earth station owner has some flexibility in where the site is to be located. From a practical standpoint, this is not true in many cases. Most of us will be limited to a site in a backyard or a short distance from the home. In these situations, you do the best you can. In some areas, it simply will not be practical or possible to efficiently receive the signals from all of the satellites which have open windows to your geographical area. Here, you will have to choose the location within the area alloted to you which provides the best reception of the largest number of potential satellites. Compromises and tradeoffs will certainly be mandated here. Of course, it may be possible to get a neighbor involved who might allow you to use a small portion of his property if this location provides a better mounting site. A small rental fee could be paid each month for this privilege, but more often, your Earth station's landlord will want a free tap-in. A signal splitter would be required and might be the cheapest way to go when the total cost is spread out over a few years of operation.

Some homeowners find that they must locate their stations in

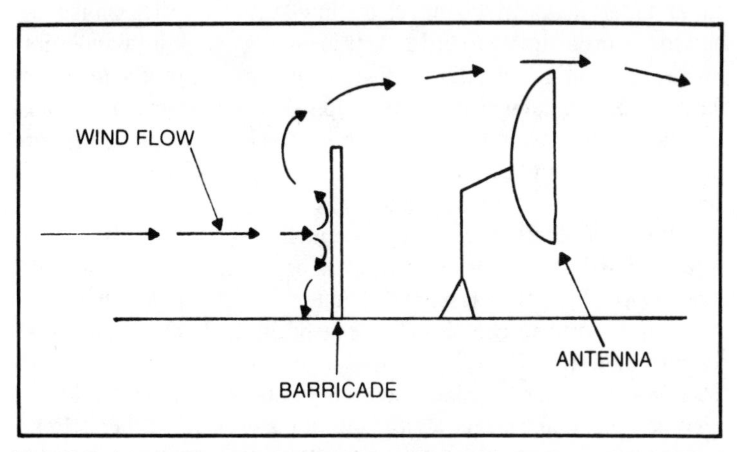

Fig. 10-8. A protective structure such as this would provide a wind breaker in areas where winds tend to be severe.

the front yard rather than in the relatively hidden recesses of the backyard. If you must do this, be certain to provide some means of barricading the station from prying eyes and hands. A sturdy chain link fence with barbed wire at the top is rather costly but very effective, although it does little for the appearance of your home. If your entire front yard is fenced in, a locked gate may serve almost as well. Some owners who find themselves in a situation which requires a front yard installation buy large tarpaulins which are used to cover the antenna and other components during times when no one is home. When vacations are planned, it's a good idea to have a neighbor keep an eye on things for you. It is also wise to partially disassemble the station by removing the LNA and any antenna-mounted electronics. This effectively leaves only the dish accessible. It's pretty hard to steal one of these without being seen, and the antenna is the most rugged portion of the station, so it is not as easily damaged by vandals.

Further protection of an Earth station can be provided by installing some sort of an alarm system. These can be built from parts obtained from your local hobby store or installed by professionals. Proximity alarms are triggered whenever anything enters the protected area. No physical contact with any portion of the alarm is necessary. These types of systems have excellent applications when a physical barrier is placed around the TVRO Earth station, but when the barrier is not present, dogs, cats, and other stray animals can keep the alarm sounding all night. Many TVRO Earth station receivers are now equipped with warning circuits which will trigger an external alarm whenever the low noise amplifier is disconnected. In order to steal an LNA, it is necessary to disconnect its power supply and cable. This will automatically trigger the receiver alarm in those units which are so equipped. The low noise amplifier and the down converter which often is attached represent a great deal of expense and could be potentially high-priority rip-off items for persons who are so inclined.

Of course, the best way to prevent theft and vandalism is to hide the Earth station as effectively as possible. This is not easily done, especially with that large dish antenna sticking out into space for all the world to see. For this reason, a backyard location is always preferable.

For some readers, this discussion will be mostly academic, as most persons don't select a site so much as it selects them. Fortunately, there are enough orbiting satellites, each carrying a significant number of program services, to make it worthwhile to install a station which receives only one or two. Most owners home in on one prime satellite for 95 percent of all viewing. The other 5 percent of the time, the system is used for experimental purposes to see just how many satellites can be received. Of course, more and more satellites are planned for the near future, so it will behoove the Earth station owner to choose as flexible a site as possible.

The selection of an appropriate TVRO Earth station site will be the prime criteria for all future reception uses. A poor site choice will result in poor selection and flexibility, regardless of the quality of the Earth station components. It is for this reason that all potential Earth station owners should spend the hours required in researching this basic aspect of satellite TV reception well before any equipment is purchased and any construction is attempted.

A very few individuals will find that they are just not in a position to install an Earth station which will provide satisfactory performance. There is no one quite so miserable as the person who spends many thousands of dollars for a piece of equipment which cannot do what he wants it to do because of location. This will be the rare exception rather than the rule. If you can receive only one or two satellites from your location (and receive them well) then the work and expense involved in owning your own TVRO Earth station will be worthwhile. If you are contemplating a move in the near future and your available site is not really appropriate, I suggest that you wait awhile. Maybe your new location will provide several prime sites for the installation of an efficient receiving system.

Don't make a decision until you know all the facts. You can start your research by purchasing a computer printout of windows in your area. Then, using the homemade sight and a few other inexpensive instruments, you can determine very accurately exactly what your location has to offer. Too many persons have become so avid in their quest for a TVRO Earth station that they have neglected this prerequisite. Most have gotten away with it very handily, but a few have been burned. Don't be one of the latter. The thirty or forty dollars you spend for the printout and the simple instruments, added to the couple of hours time spent in recording measurements, will pay off very handsomely for you when the time comes to purchase your personal TVRO Earth station equipment.

Chapter 11

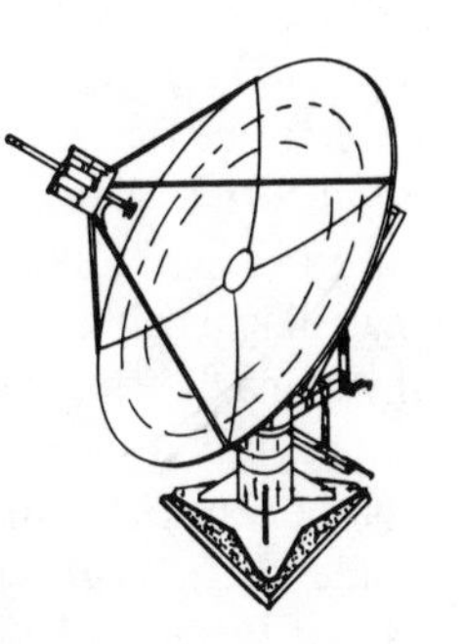

Future Applications

WHAT IS THE FUTURE OF SATELLITE TELEVISION? THIS IS a question that literally millions of people have asked and thousands have tried to answer . . . and with a thousand different answers. The long-term, purely entertainment benefits of a personally owned TVRO Earth station are certainly a question. This is due to plans of present major satellite broadcasters to switch at some time in the future to Direct Broadcast Satellite (DBS) television. The fact of the matter is that all of the signals we presently receive from satellites as individuals are really not intended for the individual, but for commercial services, such as cable television companies. There has been much controversy since the personally owned Earth station became a reality over the legal reception of satellite broadcasts. A cable service pays a satellite broadcaster a specific rate based on the number of cable subscribers. However, when an individual receives the service direct without any cable intervention, a more and more frequent occurrence, how is the satellite broadcaster reimbursed?

The fact is that the broadcaster is not reimbursed in the great majority of cases. This matter has been discussed in Chapter 9, but it is becoming more and more important to the broadcasters themselves. A few years ago, the number of TVRO Earth stations in service to individuals was minimal, and most broadcasters were not really concerned about the reception of their signals by this handful of receiving stations. Many of the satellite services simply made it a policy not to charge an individual with his/her own Earth station, while others charged a minimal fee based upon the fee charged to cable companies. This can be on the order of $.10 or so per month per subscriber. Assuming the charge is this low, it would cost the satellite broadcaster more to administrate and collect fees from individual users than they would actually take in by doing so.

Recently, however, it has become apparent that the ownership of TVRO Earth stations is not a passing fad, but rather an established practice that is becoming more widespread than was originally anticipated. Admittedly, there have been abuses. Most involve enterprising individuals of questionable scruples who set up their own illegal cable services within small communities or isolated subdivisions. These persons installed a single Earth station and then fed several different homes the received signal for a fee. Surplus components dredged from an ever-changing satellite TV industry made the job far simpler and less expensive as the science progressed. Naturally, the satellite broadcasters became very alarmed over what was seeming to take shape.

In response to the illegal reception and use of their satellite transmissions, broadcasters began examining antitheft methods that involved scrambling devices. Currently, very few satellite transmissions are scrambled. Most of those that are scrambled offer formats comprised mostly of adult entertainment. However, this may change in the very near future as more and more services protect their signals by scrambling. This would mean that only a subscriber who possessed the descrambler supplied by the broadcaster would be able to use their broadcast. In many ways, this would remove the individual TVRO Earth station owner from the picture entirely. It's one thing for a cable company to spend several thousand dollars for a descrambler and another for an individual to do so when the price of the descrambler may be more than the entire Earth station. Some see a tremendous black market in descramblers to beat such a system, but I do not agree. Due to the complexity of the scrambling system, cheap, black market descramblers are not a likely event.

This is especially true when one considers the new technologies involved in building scramblers and descramblers. The capability is presently available to design a scrambling system whereby each descrambler located at the receiving end can be controlled via the satellite transmission itself. This involves a subcarrier or subchannel that rides in on the video signal, but does not interfere with it in any way. The subchannel carries information that can constantly reprogram the descrambler to match the scrambling pattern, which is also constantly changing at the transmitting end. Going even further, the technology is even available to disconnect a descrambler (presumably owned by some person who hasn't paid his monthly charge to the broadcaster) from the transmitting station end. Each descrambler would be given a specific code sequence for access, and this sequence would only be known to the broadcaster.

DIRECT BROADCAST SATELLITE TELEVISION

The Direct Broadcast Satellite television probably provides the biggest threat to current TVRO operations. Unlike the present service, DBS is aimed directly at the individual, or more specifically, the home. DBS would feed signals from the satellite directly to a home antenna. This closely parallels the operation of our present Earth-based television stations. Each home would pay a subscriber service and would undoubtedly be equipped with a rental descrambler that would be individually programmed by the broadcaster through the satellite transmissions. One can imagine a satellite broadcaster having offices in every major area in the United States and Canada that would in many ways resemble the present telephone company. If you wanted to subscribe to the service, you would simply call the closest office, pay an installation fee, and their workmen would install the station for you. Monthly fees would then be paid for the service and rental of the equipment.

At present, most satellite broadcasts received by TVRO Earth stations are transmitted from the satellite transponders with an output power of about 5 watts. This is very low power; thus, a large receiving dish is required at the receiving end. As power is increased, the size of the antenna decreases to maintain the same level of signal to the receiver. The antenna dimensions do not shrink in direct proportion to the increase in signal. For example, if power is doubled, the antenna's equivalent size cannot be halved.

The receiving antenna is often the most expensive element in a TVRO Earth station. DBS, however, is designed along the lines of small antennas (maybe three feet in diameter) and high-powered satellite transmission in the neighborhood of 175 watts. DBS would use the frequencies in the K band, whereas present satellite broadcasts originate in the C band. These later frequencies are transmitted at around 4 gigahertz, whereas the K band spectrum lies in the vicinity of 12 gigahertz and tends to be more interference-free than the lower frequency.

There has been a tremendous amount of controversy surrounding Direct Broadcast Satellite television. Obviously, this service eliminates the middleman in the present link, the cable television company. If DBS could be offered at a rate that is competitive with the present cable services, most of these would probably go under, or at least be very limited in what they could offer. Also, by nature, cable companies cannot serve everyone. Typically, towns and cities have cable services, whereas rural and isolated areas do not.

In any event, DBS would seem to be a long way off as of this writing, although I could not have made this statement a few months ago. DBS is not some system that is being developed for the future; it has already been developed, and indeed, satellites are already in orbit and leases have been taken by many satellite broadcasters. Politics has entered the picture recently, and it is my understanding from people within the industry that DBS has been tied up in red tape (governmental) and is relatively dead for the next few years. One major distributor of TVRO Earth station components did offer the opinion that in about five years (1990), DBS may become a very popular reality.

NEW SATELLITE INDUSTRIES

With the coming of DBS, universal scrambling of satellite broadcasts will be commonplace. This could strike a severe blow to the present TVRO Earth station market, although it is possible that a new trend could take place similar to the one that affected uhf television broadcasts (channels 14-83) a decade ago. In many areas, there were no uhf channels to be received because the whole industry was geared toward vhf (channels 2-13). For the most part, uhf broadcasts were aimed at schools and contained dry, educational programs. However, ten to fifteen years ago, uhf stations began to make a dent in the television market by offering popular local and regional services and by broadcasting highly original and entertaining programs. Many of these are known as PBS, noncommercial programs, but still other uhf broadcasters went the commercial route. Some of them have cleaned up by broadcasting popular reruns and movies that they were able to purchase rather inexpensively.

In Washington, D.C., WDCA Channel 20 made great inroads into uhf broadcasting. Transmitting with a power of 4 million watts, WDCA seemed to fill a gap by broadcasting popular old movies, late-night horror flicks, and reruns of everything from Gilligan's Island to Hogan's Heroes. It wasn't the kind of station that you would want to watch continuously, but then neither is any other channel, vhf or uhf. WDCA became known as the station that always had something mildly interesting going on, and it attracted a tremendously large viewership.

The same thing could happen to the current C band satellite channels as the major broadcasters leave them for DBS. Of course, all satellites have a limited useful life, as orbits tend to decay after a period of years and equipment malfunctions take place. When the grade of a satellite signal drops below a certain level, its video broadcasters generally switch to a newly-launched satellite. The vacated satellite is then leased to other services that don't require the same signal quality as television broadcasting. The degraded satellite might still be effectively used for some types of video broadcasts and might be leased by small operators who could afford a fairly low satellite rental rate. What these persons choose to broadcast might be as big an innovation as uhf television was at the beginning of its popularity.

However, these satellites might also be used to broadcast telemetry or control signals and provide an extremely futuristic and unique service. One such service that comes to mind is supplying information directly to home computers that are tied in with a TVRO Earth station. Of a more exotic nature is robotics, or more specifically, the control or programming of personal robots.

Robots

Undoubtedly, you may be thinking that I have left the world of nonfiction, the environment in which this book is intended, and moved into the world of science fiction. However, robotics may be the next great scientific technology that can be directly utilized by the individual. Ten years ago, the idea of inexpensive computers being put to practical use by every member of a typical family would have been considered by most to be pure science fiction. Today, this is a reality in many homes, and the use of microcomputers continues to grow by leaps and bounds.

What's next? To my way of thinking, the next great scientific fad that will be converted to practical necessity is the robot. If you think about it in the right way, a robot is simply a computer that has some mechanical abilities that emulate in many ways the mechanical workings of the human body. Today, computers are mimicking some of the minor functionings of the human brain and human thought processes, so it seems only natural that the far simpler mechanical processes should follow suit. For that matter, a robot could simply be a "dumb" remote-controlled mechanical device whose "brain" is a home computer located elsewhere that transmits its signal to the mechanical "idiot."

Nor am I looking to the distant future for such technologies. The home robot is already a reality, although most are extremely limited in their mechanical capabilities. Recently, however, many companies have begun to offer home robots that are more or less intended for experimentation in the field of robotics. One of the more promising competitors is manufactured by Androbot of San Jose, California. They previously offered a very simple, but attractive robot called Topo, but have recently released their B.O.A./XA series, which offer far more. Figure 11-1 shows the Topo model on the left, with the B.O.B./XA model on the right. The optional head assembly for the latter is shown in the foreground.

The B.O.B./XA series from Androbot is probably the most sophisticated "personal robot" offered today at other than an impossible price. Right now, the starting price of this unit is $7,500 (retail). However, like the personal computer, if robots become more popular, this price could drop substantially. The thing that is most

Fig. 11-1. Two home robots from Androbot include Topo at the left and the eminently smarter B.O.B./XA at the right. Courtesy of Androbot Inc.

impressive about this robot is that it contains its own IBM PC bus-compatible computer on-board. It also offers infra-red sensors, a fluxgate compass, navcode readers, a lift arm and virtually unlimited expansion capabilities. B.O.B./XA is designed to be a moving development system. (Incidentally, B.O.B. stands for "brains on board", referring to the self-contained computer.) Current applications being studied include production, materials handling, security patrol and artificial intelligence research. B.O.B./XA is the robot of choice for companies and groups interested in state-of-the-art robotic applications.

The robot stands 24 inches high, 22 inches wide and 22 inches deep. A small number of units are being farmed out to computer hardware and software manufacturers to allow them to come up with their own specialized implementations. This could result in the first highly popular personal robot.

Power System. The robot is powered by two 12 V GelCell batteries in series to provide a 24 V power system with about four to five hours of continuous use. It contains a battery-charging circuit with an external charging unit that is plugged into a 120 V outlet and a jack on the cardcage. Recharge time is approximately six hours. All motors are driven directly off the batteries, and a dc/dc converter provides ±5 V and ±12 V power for the electronics. Power can be shut off under software control and electronically turned by any of a number of on-board devices. Batteries can be charged while the power is on. A monitor circuit will flash an LED indicator to inform a user of a low battery-level condition.

Motion System. The motion system rides on a stable, four-wheel base. Two drive wheels with a 7-inch diameter, dc motor, and gear box with quadrature encoder feedback for direction and speed control are located in the front of the robot. Two 2 1/2-inch casters in the rear provide stability. Also, there is a 1 1/2-inch clearance on the bottom of the robot. The motor drive circuitry is located on the power board.

CPU Section. The CPU board is located in the card cage and is IBM PC bus-compatible with extensions for additional interrupt, as well as shared memory and multiprocessor capability. The board consists of the following standard items:

- Intel 8088 microprocessor.
- 4 IBM PC-compatible expansion card slots.
- 4 Androbot expansion card slots.
- DMA controller.

- 2 interrupt controllers (12 intervals).
- 2 triple interval timers.
- Serial interface controller RS-232 connector, and 30-foot cable.
- 128 K dynamic RAM (expandable to 256 K).
- 8 cartridge slots, each for up to 16 K × 8 ROM or EPROM; 2 slots can be used for up to 2 K × 8 to 16 K × 8 EPROMs.
- Motor, keypad and Androlift control connectors.
- Joystick connector.
- Standard socket for IBM expansion chassis (for external use only, not fitted in standard robot).
- Switch matrix sensor connectors for 8 switches.
- Emergency stop switch connector.

Sensors. The sensor board plugs into one of the 4 Androbot card slots. The sensors all come in identical 3-inch by 3-inch boxes that are mountable in any of 4 vertical sensor columns. The sensor columns can each hold 4 sensor boxes, each in one of 4 different vertical locations. The 2 front sensor columns can hold sensors pointing either directly forward or directly to the side. The side columns can hold sensors pointing only to the side.

Standard Sensors:

1. 4 ultrasonic rangefinders.
2. 1 IR nav code reader.
3. 1 flux gate electronic compass.
4. 1 battery monitor.

Optional Sensors:

1. Up to 4 additional ultrasonic rangefinders (8 total).
2. Up to 3 additional IR transceiver Nav Code Readers (4 total).
3. Audio direction/level detection sensor with 3 microphones.
4. Analog joystick.
5. Light level sensor.
6. Temperature sensor.
7. Floor stripe sensor.
8. Up to 5 additional general purpose A to D inputs.
9. 24 general purpose, user-programmable digital I/O lines.
10. Up to 3 additional interrupt lines (7 total).

Speech. Two types of speech capabilities are standard, using a speech synthesis board with a TI 5220 speech synthesis chip. Both use the single, 3-inch speaker mounted behind the front panel immediately above wheel level. Speech synthesis is accomplished through use of a linear predictive coding (LPC) technique under direct control from the 8088 microprocessor on the CPU board.

Androlift. The lift is capable of lifting up to 20 pounds of weight. The standard lifting element is a curved fork that is 3/4 of an inch thick, forming a 2/3 circle with 4-inch inner diameter and 3-inch fork spacing. The center of the fork is about 6 inches in front of the robot. It is mounted to the front cover plate in the center front of the robot immediately between the two front sensor columns. There are up to eight levels of lift height possible, each adjustable from 16 inches to 20 inches above the ground. The fork element automatically retracts flush to the front of the robot in the lower one inch of lift travel and extends parallel to the ground above that level. Lifting time is about 4 seconds from the extreme points of lift travel.

The lift is designed to lift an 18-inch tall round table 12 inches in diameter with a single-centered leg 2 1/2 inches in diameter. Through readjustment of the lift heights and replacement of the removable fork element, different types of articles can be lifted. Custom fork elements are available.

Control Panel. The control panel is located on the top of the robot. It includes a power switch, power status indicator LED, panic stop button, and a 16-key keypad. The power switch is a three-position switch, giving the possibility of full off, full on, or automatic mode. Automatic mode allows the CPU to turn off the power and selected battery driven electronics (on the expansion board) and turn the power back on upon the occurrence of certain program-selectable events. The power status indicator LED will be on when the power is on and will blink at various rates to indicate battery level condition. The panic stop button can be used at any time to immediately stop the motion of the robot to prevent it from damaging itself or another object. The 16-key keypad is user-reprogrammable to provide easy selection of programs and input of information into the robot.

Software Features. The BOSS System Cartridge includes a command line interpreter (CLI), cartridge incorporation, dynamic memory allocation, communication controller for dumb terminals (keyboard and display) and host computers, programmable event timer, programmable keypad driver, and switch matrix drivers. The

Robot System Cartridge is for drive motor control, sensor control, lift control, and mathematical functions.

Drive motor control software uses a proportional-integral (PI) speed control algorithm utilizing encoder count quadrature feedback for speed and direction determination and pulsewidth modulation (PWM) for speed control. It provides for three different modes of operation. Joystick mode will provide user control of the robot with a 9-pin, digital joystick. Exact mode will provide very accurate position control for straight lines, circular arcs, and rotations about a point. It requires stopping of the robot between motion commands and has the capability to queue up to ten commands. Speed mode will produce smooth operation between straight line, circular and rotational motion, but will not provide accurate position control. It is designed to be used primarily with sensor feedback updates. The motion system can maintain accuracy at any battery level, but will have reduced maximum speed at lower battery levels. It also will automatically halt the motors if it detects the motors are stuck on an obstacle. Calibration routines provide improved accuracy as well as the use of other sizes and configurations of wheels with no change in software control routines.

Sensor control software provides all the necessary drive routines for the current sensor set, including ultrasonic rangefinders, IR navcode readers, flux-gate electronic compass, and battery monitor. The ultrasonic rangefinders may be fired and read as well as maximum echo wait time set. The electronic compass routines provide for calibration of the sensor as well as bearing determination. The navcode readers may be configured to either simply note the presence of an IR reflection from some surface, or to read and validate the presence of one of sixteen unique combinations of retroreflective material (a navcode). The battery monitor can return the actual value of the batteries to a resolution of 120 millivolts, either the total of the two 12-volt batteries, or the level of the lower battery only.

Lift control software provides for raising and lowering of the lift, as well as monitoring the status of those operations.

Mathematics routines include square root of a 16- or 32-bit integer, sine/cosine of an angle, arcsine/arccosine of a 16-bit number, and arctangent of the ratio of two 16-bit numbers.

Forth Language Cartridge. These two cartridges contain the BOBFORTH language interpreter, which is based on the MVP FORTH interpreter and has been extended to include all the words required to control all the software functions available onboard the robot. It has the MVP FORTH EDITOR capability and also has an 8088 assembler for the generation of efficient code in critical areas.

Speech Cartridge. The speech cartridge contains all the speech rules and control software for text-to-speech capability. It also has the capability to select prerecorded words from optional vocabulary cartridges in an easy-to-use fashion. Words that are not available in the vocabulary cartridges are spoken using the text-to-speech algorithm, with volume and pitch matched to the vocabulary cartridges installed. A help function to determine what vocabulary words are installed is also included.

Vocabulary Cartridges. Prerecorded words stored in cartridges for use by the speech control routines. There are about 140 words per cartridge. There are currently two cartridges available.

Demo Cartridge. Provides the following FORTH language demo routines, available either in cartridge form or on a host computer diskette: Table Fetch—locates and picks up the special table designed for the lift; Wall Hug—will follow parallel to a wall and indicate when the wall is gone, Follow Me—(with head and shoulder option only) will wait for a person to stand within 150 centimeters in front of the robot and will follow that person when he/she moves.

Host Computer File Support. Support routines to make host computer act like a dumb terminal to BOB and provide access to the host file system. Currently supported on Apple II, II+, *II*e, and IBM PC. Specifications are available for support for other hosts.

Options. The following options are available upon request:

- Remote communications (RF or IR).
- Speech Recognition.
- Multitasking Operating System.
- Additional Languages.
- General purpose expansion board, including time of day with battery backup, battery backed up RAM, cartridge/RAM expansion, and auto turn on monitor circuits.
- IBM PC compatibility board, including keyboard driver circuit and bus clock driver.
- IBM PC compatibility (BIOS).
- Head and shoulder assembly with ultrasonic eyes.

This offering from Androbot, to say the least, is most impressive. What may be its most impressive feature is its open-ended design. Unlike many previous attempts at developing a personal robot, the model described seems to be infinitely expandable and therefore may be pressed into service in many, many different applications.

Many robots that have gone before and were targeted to a general audience were little more than interesting toys. Most demonstrated the general principle of robotics, but few had the capability of doing much more than perform a few interesting tricks. The B.O.B./XA series would seem to be the first attempt to directly interface a mechanical robotic system with a fairly powerful onboard computer. The fact that Androbot is opening the architecture of the brain and its controlling functions to selected researchers could mean that a great deal of third-party support will be forthcoming. Third party support is part of what makes the IBM Personal Computer series so popular today. A well-learned lesson in the microcomputer field has caused many computer manufacturers to freely reveal the workings of their computers so that other companies may make add-on products. Companies that have kept their computer architecture secret in an attempt to be the only supplier of products have often met with failure.

Robotics is definitely a technology that will result in products that will be used in many homes in the very near future. It is conceivable that one day, satellite transmissions may be used to affect these devices. I used the word affect rather than control, although control may be an aspect of it. Robots may one day be taught by information transmitted from satellites and received via a TVRO Earth station.

Satellite Bulletin Boards

We may see in the not too distant future a kind of satellite bulletin board much like computer bulletin boards of today. One could speculate that a satellite whose quality of transmission has decayed below the levels necessary for television broadcasts might be rented by a firm that would accept short personal advertisements. These would be broadcast in computer bulletin board fashion and could include everything from help wanted ads to flea market items for sale.

Admittedly, the ideas expressed in this chapter may be somewhat far-fetched, but the fact of the matter is that the C band satellites currently used for broadcasting won't be totally abandoned should a switch be made to DBS. Besides being a waste of available satellites, there are all those personal Earth station owners who could make someone a very lucrative market. As the major broadcasters go to different satellites and different frequencies, some of the present "minor league players" may move into the major leagues, using what the big boys left behind.

UPGRADING TO DBS

A question I'm often asked is "Would it be possible to convert my present TVRO Earth station to DBS should the whole industry move in that direction?" The answer is yes. The extent of the modifications, however, would depend on the specific system currently in use. The present TVRO antenna would offer exceptional gain at 12 GHz instead of its designed 4 GHz. It would be necessary to change the feed horn assembly, along with the LNA and down converter. In some situations, the current receiver could be used, but this is a matter for further discussion.

From the practical standpoint, there would probably be no advantage to switching to DBS, since it's my understanding that all DBS channels will be scrambled anyway. To put it bluntly, DBS places satellite reception in the category of "purchasing a service" rather than striking out on your own. The DBS equipment will have to be either rented or purchased from the company supplying the service.

SATELLITE SIGNAL SCRAMBLING

The subject of broadcasters scrambling their video and audio signals before beaming them to a satellite for retransmission is one that is often discussed, argued, and in general, bandied about by any crowd of Earth station owners or dealers. As soon as personal Earth stations began to catch on, the rumor began to spread that all of the major users of satellites would begin scrambling their signals to prevent nonpaying users from gaining access. Scrambling involves a complex rearranging of the television signals from the transmit end. At the receiving station, a descrambler is used to put the signal back in shape again so that it may be properly detected by the television receiver. When a signal is scrambled, you can still receive it, but it's simply useless garbage on the television screen. For many years, the FCC (Federal Communications Commission) strictly prohibited the scrambling of any and all standard broadcasts. In recent years, however, exceptions have been made.

Currently, the great majority of satellite transmissions are not scrambled in any way. But even as I write these words, the rumor is again circulating that most of the major services will be going to scramblers in the very near future. While I don't necessarily agree with this, there certainly will come a time (as personal Earth stations become more and more popular) when broadcasters may deem it necessary to protect their billion-dollar investments in what is accurately called "pay television."

Even with the current popularity of personal Earth stations, most satellite broadcasters do not really feel threatened by what is still only a miniscule loss of revenue. Many of these broadcasters do not even make provisions for the personal Earth station owner, since their targeted buyers are local cable companies from coast to coast. This is where the main portion of their revenue is derived. I have noticed that even those companies that do instigate charges for personal reception have very low rates. This is really based on the per-user rate that is charged commercial cable companies that receive the broadcasts via an Earth station and then transmit it over cable to their subscribers. The rate, then, that is charged the owner of a personal Earth station is generally based on this same rate, which is extremely low. One must realize that the cable companies figure the cost of their Earth station, converters, service, maintenance, and so on, along with the per-user rate in arriving at their own per-user rate. Since personal Earth station users have effectively cut out the middleman (the cable company), the rate that is charged by the broadcast service itself is extremely low. In many instances, a fee of $10.00 will pay you up for several years.

There are many "free" broadcast services on satellite, but for the most part, satellite television can be thought of as true "pay television." In North America, most of us are accustomed to all broadcast services being free of charge. In other words, if we have the equipment to receive it, we have the right to receive anything we want.

Don't be so sure! Just because something has always been does not mean that it will always remain. The major reason that the commercial broadcast airwaves have been relatively free of charge to users at the receiving end lies in the fact that the broadcasters have traditionally paid for their overhead and received a profit from those advertisers whose commercials are broadcast along with the programs. It is a worldwide tradition to complain about commercials, but this is what has kept us from having to pay for these broadcasts. However, pay television is often commercial-free. Let's take one of the movie networks, for instance. You will never see a commercial on HBO because it is an entertainment network. Its only purpose is to broadcast programs, not commercials. We all know that vast amounts of money are involved in standard, Earth-based television broadcasts, but even more is involved in satellite broadcasts. Since there are no commercials on HBO, the corporation that runs it must recoup its investment and make a profit by charging users. Ninety-nine percent of the time, these users are cable companies. What this amounts to is the vast majority of pay television subscribers breaks down to only a handful of actual receiving stations when compared with the overall populace that eventually receives the broadcasts via cable. This greatly simplifies the business of enforcing payment for services, in that it's much easier to monitor the broadcasts of a cable company than to monitor the broadcasts to millions of persons, as is the case with standard television. It's not very likely that a cable company would take the chance of stealing a broadcast from a satellite and displaying this theft by broadcasting the stolen program to hundreds of thousands of subscribers.

However, as TVRO Earth station owners increase in number, the potential for lost revenues through theft increases tremendously. There is no practical way for the satellite broadcaster to determine just what a personal viewer is watching. When the predicted revenue loss through personal theft is significant enough to cover the cost of adding scramblers at the broadcast end and descramblers at the reception end, then there certainly will be a move to do something about these lost revenues. Again, this applies only to those satellite broadcast services that charge for their programs. Others, like Superstation, have paid advertising, and this can sometimes mean that there is no user charge.

It's not too difficult for a satellite broadcast company to determine the amount of theft that is taking place. All that is necessary is to amass the figures on the number of personal Earth stations in existence and compare them with the number that have contacted any one service and paid to receive the broadcast. These figures would be compared with nationwide ratings that determine the percentage of people who watch these programs via cable.

In recent years, the entire television industry has been complaining about theft of programs through the use of videocassette recorders. The same objections have been voiced by the movie industry. Regarding satellite transmissions, the simplest method for the broadcasters is to scramble their signals and provide descramblers to the cable companies. The descrambled signals would then be sent to their subscribers. No descrambler would be required by each subscriber because the signal has been straightened out at the cable company's Earth station. But what does this do to the Earth station owner?

It may literally put him out of contention altogether. Scramblers and descramblers can be simple or highly complex. You may be able to pick up a simple descrambler for $100 or so, but a complex descrambler can cost well over $1,000, and indeed, many of the more sophisticated types could cost $10,000 to $20,000 or even more. Undoubtedly, there will be cries of "Hey! This isn't fair!" While my opinion can be argued, it probably is fair, especially if you stop to consider that unlike standard Earth-based transmitting stations which are required by the FCC to provide a public service (i.e., community broadcasts, nonprofit public service announcements, etc.), most satellite broadcasters are purely commercial endeavors that do not fall under the same regulators. In short, the broadcasters are there to make a buck, and their tremendous investment should certainty entitle them to do so.

The satellite broadcasters are not prohibited, through regulations, from scrambling their signals. The main reason this has not generally been done so far lies in the initial cost outlays involved and the fact that most of these services are not really threatened by the small percentage of theft that can and does occur. However, all of the major broadcasters are now preparing for scrambling. This will not apply to the present TVRO frequencies.

It would seem to many TVRO Earth station owners that the broadcasters have ignored their existence completely. This may be basically true, because the current broadcast frequencies and equipment required for reception was designed along the principle of commercial use only. These satellite systems were designed to broadcast signals to cable companies and other commercial installations. I don't think anyone dreamed at the time that technology would advance to a point where personally owned Earth stations could be abundant.

However, these same broadcasters were also looking toward the not-too-distant future and a more personalized service called DBS. The Direct Broadcast Service has been operational for quite some time, but it does not presently boast the multitude of program offerings the current system used by cable companies and Earth station owners does. DBS was designed as a system that would allow every home to receive satellite broadcasts at a price within the range of a color television receiver. For this reason, present TVRO Earth stations may suffer.

The Direct Broadcast Service transmits from space at a frequency three times that of the current 4 GHz used by present services. The 12 GHz frequency was chosen for many reasons. One of the most important of these is that an antenna for 12 GHz can be one-third the size of an antenna for 4 GHz and still boast the same signal pickup efficiency. The size or diameter of any antenna is directly related to the frequency or frequencies it is designed to receive. The higher the frequency, the smaller the physical size of the antenna. Of course, this also means that antenna dimension deviations become far more critical. A deviation of one-sixteenth of an inch might be negligible at one frequency, while at another higher frequency, it can mean the difference between successful reception and no reception at all. In other words, a 12 GHz antenna will be much smaller than one designed for 4 GHz, but its design will be far more critical.

With the coming of DBS will also come inexpensive satellite receiving stations for every home. Whereas a 10-foot dish is the standard for most present Earth stations, DBS antennas will measure 3 feet or so in diameter. Just an antenna design is more critical at higher frequencies, so is the design of the electronics involved in detecting these signals. However, mass market potential is anticipated with DBS, and this will help keep costs down to a very reasonable level. DBS may indeed be the service that will bring satellite television into every home.

Let's return to the subject of scrambling, which directly relates to DBS. The official word is that HBO, one of the most popular entertainment channels in North America, will be fully scrambled by 1985. However, if this comes about, it will have little effect over the short term on TVRO Earth station owners who are presently receiving this channel. As of this writing, HBO is broadcast in the East via geostationary satellite F3. However, satellites have a life expectancy of only eight years or so in this type of service. After this period of time, signal quality falls below acceptable levels, and the services must move to another satellite. Generally, the satellite they vacate is leased to other services that do not require the higher levels of quality. It has been stated that when HBO makes the switch to a new satellite, the service will still be broadcast over F3. The new satellite will use scrambling, whereas F3 will not. It is conceivable that several more years of HBO service can be expected from F3 before the signal becomes unusable. The scrambling system that HBO will use is said to have the capability of changing the scrambling pattern at a rate of up to ten times per second. The descramblers will be sold or leased to the cable companies.

However, the main reason for the switch to scramblers is probably not due to theft of signals under the current service, but the anticipated move to DBS. There is some question as to exactly what the future of the 4 GHz service will be, but I imagine that it will be active for many years to come. When pay television is offered directly to the consumer by cutting out the middleman, the cable company, scrambling becomes an absolute necessity from a monetary standpoint. The DBS system should be fairly inexpensive and will probably start a widespread trend in television reception. More people will be able to take advantage of DBS because of its lower cost and less stringent installation requirements. Since the broadcast is now selling direct to the public, steps must be taken to avoid what could constitute widespread theft amounting to revenues in the billions of dollars. Here is where each home that uses DBS must be equipped with a descrambler. The descrambling units will probably be sold as a part of the subscription package to the various DBS broadcast stations.

The current projected cost of these descramblers is $250. This doesn't sound too bad until you stop and think that such a descrambler will be good for only one channel. This would be the channel you subscribe to. If you want to pick up another channel, another descrambler is required. Obviously, DBS will not take the place of the standard Earth-based broadcast service, since to pick up five channels, you would have to spend $1,000 on descramblers. It is more likely that owners will not be required to buy descramblers, but will simply lease them in the price of the subscription.

Undoubtedly, a lucrative black market will spring up to sell descramblers. Such a situation could make what is currently known as computer crime look like small potatoes. I'm not saying that the descrambling system I have outlined is what will eventually come to be, but that these would seem to be the current industry trends.

Some people have suggested that an alternative will be used whereby a single type of descrambler will be approved. This more or less points to a universal descrambler that can be used with all services. With a subscription to the service, you would receive a special code number to punch into your programmable descrambler to allow you to access any service to which you have subscribed. These codes would change on a regular basis, and only those paid-up subscribers would receive the current access codes.

I don't think this will come to pass because there's too little security in this type of system. Certainly, there would be a large amount of swapping of current code numbers, meaning theft could easily run rampant. The broadcast industry is trying to come up with a means of protecting the unauthorized reception of their broadcasts that is foolproof and inexpensive.

Fortunately, technology may have solved this problem to a high degree. Descramblers are now available that can be individually programmed at the point of transmission. These are called addressable descramblers, and they work in the following manner. A subscriber is supplied with an addressable descrambler upon taking a particular service. The subscriber's identification number is input to a computer. During the television broadcast, the computer constantly outputs the coding sequences that cause this subscriber's descrambler to allow access. This would mean that millions of descramblers would be individually programmed on an ongoing basis by a computer tied into the transmitter that initially broadcasts a television signal. The programming information would be placed on the signal in such a manner that it would not interfere with picture reception.

If a universal addressable descrambler design could be arrived at, it could be produced by many different manufacturers, and this would mean that subscribers would need to have only one in order to pick up any service they wished to access. In each case, a subscription to each service would be required. This would allow the subscriber's number to go into the individual computers of the various services to be accessed.

This would give the broadcast services the option of disconnecting you anytime your subscription is not paid up by simply removing your identification number from their computer. This could be the trend of the future, since the technology is readily available today.

While I have theorized that this addressable descrambling system might be the logical approach, I can argue the point as well. The FCC, which has traditionally controlled all broadcast services with an iron hand, seems to be advocating deregulation of the satellite broadcast services. If this deregulation continues, it may not be possible to arrive at a universal standard for descramblers, and each company may elect to use its own type. We have seen a sample of this in the videocassette industry, where Beta is used by some and VHS by others. A third standard has even cropped up, and none of the three are compatible. One must realize that the satellite broadcast services are interested only in selling their own product and not with whether or not their descrambling equipment is compatible with that of a competitor. Certainly, it's possible that a pact could be agreed upon within these services, but this would not be consistent with the much-reputed network battles that are ever ongoing in Earth-based television services today.

I certainly would not advise anyone interested in owning a TVRO Earth station to forget the idea just because of the scrambling that is rumored to be taking place. Certainly, if there is a massive switch to DBS, new equipment will be necessary, but this will not necessarily make your present Earth station obsolete. As a matter of fact, it would probably be quite easy to convert a current 4 GHz Earth station to 12 GHz by simply replacing the LNA and down converter (and possible the feedhorn) with units designed for the higher frequency. This could represent a cost of approximately $1,000 today, but might be far less expensive two years from now. The present receiver, which detects the 70 MHz output from the down converter, could still be used, and the larger dish size would simply boost the signal strength. This is only conjecture, and other problems could crop up, but I think it's fair to assume that a TVRO Earth station purchased today would still be in use five years from now.

This section has addressed not only scrambling of satellite signals but also the reasons for scrambling, many of which relate to future changes in satellite broadcasting. To be honest, no one knows for sure exactly where this service is headed. While it is possible that the face of TVRO Earth station ownership could change drastically in the next half decade, it may also be as accurate to guess that it will change little. For my money, it's anybody's guess.

The gist of this entire discussion is that the area of TVRO Earth stations owned by individuals is certainly in for some changes in the next five years. However, with the recent blockage of DBS for the time being and the greatly advanced technology that has been made available, some people in the field feel that the personal Earth station market is in its golden era. Manufacturers who have tended to be quite conservative have suddenly sprung their new innovations on a receptive market. Most of this conservatism was due to a fear of possible obsolescence by the coming of DBS. Now that DBS has been forestalled, at least for several years, Earth station technology is moving forward and the individual is in a perfect position of being able to choose from the most advanced equipment ever and at some of the lowest prices ever. It seems like every week, newer and better equipment is being offered. Since TVRO Earth station ownership is on the increase, quantity sales have bought prices to the lowest point in the entire history of this field. Now may be the best time ever to make that purchase you've been dreaming about for years.

Chapter 12

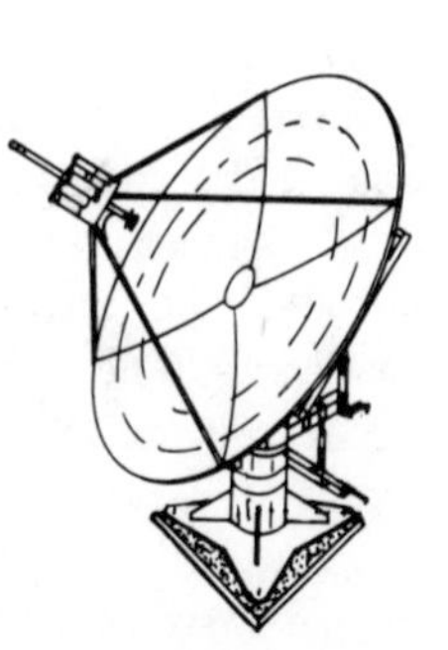

The Earth Station Assembly

THE PURPOSE OF THIS BOOK HAS BEEN TO PROVIDE THE READer with the information necessary to make intelligent decisions regarding what equipment to consider and ultimately, to buy for assembling a complete TVRO Earth station. This last chapter will serve as a summary of all of the information contained in this text and to describe what the assembly of a complete TVRO Earth station will involve. The appendices are filled with charts and other materials which will aid you in the initial planning stages and which may be used in the years to come as a convenient reference source.

SITE DETERMINATION

Your first duty as one who is interested in installing a personal TVRO Earth station is to select a site. This will involve obtaining a computer printout for your geographical area to determine the number of satellites which you have a reasonable chance of accessing from your area. When you have this information, you will be in an excellent position to determine how much use you will get from an Earth station.

If the computer scan looks favorable for your area, you will then select a specific site near your home a which the antenna and other TVRO equipment is to be mounted. Many criteria for selecting a favorable site have been provided in Chapter 10, and you may want to refer to it when choosing between several possible TVRO locations.

As is outlined, you will want to run an "open window" check, which will allow you to detect any interfering objects which may be situated between the proposed antenna location and the satellites you wish to receive. This check will provide the necessary information as to which satellites generally available to your area may be specifically received by your station.

During this entire process, you will also want to refer to the relative signal strength charts provided in the appendices of this book to determine how strong the satellite transmissions are in your area. Chances are, the manufacturers of specific Earth station antennas and other components will also supply you with the same basic charts. The later are often coded with specific antennas and LNAs recommended by these manufacturers for stations in different areas of the United States.

Having done all this, it is now necessary to determine the physical nature of your site. Referring to the previous chapter on this subject, we know that the site has to be accessible and fairly stable in regard to soil conditions in order to provide a firm base for the antenna. From the information contained in this section of the book, you will probably be able to determine whether or not additional site preparation will be necessary before beginning construction of the antenna mounting base.

EQUIPMENT SELECTION

After you have determined that you are indeed in a situation to make full use of a TVRO Earth station, the fun part begins. You will have to decide on which manufacturer or manfuacturers you will select to supply the TVRO Earth station components. You may even elect to build some components yourself. In any event if you are like most TVRO enthusiasts in the country, you will probably purchase all of your equipment from one manufacturer. Many products from a variety of companies which sell TVRO Earth station equipment have been mentioned in this book. Additionally, there is a listing of manufacturers in the appendices. Before the selection is made, you should obtain complete, up-to-date information from a number of these companies who will most likely supply you with a chart of which equipment they recommend for your specific situation and requirements. From this information, you will be able to effectively compare overall pricing and the abilities of several different Earth station packages. This will make the selection process much easier.

Once you have ordered the equipment, you can immediately begin to prepare the mounting site for the antenna. You should request the installation manual from your selected manufacturer in advance to allow you to utilize the time between equipment order and arrival to complete this phase of assembly. The sinking of the footers for the antenna base will require drilling three or more holes into the earth and pouring concrete. You can do this yourself or have a local contractor do the job for you. Either way, it will be necessary to allow about one week for the concrete to set. This time cannot be hurried and may even be lengthened, depending upon weather conditions in your area.

THE EQUIPMENT ARRIVES!

Finally, the glorious day is here, and your complete Earth station package arrives in several large boxes. Assuming that your site has been properly prepared in advance, your first job will be to examine each component carefully, looking for signs of damage. Since most antennas will arrive in sections that must be assembled, you will also want to check every bolt and bracket to make sure that you have not been accidentally shorted by the manufacturer. Most companies will include a packing list and a chart which will help you in taking this inventory. Damaged equipment should be reported immediately to the trucking company which made the delivery and simultaneously to the manufacturer. In most instances, it's best to wait until all equipment is in hand before beginning the construction. In a few instances, this may require waiting a few extra days; and while this is nervewracking, it will most likely assure a trouble-free installation which will serve you for many years to come.

You can probably assemble the antenna by yourself or with the aid of a friend. From a practical standpoint, however, three adults will make the assembly go much faster and more smoothly. These same three persons will definitely be needed to lift the antenna to its mounting position. You will also have to construct the base and bolt it to the concrete footers, and this is another job which is best done with some extra hands.

Before enlisting the aid of some fellow workers, sit down and re-read the entire assembly from cover to cover. Chances are, the manufacturer supplied you with this information as soon as you placed your order and you have already pored through it once. By rereading the information, you will obtain a better understanding of just what is to be done and why certain specific procedures are necessary. A rebriefing the night before or the day of construction will put all of the assembly information fresh in your mind and you should be able to proceed without constant reference to the manual during each and every little step. After individual phases of construction are completed, mark should be made in the manual indicating that this phase has been properly completed and tested.

Be sure to allot an adequate amount of time to finish this project. Hurried installations often result in improper performance or a structure which may not be mechanically sound. This could mean that your antenna is blown into the next country by the first strong wind that comes along. Even though the manufacturer may state that the entire station may be built in a few hours, this may not be true in your case. It's a good idea to double the stated installation time to allow for any problems or inconsistencies which might crop up. For this reason I recommend that you begin your project early in the morning (at first light). If you start too late in the day, darkness may overtake you midway in the project and you will be forced to leave your partially assembled Earth station exposed to the elements overnight. If bad weather occurs the following day, your installation could be in real trouble. Anything partially assembled is more prone to accidental breakage and environmental damages than when completely installed. Also, if you have to stop midway, you can easily lose track of just where you left off when construction was halted. Most of the Earth stations discussed in this book are intended for one-day owner-installation.

Moving along in the assembly, the base is firmly anchored to the concrete footers in strict accordance with the manufacturer's instructions. This is most important, because most companies will not warranty their antennas against damage which is incurred through improper mounting. The manufacturers know best how to install their products for maximum stability and safety. This is an area where experimentation is not recommended on the part of the owner.

Now that the base is up, the antenna may be completely assembled. Often, this will involve the piecing together of the dish segments, which are sometimes referred to as petals. You will also have to properly align the pieces to make certain the dish follows the approval contours which will result in maximum efficiency. The manufacturer's instruction manual will completely explain how to perform this precise but relatively simple procedure.

After the dish portion is complete, the structure to which the feed horn is mounted will be attached. This will often involve three struts which are attached to the sides of the dish and which meet at the focal point of the antenna, forming a sort of tripod.

Before mounting the dish to the base, most installations call for the feed horn, LNA and polarity rotor to be mounted as the apex of the triangle mount. Associated coaxial cable and control wiring is also installed at this time. Some antenna designs will not use the tripod arrangement but instead will mount the LNA and feed horn at the end of a shaft which is connected at the center of the dish.

Once you have completed the antenna assembly, go back over all of your work, checking carefully with the specified installation procedures. If everything checks out, the dish is then aligned if this procedure is called for during this phase of construction. Now, it will be necessary for you and one or two other persons to lift the antenna and set it into place on its mount. Usually, two persons hold the antenna in place while the third inserts the connecting bolts at the back of the dish. These hold the entire assembly firmly to the mount.

Once the antenna is properly set up, it's a good idea to take a halfhour break. During this period, you can go over the next stage of construction with your fellow workers. Make certain that you and your assistants understand exactly what is to be done. Good intentions coupled with a lack of knowledge have often been the ruin of many a project.

After everyone is refreshed, you can begin again. This next stage of construction is relatively simple and can probably be completed within an hour. It will vary slightly according to the particular components you are using, but it basically involves bolting the receiver to the antenna base, interconnecting the proper cables between this component and the LNA, and then running the feed lines and control wiring to the television set. If your receiver does not contain built-in modulator circuitry, a discrete unit will also be installed. Sometimes, this is done at the antenna site; but more often, the modulator will be mounted indoors near the television receiver and remote channel selector. You will want to be especially certain that all wiring connections are tight. This means tightening the collars of the coaxial connectors, tightening of all screw-in terminals, and the tight insertion of other types of control wiring connecting devices. If you live near the ocean, it would be a good idea to apply some waterproofing compound to connector surfaces in order to prevent corrosion. Be sure that there are no nicks in the protective outer cover of the coaxial cable. This can lead to water infiltration of the line. It's also a good idea to carefully examine the waterproof housings which are used to enclose receivers, LNAs, and other associated equipment which is to be mounted outdoors.

THE CHECKOUT

Once your TVRO Earth station is complete, go back through every major step of assembly in the manual and check to make certain all has gone according to the instructions. Sometimes, you will find that additional work is needed, such as tightening bolts, redoing connectings, and readjusting coaxial cable. This procedure will also point out some areas which need a little more work, although not mentioned in the instruction manual. This could take the form of tying down coaxial cable and other conductors which could swing back and forth in a strong wind. It's usually possible, using electrical tape to tie these loose ends down in a few minutes. When cable is allowed to swing with the wind, mechanical breakdown can occur at the connectors. When the internal conductors are severed, reception will be lost.

After this close examination and any additional small jobs, you are ready to activate the system and check its performance. Antenna aiming information is included throughout this book and in the appendices. Direct your antenna toward a satellite. Turn your television receiver to the appropriate channel which will receive the output of the modulator (this is often channel 2 or 3 but can vary with modulator design). Using the satellite guide, select either vertical or horizontal polarization and switch the satellite receiver remote control unit to the appropriate channel. Now, activate the entire TVRO station, and you should receive a picture. It will probably be necessary to fine tune the audio control on the remote station for clearest reception. Ideally, the picture will be crisp, the audio clear, and the overall reception identical to what you can obtain from the standard television broadcast service.

Rarely, however, is ideal reception obtained in the first try. It will probably be necessary to do some minor realignment of the dish position at the antenna site using the manual controls. If you have motorized antenna controls, this can be done from your viewing site by utilizing the remote antenna control unit. Chances are, your first picture may be fuzzy or even nonexistent. Maneuvering the antenna just a bit should serve to clear up reception. Eventually, through trial and error adjustments, perfect reception will be obtained. As you become more and more familiar with your station and its various settings, tuning will be a simple matter which can be accomplished by any member of your family.

PROBLEMS

Let's assume that you try every antenna setting imaginable and still don't obtain adequate reception. Actually, if any portion of the TVRO Earth station is not operating properly, you will probably receive no picture at all. The first thing to do is to check the electrical connections to the station as a whole. In other words, make sure it's "plugged in". Most Earth station equipment contains lighted indicators which will tell you if individual components are receiving power. Start from the antenna and work back to the television receiver. Is the LNA receiving power? Chances are, this voltage source will be derived from the satellite receiver. Is the receiver activated? Is the modulator receiving power? A complete checkout procedure for each piece of equipment will most likely be provided by the manufacturer.

If you're sure that each component in the Earth station is receiving power, then additional checks are necessary. However, if one component is not being activated, it will be necessary to determine the cause of the problem. Are all cables which supply operating current properly connected? Have you observed connection polarity when separate lines are used for powering the LNA? A reversed connection here may possibly damage the internal circuitry of this device. An ac voltmeter will help you to determine if house current is available at the equipment which operates from this source. Check the fuses. Sometimes, receivers and other devices will blow a fuse for a variety of reasons. Replace any open fuses (making sure the power is off) and then try again. If the fuse blows again, then there may be internal circuit problems and the manufacturer should be consulted.

As was mentioned above, if your station is receiving power and you still aren't receiving a picture, you must follow another path of inspection. First of all, check all coaxial cable connections. These often are the cause of many TVRO problems especially if the cables are abused. Then too, coax often serves as the dc power line for the low noise amplifier. If all other equipment is properly activated and the LNA does not receive power, then this line should be highly suspect. In continuing your trace, inspect all other coaxial cables and any appropriate lines. If you're in doubt about the condition of any of these, check them out for continuity with an ohmmeter. Better yet, replace them entirely.

If a picture still is not received, check antenna alignment again and also make certain that the feed horn is positioned for the type of polarization used in the satellite transmission. If it's set to vertical polarization, switch it to horizontal just as a test. Remember, you will probably receive absolutely nothing if the wrong polarization is chosen.

If you still don't have reception, turn your antenna to another satellite and try again. It is conceivable that the first one is not operational for some reason. Also, try several different positions on the remote channel select unit. If your system is operational, you will certainly get something at the television receiver by trying a different satellite.

If all of this fails, then consult your buyer's manual once again to see if the manufacturer recommends any additional procedures. The checkout mentioned in this discussion is very broad in nature and specific Earth stations will come with manuals that go into a great deal more detail. If you can't determine what the problem is, then a call to the manufacturer will be necessary. Before doing this, however, go through your antenna alignment procedure one more time from the start. An error way back when you were choosing the site could mean that your dish is not aimed to the correct part of the sky. This can be a costly mistake, especially at this point in the setup, because your site location was probably chosen based upon these initial calculations. If you find you have made an error, point your antenna in the correct direction and start over again.

CALLING THE MANUFACTURER

Before calling the technical assistance department of the company from which you purchased your TVRO Earth station equipment, jot down all of the results of your troubleshooting checks. List which components seem to be receiving power or appear operational and those that aren't. Write down any specific questions you have about areas where you suspect the problem may be occurring. Check out your television receiver to make sure it is operational and note any unusual occurrences. Also, add to your list a description of any types of video or audio information you may be receiving on the satellite channels.

Before placing the call, have an assistant nearby. The technician may request that you try a certain procedure and report the results to him while on the phone. Your assistant may be able to help perform these additional checks or simply communicate what you find at the TVRO site, acting as a go-between. The companies I have been in contact with offer excellent technical assistance departments which are designed to take care of a multitude of problems by talking to the TVRO Earth station owner by telephone.

It should be pointed out that reputable companies check all of their equipment before it is sent out to the customer. Earth stations are not selling by the millions at the present time, so it is easier for the manufacturer to run comprehensive checks and to assure quality performance with a relatively small staff. Chances are, the equipment you receive (if not damaged in transit) will be fully operational and will require only minor adjustments after initial installation if any adjustments at all are actually necessary.

If from your report, the technician decides that a component is malfunctioning, it will probably be necessary for you to return it to them for service or replacement. Many companies will simply send out a replacement immediately in order to get your station operational in the shortest period of time.

In speaking with many TVRO Earth station owners, I have come across only one situation where a piece of defective equipment was the cause of a receiving failure upon initial installation and checkout. The manufacturer immediately sent out a replacement which arrived within 24 hours and was most apologetic about the inconvenience to their customer. Certainly, these situations can occur, but they are few and far between.

The vast majority of TVRO Earth station owners install their systems and receive some sort of reception on the first try. it is usually necessary to make some minor antenna adjustments and perhaps rearrange a cable or two to obtain perfect reception. The checkout procedure usually takes less than an hour for the first satellite received and just a few minutes for all those that follow.

INTERFERENCE

Some TVRO Earth station owners have experienced problems with interference to their reception by other communications companies which operate equipment that transmits a signal which can enter the TVRO system. Several manufacturers contacted mentioned that as a part of a normal sale, they will check to see what possible interference problems may exist for a particular location. The owner is then advised of the potential problems and supplied with information and/or filters which will serve to eliminate them. Of course, your television receiver may be subject to interference from CB radios, ignition noise from automobiles, and the other routine problems which have existed for a long time. A TVRO Earth station will often do nothing to prevent these problems from continuing and it will be necessary for the television owner to install the appropriate filters at the television receiver proper to clean this up. These can often be purchased at local electronic hobby stores.

SUMMARY

It is important to mention that most manufacturers of personal TVRO Earth stations designed for home assembly realize that the great majority of their customers are not scientists or technicians who have a great deal of expertise in microwave and space communications. The success of these manufacturers in marketing their products depends directly upon offering a package which can be installed in a short period of time by customers with average skills. When the installation is complete, the station should work perfectly. Should a package be offered which requires a great deal of technical skill to set up and align, it would not be marketable to the vast majority of average persons who are the prime targets of the TVRO Earth station industry.

By looking at installation plans from various manufacturers before ordering, the potential buyer can get a good idea of whether or not he feels qualified to perform the assembly and checkout procedures. These manuals are as important to the customer as is the station itself. Without a properly written manual, the joys of building your own TVRO Earth station can quickly turn into a massive headache. The step-by-step plans from the companies mentioned in this book can only be described as excellent.

When assembling your station, pay attention to every little detail. Strictly follow the manufacturer's instructions and don't take any shortcuts. If you do all of this, you can be 99 percent sure that you will quickly be the proud owner of a personal TVRO Earth station that will provide you with many years of information, entertainment, and untold satisfaction.

Appendix A

Listing of Satellite Video Programming Sources

RCA SATCOM 1

TR-1:	NICKELODEON—Premium Children's Programming
TR-2:	PTL (Praise the Lord)—Religious
TR-3:	WGN-TV, Chicago
TR-5:	THE MOVIE CHANNEL—24 hr./day new movies
TR-6:	WTBS, Atlanta—Ted Turner's Superstation
TR-7:	ESPN (Entertainment & Sports Network)—24 hr/day sports
TR-8:	CBN (Christian Broadcasting Network)—Religious
TR-9:	C-SPAN—Live coverage from the House of Representatives
	USA NETWORK: Madison Square Garden Sports, Calliope, and Black Entertainment Network
TR-10:	SHOWTIME (west)—first-run movies, entertainment specials
TR-11:	NICKELODEON—Premium Children's Programming
TR-12:	SHOWTIME (east)—first-run movies, entertainment specials
	ABC Network Special Programming
TR-13:	TBN (Trinity Broadcasting Network)—Religious
TR-14:	CNN (Cable News Network)—24 hr/day news
TR-16:	SHOWTIME (spare)
	Occasional network, remote and sports events feeds
	ACSN (Appalachian Community Service Network)
	WWS (Window on Wall Street)—financial
TR-17:	WOR-TV, New York
TR-18:	REUTERS NEWS SERVICE
	GALAVISION—The best in Spanish-oriented programming

TR-20:	HOME BOX OFFICE CINEMAX (east)—time-structured HBO
TR-21:	HTN (Home Theater Network)—quality G & PG movies
TR-22:	HBO (Home Box Office) (west)—first-run movies, entertainment specials MSN (Modern Satellite Network)—general entertainment
TR-23:	HBO CINEMAX (west)—time-structured HBO
TR-24:	HBO (east)—first-run movies, entertainment specials

Audio Services on SATCOM 1

TR-2:	WAME (AM), Charlotte, North Carolina (6.2)
TR-3:	WFMT (FM), Chicago, Illinois (5.8) Seeburg Easy Listening Music (7.6)

ATT/GTE COMSTAR 2

TR-2:	Occassional transmissions: teleconferencing, sporting events, news & network feeds
TR-4:	Occasional transmissions: teleconferencing, sporting events, news & network feeds
TR-6:	Occasional transmissions: teleconferencing, sporting events, news & network feeds
TR-7:	NCN (National Christian Network)—religious BRAVO—Performing and Cultural Arts Programming ESCAPADE—"R" rated movies only
TR-10:	Occasional transmissions: teleconferencing, sporting events, news & network feeds
TR-13:	TBN (Trinity Broadcasting Network)—religious
TR-15:	Occasional transmissions: teleconferencing, news, and network feeds
TR-17:	HOME BOX OFFICE CINEMAX (east)—time-structured HBO
TR-18:	HOME BOX OFFICE (east)—first-run movies, entertainment specials
TR-19:	LVEN (Las Vegas Entertainment Network)—Live on-stage programming from Las Vegas CINEAMERICA—entertainment programming directed for ages over 50
TR-21:	Occasional transmissions: teleconferencing, sporting events, news & network feeds
TR-22:	Occasional transmissions: teleconferencing, sporting events, news & network feeds

Audio Services on COMSTAR 2

TR-7:	Family Radio Network (east) (5.8) Family Radio Network (west) (7.7)

WU WESTAR 3

TR-2: HUGES SPORTS NETWORK
Occasional transmissions: sporting events, news & network feeds

TR-3: XEW-TV, Mexico City

TR-5: WOLD COMMUNICATIONS—occasional transmissions: sporting events, news & network feeds
PRIVATE SCREENINGS—Hardcore, sexploitation "R"

TR-6: CBS Network Contract Channel—live network feeds
Occasional transmissions: sporting events, news news feeds

TRS-8: SIN (Spanish International Network)

TR-9: SPN (Satellite Program Network)—classic movies

TR-10: ABC Network Contract Channel—live network feeds

TR-11: CNN (Cable News Network) Contract Channel—news feeds
Occasional transmissions: sporting events

TR-12: Occasional transmissions: sporting events, news & network feeds

WU WESTAR 1

TR-1: Occasional transmissions: sporting events, news & network feeds

TR-3: Occasional transmissions: sporting events, news & network feeds

TR-5: Occasional transmissions: sporting events, news & network feeds
VEU (Video Entertainment Unlimited)—STV feed: first-run movies, concert specials, sporting events

TR-6: Occasional transmissions: sporting events, news & network feeds

TR-8: PBS (Public Broadcasting) Schedule A Programming

TR-9: PBS (Public Broadcasting) Schedule B Programming

TR-11: PBS (Public Broadcasting) Schedule C Programming

TR-12: PBS (Public Broadcasting) Occasional Feeds
CENTRAL EDUCATIONAL NETWORK
Occasional transmissions: sporting events, news & network feeds

WU WESTAR 2

TR-2: Occasional transmissions: sporting events, news & network feeds

TR-9: Occasional transmissions: sporting events, news & network feeds

RCA SATCOM 2

TR-2:	Occasional transmissions: sporting events, news & network feeds
TR-5:	Occasional transmissions: sporting events, news & network feeds
TR-8:	NBC Network Contract Channel—live & taped network feeds
TR-9:	ALASKAN FORCES SATELLITE NETWORK—assorted independent & network programming
TR-13:	NASA CONTRACT CHANNEL
TR-23:	ALASKA SATELLITE TELEVISION PROJECT—assorted network and independent programming

ATT/GTE COMSTAR 1

TR-20:	Occasional transmissions: sporting events, news & network feeds

ANIK B(2) (CAMADOAM)

TR-4:	Occasional transmissions: sporting events, news & CBC/CTV network feeds
TR-6:	CBC NORTH—assorted CBC network programming
TR-7:	Occasional Transmissions: sporting events, news — CBC/CYV network feeds
TR-8:	CBC (French Channel)—French language CBC programming
TR-9:	CBC (English Channel-1) English CBC programming
TR-10:	CBC (English Channel-2) English CBC programming

ANIK 3 (CANADIAN)

TR-1:	Occasional Transmissions: sporting events, news & CBC/CTV network feeds
TR-4:	Daily Live Coverage of Canadian House of Commons from Ottawa (with French translation)
TR-7:	Daily Live Coverage of the Canadian House of Commons from Ottawa (standard English)
TR-12:	CTV NORTH—assorted CTV network programming

Appendix B

List of Geostationary Satellites by Orbital Positions

The following list includes both satellites already in orbit and those planned for future launching into the geostationary satellite orbit. The list is based on, and limited to, information supplied to the International Frequency Registration Board (IFRB) by ITU Member administrations under the provisions of the Radio Regulations paragraphs 639AA, 639AJ, 639BA. The designations of the satellites are those officially notified and may not always correspond to the name in general use.

Orbital Position			Space station		Frequency Bands								
					GHz <1	<3	4	6	7	11	12	14	>15
0	E		F/GEO	GEOS-2	1	3							
0	E		F/MET	METEOSAT	1	3							
10	E		F/OTS	OTS	1					11		14	
13	E	#	F	EUTELSAT 1-2	1					11		14	
17	E	#	ARS	SABS						11	12	14	
19	E	•	ARS	ARABSAT I		3	4	6					
26	E	•	ARS	ARABSAT II		3	4	6					
26	E	•	IRN	ZOHREH-2						11		14	
29	E		F/GEO	GEOS-2	1	3							
34	E	•	IRN	ZOHREH-1						11	12	14	
35	E		URS	STATSIONAR-2			4	6					
40	E	•	F/MRS	MARECS-D	1	3	4	6					
40	E		F/MTS	MAROTS	1	3							
40	E	#	URS	STATSIONAR-12			4	6					
45	E	•	URS	GALS-2					7				
45	E	•	URS	LOUTCH P2						11		14	
45	E		URS	STATSIONAR-9			4	6					
45	E	•	URS	VOLNA-3	1	3							
47	E	•	IRN	ZOHREH-3						11		14	
53	E		URS	LOUTCH-2						11		14	
53	E		URS/IK	STATSIONAR-5			4	6					
53	E		URS	VOLNA-4		3							
56.5	E		USA/IT	INTELSAT3 INDN1			4	6					
57	E	•	USA/IT	INTELSAT4 INDN2			4	6					
57	E	•	USA/IT	INTELSAT4A INDN2			4	6					

Orbital Position			Space station		Frequency Bands GHz <1	<3	4	6	7	11	12	14	>15
57	E	#	USA	INTELSAT5 INDN3			4	6		11		14	
57	E	#	USA	INTELSAT MCS INDN C 3			4	6					
60	E		USA/IT	INTELSAT4 INDN2			4	6					
60	E		USA/IT	INTELSAT4A INDN2			4	6					
60	E	*	USA/IT	INTELSAT5 INDN2			4	6		11		14	
60	E	*	USA/IT	INTELSAT MCS INDN B 3			4	6					
60	E		USA	USGCSS PHASE2 INDN					7				
60	E	#	USA	USGCSS PHASE3 INDN					7				
63	E		USA/IT	INTELSAT4 INDN1			4	6					
63	E		USA/IT	INTELSAT4A INDN1			4	6					
63	E	*	USA/IT	INTELSAT5 INDN1			4	6		11		14	
63	E	*	USA/IT	INTELSAT MCS INDN A 3			4	6					
64.5	E	*	F/MRS	MARECS-C	1	3	4	6					
66	E	*	USA/IT	INTELSAT4 INDN1			4	6					
66	E	*	USA/IT	INTELSAT4A INDN1			4	6					
66	E	#	USA	INTELSAT5 INDN4			4	6		11		14	
66	E	#	USA	INTELSAT MCS INDN D 3			4	6					
73	E		USA	MARISAT-INDN	1	3*	4*	6*					
74	E		IND	INSAT-1A	1	3	4	6					
75	E		USA	FLTSATC INDN	1				7				
77	E		INS	PALAPA-2			4	6					
80	E		URS	STATSIONAR-1			4	6					
80	E	#	URS	STATSIONAR-13			4	6					
83	E		INS	PALAPA-1			4	6					
85	E	*	URS	GALS-3					7				
85	E	*	URS	LOUTCH P3						11		14	
85	E		URS	STATSIONAR-3			4	6					
85	E	*	URS	VOLNA-5	1	3							
90	E		URS	LOUTCH-3						11		14	
90	E		URS	STATSIONAR-6			4	6					
90	E	#	URS	VOLNA-8		3							
94	E	*	IND	INSAT-1B	1	3	4	6					
95	E	#	URS	STATSIONAR-14			4	6					
99	E		URS	STATSIONAR-T	1			6					
99	E	#	URS	STATSIONAR-T2	1			6					
102	E	*	IND	ISCOM	1		4	6					
108	E	*	INS	PALAPA-B1			4	6					
110	E		J	BSE		3						14	
113	E	*	INS	PALAPA-B2			4	6					
118	E	*	INS	PALAPA-B3			4	6					
125	E	*	CHN	STW-1			4	6					
130	E		J	ETS-2	1	3				11			34
130	E	#	J	CS-2A		3	4	6					20/30
130	E	#	URS	STATSIONAR-15			4	6					
135	E	#	J	CS-2B		3	4	6					20/30
135	E		J	CSE		3	4	6					18/29
140	E	*	J	GMS	1	3							
140	E		URS	LOUTCH-4						11		14	
140	E		URS	STATSIONAR-7			4	6					
140	E		URS	VOLNA-6		3							
172	E		USA	FLTSATC W PAC	1				7				
174	E		USA/IT	INTELSAT4 PAC1			4	6					
174	E	*	USA/IT	INTELSAT4A PAC1			4	6					
174	E	#	USA	INTELSAT5 PAC1			4	6		11		14	
175	E		USA	USGCSS PHASE2 W PAC					7				
175	E	#	USA	USGCSS PHASE3 W PAC					7				
176.5	E		USA	MARISAT-PAC	1	3	4	6					
179	E		USA/IT	INTELSAT4 PAC2			4	6					
179	E	*	USA/IT	INTELSAT4A PAC2			4	6					
179	E	#	USA	INTELSAT5 PAC2			4	6		11		14	
172	W	*	F/MRS	MARECS-B	1	3	4	6					
170	W	*	URS	GALS-4					7				
170	W	*	URS	LOUTCH P4						11		14	
170	W		URS	STATSIONAR-10			4	6					
170	W	*	URS	VOLNA-7	1	3							

Orbital Position			Space station		Frequency Bands							
				GHz	<1	<3	4	6	7	11	12	14
149	W		USA	ATS-1	1		4	6				
136	W		USA	US SATCOM-1			4	6				
135	W		USA	GOES WEST	1	3						
135	W	#	USA	USGCSS PHASE3 E PAC					7			
135	W		USA	USGCSS PHASE2 E PAC					7			
132	W	#	USA	US SATCOM-3			4	6				
128	W		USA	COMSTAR D1			4	6				
123.5	W		USA	WESTAR-2			4	6				
122	W		USA	USASAT-6A						11		14
119	W		USA	US SATCOM-2			4	6				
116	W	•	CAN	ANIK-C2						11		14
114	W		CAN	ANIK-A3			4	6				
112.5	W	•	CAN	ANIK-C1						11		14
109	W		CAN	ANIK-A2			4	6				
109	W		CAN	ANIK-B1			4	6		11		14
106	W		USA	USASAT-6B						11		14
105	W		USA	ATS-5	1	3						
104	W	#	CAN	TELESAT D-1			4	6				
104	W		CAN	ANIK-A1			4	6				
102	W	#	MEX	SATMEX-1			4	6				
100	W		USA	FLTSATC E PAC	1				7			
99	W		USA	WESTAR-1			4	6				
95	W		USA	COMSTAR D2			4	6				
91	W	•	USA	WESTAR-3			4	6				
87	W		USA	COMSTAR D3			4	6				
86	W		USA	ATS-3	1							
75.4	W		CLM	SATCOL-2			4	6				
75	W		USA	GOES EAST	1	3						
75	W		CLM	SATCOL-1			4	6				
34.5	W	#	USA	INTELSAT MCS ATL E		3	4	6				
34.5	W		USA/IT	INTELSAT4 ATL5			4	6				
34.5	W		USA/IT	INTELSAT4A ATL4			4	6				
34.5	W	•	USA/IT	INTELSAT5 ATL4			4	6		11		14
31	W	•	USA/IT	INTELSAT4A ATL4			4	6				
29.5	W		USA/IT	INTELSAT4 ATL2			4	6				
29.5	W		USA/IT	INTELSAT4A ATL3			4	6				
29.5	W		USA/IT	INTELSAT5 ATL3			4	6		11		14
27.5	W	•	USA/IT	INTELSAT4A ATL3			4	6				
27.5	W	•	USA/IT	INTELSAT5 ATL3			4	6		11		14
27.5	W	•	USA/IT	INTELSAT MCS ATL B	3		4	6				
25	W	•	URS	GALS-1					7			
25	W	•	URS	LOUTCH P1						11		14
25	W		URS	STATSIONAR-8			4	6				
25	W	•	URS	VOLNA-1	1	3						
24.5	W	#	USA	INTELSAT MCS ATL D		3	4	6				
24.5	W		USA/IT	INTELSAT4A ATL1			4	6				
24.5	W		USA/IT	INTELSAT5 ATL1			4	6		11		14
23	W		USA	FLTSATC ATL	1				7			
21.5	W	#	USA	INTELSAT MCS ATL C	3		4	6				
21.5	W	#	USA	INTELSAT5 ATL5			4	6		11		14
21.5	W	•	USA/IT	INTELSAT4 ATL2			4	6				
21.5	W	•	USA/IT	INTELSAT4A ATL1			4	6				
19.5	W		USA/IT	INTELSAT4 ATL3			4	6				
19.5	W		USA/IT	INTELSAT4A ATL2			4	6				
18.5	W	•	USA/IT	INTELSAT4 ATL3			4	6				
18.5	W	•	USA/IT	INTELSAT4A ATL2			4	6				
18.5	W	•	USA/IT	INTELSAT5 ATL2			4	6		11		14
18.5	W	•	USA/IT	INTELSAT MCS ATL A		3	4	6				
18	W	•	BEL	SATCOM III ATL					7			
18	W	•	BEL	SATCOM-III					7			
18	W		BEL	SATCOM-II					7			
15	W	•	F/MRS	MARECS-A	1	3	4	6				
15	W		I	SIRIO	1					11		
15	W		USA	MARISAT-ATL	1	3	4	6				

Orbital Position			Space station		GHz <1	<3	4	6	7	11	12	14
14	W		URS	LOUTCH-1						11		14
14	W		URS/IK	STATSIONAR-4			4	6				
14	W		URS	VOLNA-2		3						
13	W		USA	USGCSS PHASE2 ATL					7			
12	W	*	USA	USGCSS PHASE2 ATL					7			
12	W	#	USA	USGCSS PHASE2 ATL					7			
11.5	W		F/SYM	SYMPHONIE-2	1		4	6				
11.5	W		F/SYM	SYMPHONIE-3	1		4	6				
10	W	#	F	TELECOM-1A		3	4	6	7		12	14
8.5	W	#	URS	STATSIONAR-11			4	6				
7	W	#	F	TELECOM-1B		3	4	6	7		12	14
4	W		USA/IT	INTELSAT4 ATL1			4	6				
1	W		USA/IT	INTELSAT4 ATL4			4	6				

* Under co-ordination RR639AJ

Advanced publication only under RR639AA

Appendix C

List of Artificial Satellites Launched in 1980

Code name Spacecraft description	International number	Country Organization Site of launching	Date	Perigee Apogee	Period Inclination	Frequencies Transmitter power	Observations
Cosmos-1149	1980-1-A	USSR (PLE)	9 Jan	208 km 414 km	90.4 min 70.29°		Photographic reconnaissance satellite Recovered on 23 January 1980
46th Molnya-1 hermetically sealed cylinder with conical ends, mass: 1000 kg, 6 solar panels	1980-2-A	USSR (PLE)	11 Jan	478 km 40 830 km	737 min 62.8°	800 MHz band 40 W (emission) 1000 MHz band (reception) 3400-4100 MHz (retransmission of television)	Apparatus for transmitting television programmes and multichannel radiocommunications
Cosmos-1150	1980-3-A	USSR (PLE)	14 Jan	989 km 1028 km	105.0 min 83.0°		Navigation satellite
FLTSATCOM-3 3-axis stabilized hexagonal satellite, width: 2.44 m, overall height: 6.70 m, mass at launch: 1875 kg, mass in orbit: 1005 kg	1980-4-A	United States USDOD (ETR)	18 Jan	35 745 km 35 829 km in geostationary orbit at 23° W	1436.1 min 2.6°	240-400 MHz band (communications) 2252.5, 2262.5 MHz 2.4 W (telemetry)	Government communications satellite. Replaces *FLTSATCOM-2* which is being moved to 75° E. Nine 25 kHz channels and twelve 5 kHz channels for small mobile users. A 25 kHz broadcast channel and a 500 kHz channel
Cosmos-1151	1980-5-A	USSR (PLE)	23 Jan	650 km 678 km	97.8 min 82.5°		Ocean monitoring satellite
Cosmos-1152	1980-6-A	USSR (PLE)	24 Jan	181 km 370 km	89.7 min 67.1°		High-resolution, reconnaissance satellite Recovered on 6 February 1980
Cosmos-1153	1980-7-A	USSR (PLE)	25 Jan	983 km 1031 km	105 min 83°		Navigation satellite
Cosmos-1154	1980-8-A	USSR (PLE)	30 Jan	634 km 671 km	97.3 min 81.3°		Electronic monitoring satellite
Cosmos-1155	1980-9-A	USSR (PLE)	7 Feb	206 km 422 km	90.4 min 72.9°		Medium-resolution reconnaissance satellite Recovered on 21 February 1980
KH-11	1980-10-A	United States USAF (WTR)	7 Feb	220 km 498 km	91.7 min 97.0°		Digital imaging reconnaissance satellite

Code name Spacecraft description	International number	Country Organization Site of launching	Date	Perigee Apogee	Period Inclination	Frequencies Transmitter power	Observations
Navstar-5	1980-11-A	United States (WTR)	9 Feb.	20 095 km 20 165 km	715.9 min 63.7°		Global positioning system navigation satellite. Replaces *Navstar-1*
Cosmos-1156 **to** **Cosmos-1163** mass: 40 kg each	1980-12-A to 1980-12-H	USSR (PLE)	12 Feb.	1450 km 1528 km	115.4 min 74.0°		Government communication satellites
Cosmos-1164	1980-13-A	USSR (PLE)	12 Feb.	220 km 640 km	92.9 min 62.8°		
SMM 3-axis stabilized satellite; width: 1.20 m; length: 4 m; mass: 2315 kg; 2 fixed solar arrays (3 kW); Ni-Cd batteries	1980-14-A	United States NASA (ETR)	14 Feb.	571.5 km 573.5 km	96.12 min 28.5°	2287.5 MHz (tracking and telemetry)	Solar Maximum Mission. Objectives: to measure solar radiation during the period of maximum solar activity. Carries gamma ray spectrometer, hard X-ray burst spectrometer, hard X-ray imaging spectrometer, ultraviolet spectrometer and polarimeter, X-ray polychrometer, coronograph/polarimeter and solar constant monitoring package
Tansei-4 (MS-T4)	1980-15-A	Japan ISAS (KSC)	17 Feb.	517 km 672 km	96.5 min 38.7°	136.725 MHz (tracking) 400.45; 2280.5 MHz (telemetry)	Tests of new technology for future satellites and test of the new *M-3S* launcher
Raduga-5 3-axis stabilized satellite; mass: 5 tonnes; solar cells	1980-16-A	USSR (BAI)	20 Feb.	36 610 km geosynchronous orbit	24 h 38 min 0.4°	5.7-6.2 GHz (reception) 3.4-3.9 GHz (emission)	Carries apparatus for transmitting television programmes and multichannel radiocommunications
Cosmos-1165	1980-17-A	USSR (PLE)	21 Feb.	182 km 379 km	89.8 min 72.9°		High-resolution reconnaissance satellite. Recovered on 5 March 1980

Code name *Spacecraft description*	*International number*	*Country* *Organization* *Site of launching*	*Date*	*Perigee* *Apogee*	*Period* *Inclination*	*Frequencies* *Transmitter power*	*Observations*
Ayame-2 cylindrical satellite: diameter: 1 m; height: 1.50 m; mass: 260 kg	1980-18-A	Japan NSDA (TSC)	22 Feb.	206.9 km 35 512 km	625.8 min 24.6°	31.65 GHz 3.2 W 4.075; 4.080 GHz 4.7 W 3.940 GHz 3.5 W 136.112 MHz 2 or 8 W	Experimental telecommunication satellite. Was intended for geostationary orbit but contact was lost while still in transfer orbit
No name	1980-19-A to 1980-19-C	United States USN (WTR)	3 March	1053 km 1151 km	107.1 min 63.5°		Ocean surveillance satellite system. Three satellites
Cosmos-1166	1980-20-A	USSR (PLE)	4 March	208 km 406 km	90.3 min 72.9°		Medium-resolution photographic reconnaissance satellite. Recovered on 18 March 1980
Cosmos-1167	1980-21-A	USSR (PLE)	14 March	438 km 457 km	93.3 min 65°		Ocean surveillance satellite
Cosmos-1168	1980-22-A	USSR (PLE)	17 March	981 km 1026 km	104.9 min 82.9°		Navigation satellite
Cosmos-1169	1980-23-A	USSR (PLE)	27 March	478 km 521 km	94.5 min 65.8°		Satellite intercept programme
Progress-8 modified *Soyuz* spacecraft without the descent section; mass at launch: 7 tonnes	1980-24-A	USSR (BAI)	27 March	192 km 266 km	88.8 min 51.6°		Expendable supply craft. Docked with *Salyut-6* on 29 March. Separated on 25 April and was deorbited over the Pacific Ocean on 26 April 1980
Cosmos-1170	1980-25-A	USSR (BAI)	1 April	181 km 386 km	89.9 min 70.4°		High-resolution photographic reconnaissance satellite. Recovered on 13 April 1980
Cosmos-1171	1980-26-A	USSR (PLE)	3 April	976 km 1017 km	105 min 65.8		Satellite intercept programme. Target vehicle for *Cosmos-1174*
Soyuz-35 3-part spacecraft: 2 spherical habitable modules (orbital compartment and command module) connected in tandem to a cylindrical service module; diameter: 2.70 m; height: 7.10 m; mass: 6.7 tonnes; 2 solar arrays	1980-27-A	USSR (BAI)	9 April	276 km 315 km	90.3 min 51.6°		Two-man spacecraft. L. Popov, commander; V. Ryumin, flight engineer. Docked with *Salyut-6* on 10 April. Returned to Earth carrying *Soyuz-36* cosmonauts on 3 June 1980, landing some 440 km north-east of Baikonur Cosmodrome

Cosmos-1172	1980-28-A	USSR (PLE)	12 April	637 km 40 160 km	726 min 62.8°		Early warning satellite
Cosmos-1173	1980-29-A	USSR (BAI)	17 April	180 km 379 km	89.9 min 70.3°		High-resolution photographic reconnaissance satellite. Similar to *Cosmos-1170*. Recovered on 28 April 1980
Cosmos-1174	1980-30-A	USSR (BAI)	18 April	387 km 1035 km	98.6 min 65.8°		Satellite intercept programme. Interceptor vehicle for *Cosmos-1171* target vehicle. The test was a failure. *Cosmos-1174* was exploded in space on 20 April 1980
Cosmos-1175	1980-31-A	USSR (PLE)	18 April	317 km 485 km	92.3 min 62.5°		Satellite intercept programme. Interceptor vehicle employing an optical-thermal guidance system. Decayed on 28 May 1980
Navstar-6	1980-32-A	United States (WTR)	26 April	19 622 km 20 231 km	707.6 min 62.9°		Global positioning system navigation satellite
Progress-9 modified *Soyuz* spacecraft without the descent section; mass at launch: 7 tonnes	1980-33-A	USSR (BAI)	27 April	192 km 275 km	88.9 min 51.6°		Expendable supply craft. Docked with *Salyut-6* on 29 April, undocked on 20 May and was made to re-enter the Earth's atmosphere on 22 May 1980
Cosmos-1176	1980-34-A	USSR	29 April	260 km 265 km	89.6 min 65.0°		Ocean surveillance satellite. Carries a nuclear reactor. Similar to *Cosmos-954*
Cosmos-1177	1980-35-A	USSR (PLE)	29 April	181 km 365 km	89.7 min 67.2°		High-resolution photographic reconnaissance satellite. Recovered on 12 June 1980
Cosmos-1178	1980-36-A	USSR (PLE)	7 May	207 km 417 km	90.4 min 72.9°		Reconnaissance satellite. Recovered on 22 May 1980
Cosmos-1179	1980-37-A	USSR (PLE)	14 May	310 km 1570 km	103.5 min 83.0°		Navigation satellite
Cosmos-1180	1980-38-A	USSR (PLE)	15 May	240 km 296 km	89.8 min 62.8°		Satellite for geophysical observations and measurements. Recovered on 25 May 1980

Code name Spacecraft description	International number	Country Organization Site of launching	Date	Perigee Apogee	Period Inclination	Frequencies Transmitter power	Observations
Cosmos-1181	1980-39-A	USSR	20 May	992 km 1020 km	105 min 82°		Navigation satellite
Cosmos-1182	1980-40-A	USSR (PLE)	23 May	221 km 278 km	89.2 min 82.3°		Medium-resolution photographic reconnaissance satellite. Recovered on 5 June 1980
Soyuz-36 3-part spacecraft: 2 spherical habitable modules (orbital compartment and command module) connected in tandem to a cylindrical service module; diameter: 2.70 m; height: 7.10 m; mass: 6.7 tonnes; 2 solar arrays	1980-41-A	USSR (BAI)	26 May	198 km 216 km	88.0 min 51.6°		Two-man spacecraft. V. Kubasov, commander; B. Farkas (Hungary), research cosmonaut. Docked with *Salyut-6* on 28 May. *Soyuz-36* was returned to Earth with *Soyuz-37* cosmonauts aboard on 31 July 1980
Cosmos-1183	1980-42-A	USSR	28 May	208 km 414 km	90.4 min 72.9°		Medium-resolution photographic reconnaissance satellite. Recovered on 11 June 1980
NOAA-B	1980-43-A	United States NOAA (WTR)	29 May	273 km 1453 km	102.2 min 92.3°		Owing to malfunction of the launch vehicle, proper orbit was not attained and the spacecraft is considered inoperable
Cosmos-1184	1980-44-A	USSR (PLE)	4 June	621 km 662 km	97.4 min 81.2°		Electronic monitoring satellite
Soyuz-T2 solar batteries	1980-45-A	USSR (BAI)	5 June	267 km 316 km	90.25 min 51.6°		Two-man spacecraft. Y. Malishev and V. Aksenov, cosmonauts. Docked with accessories port of *Salyut-6* on 6 June. Recovered on 9 June 200 km south-east of Dzhezkazgan
Cosmos-1185	1980-46-A	USSR (PLE)	6 June	226 km 308 km	89.5 min 82.3°		Medium-resolution photographic reconnaissance satellite. Recovered on 20 June 1980
Cosmos-1186	1980-47-A	USSR (PLE)	6 June	473 km 519 km	94.5 min 74.0°		Electronic monitoring satellite

Cosmos-1187	1980-48-A	USSR (PLE)	12 June	210 km 332 km	72.9 min 89.6°		Medium-resolution photographic reconnaissance satellite. Recovered on 26 June 1980
Gorizont-4 3-axis stabilized spacecraft	1980-49-A	USSR (BAI)	14 June	36 515 km	24 h 33 min 0.8°	3.4-3.9 GHz (emission) 5.7-6.2 GHz (reception)	Communication satellite for transmission of telegraph and telephone messages and for transmission of television programmes
Cosmos-1188	1980-50-A	USSR (PLE)	14 June	628 km 40 165 km	726 min 62.8°		Early warning satellite
30th Meteor-1 3-axis stabilized cylindrical satellite mass: 2200 kg: sun-oriented solar panels	1980-51-A	USSR CAHS (PLE)	18 June	589 km 678 km	97.3 min 98.0°		Meteorological satellite
Big Bird	1980-52-A	United States USAF (WTR)	18 June	165 km 254 km	88.5 min 96.5°		Reconnaissance satellite
No name	1980-52-C	United States	18 June	1325 km 1329 km	112.2 min 96.6°		
47th Molnya-1 hermetically-sealed cylinder with conical ends: mass: 1000 kg: 6 solar panels	1980-53-A	USSR (PLE)	21 June	658 km 40 707 km	738 min 62.5°	800 MHz band 40 W (emission) 1000 MHz band (reception) 3400-4100 MHz (retransmission of television)	Television and multichannel radiocommunications
Cosmos-1189	1980-54-A	USSR (PLE)	26 June	209 km 330 km	89.5 min 72.9°		Medium-resolution photographic reconnaissance satellite. Recovered on 10 July 1980
Progress-10 modified *Soyuz* spacecraft without the descent section: mass at launch: 7 tonnes	1980-55-A	USSR	29 June	191 km 281 km	88.9 min 51.6°		Expendable supply craft. Docked with *Salyut-6* on 1 July and was made to re-enter the Earth's atmosphere on 19 July 1980

Code name *Spacecraft description*	*International number*	*Country Organization Site of launching*	*Date*	*Perigee Apogee*	*Period Inclination*	*Frequencies Transmitter power*	*Observations*
Cosmos-1190	1980-56-A	USSR (PLE)	1 July	792 km 829 km	100.8 min 74.0°		Electronic monitoring satellite
Cosmos-1191	1980-57-A	USSR (PLE)	2 July	646 km 40 165 km	726 min 62.8°		Early warning satellite
Cosmos-1192 **to** **Cosmos-1199** mass: 40 kg each	1980-58-A to 1980-58-H	USSR (PLE)	9 July	1451 km 1522 km	115.3 min 74.0°		Government communication satellites
Cosmos-1200	1980-59-A	USSR (PLE)	9 July	209 km 332 km	89.5 min 72.9°		Medium-resolution photographic reconnaissance satellite. Recovered on 23 July 1980
Ekran-5 (Statsionar) 3-axis stabilized satellite; mass: 5 tonnes; solar cells	1980-60-A	USSR (BAI)	14 July	34 474 km geostationary orbit	1420 min 0.36°	5.7-6.2 GHz (reception) 3.4-3.9 GHz (emission)	Television relay satellite
Cosmos-1201	1980-61-A	USSR (PLE)	15 July	220 km 274 km	89.1 min 82.3°		Natural resources satellite. Recovered on 28 July 1980
Rohini-1 mass: 35 kg	1980-62-A	India (SSC)	18 July				First satellite launched by Indian *SLV-3* solid propellant 4-stage rocket system
13th Molnya-3 3-axis stabilized satellite; mass: 1500 kg	1980-63-A	USSR (PLE)	18 July	467 km 40 815 km	736 min 62.8°	5.9-6.2 GHz (reception) 3.6-3.9 GHz (emission)	Television and multichannel radiocommunications
Soyuz-37 3-part spacecraft: 2 spherical habitable modules (orbital compartment and command module) connected in tandem to a cylindrical service module; diameter: 2.70 m; height: 7.10 m; mass: 6.7 tonnes; 2 solar arrays	1980-64-A	USSR (BAI)	23 July	263 km 312 km	90.0 min 51.6°		Two-man spacecraft: cosmonaut V. Gorbatko and cosmonaut-researcher Fam Tuan (Viet Nam). Docked with *Salyut-6* on 24 July. *Soyuz-37* cosmonauts returned to Earth aboard *Soyuz-36* on 31 July 1980. Soyuz-37 spacecraft was returned to Earth with *Soyuz-35* cosmonauts Popov and Ryumin on 11 October 1980

Cosmos-1202	1980-65-A	USSR (PLE)	25 July	209 km 333 km	89.6 min 72.9°		Medium-resolution photographic reconnaissance satellite. Recovered on 7 August 1980
Cosmos-1203	1980-66-A	USSR (PLE)	31 July	227 km 303 km	89.5 min 82.3°		Recovered on 14 August 1980
Cosmos-1204	1980-67-A	USSR (AKY)	31 July	346 km 546 km	93.3 min 50.7°		
Cosmos-1205	1980-68-A	USSR (PLE)	12 Aug.	208 km 332 km	89.6 min 72.8°		Photographic reconnaissance satellite. Recovered on 26 August 1980
Cosmos-1206	1980-69-A	USSR (PLE)	15 Aug.	630 km 659 km	97.4 min 81.2°		Electronic monitoring satellite
Cosmos-1207	1980-70-A	USSR (PLE)	22 Aug.	218 km 282 km	89.2 min 82.3°		Film-return earth resources satellite. Recovered on 4 September 1980
Cosmos-1208	1980-71-A	USSR	26 Aug.	181 km 362 km	89.6 min 67.1°		Long-duration reconnaissance satellite. Recovered on 24 September 1980
Cosmos-1209	1980-72-A	USSR (PLE)	3 Sept.	222 km 306 km	89.4 min 82.3°		Earth resources satellite. Recovered on 17 September 1980
6th Meteor-2	1980-73-A	USSR CAHS (PLE)	9 Sept.	868 km 906 km	102.4 min 81.2°	137.3 MHz 5 W (APT)	Meteorological satellite. Scanning telephotometer and television-type scanning equipment (0.5 to 0.7 μm), infrared scanning radiometer (8 to 12 μm)
GOES-4 cylindrical spin-stabilized satellite; diameter: 1.90 m; height: 2.30 m; mass: 397 kg	1980-74-A	United States (ETR)	9 Sept.	34 264 km 49 830 km	1767 min 0.25° in geosynchronous orbit at 95° W	2209 MHz; 2214 MHz (telemetry)	Geostationary Operational Environmental Satellite. Carries a visible and infrared spin-scan radiometer (VISSR) to provide data on the vertical structures of temperature and moisture in the atmosphere

Code name Spacecraft description	International number	Country Organization Site of launching	Date	Perigee Apogee	Period Inclination	Frequencies Transmitter power	Observations
Soyuz-38 3-part spacecraft; 2 spherical habitable modules (orbital compartment and command module) connected in tandem to a cylindrical service module; diameter: 2.70 m; height: 7.10 m; mass: 6680 kg; 2 solar arrays	1980-75-A	USSR (BAI)	18 Sept.	199 km 273 km	88.9 min 51.6°		Two-man spacecraft: Y. V. Romanenko, flight commander: A. Tomayo Méndez (Cuba). Docked with *Salyut-6* on 19 September 1980 and returned to Earth with the same crew on 26 September 1980
Cosmos-1210	1980-76-A	USSR (PLE)	19 Sept.	195 km 268 km	88.8 min 82.3°		Photographic reconnaissance satellite. Recovered on 30 September 1980
Cosmos-1211	1980-77-A	USSR (PLE)	23 Sept.	215 km 261 km	89.1 min 82.4°		Photographic reconnaissance satellite. Recovered on 4 October 1980
Cosmos-1212	1980-78-A	USSR (PLE)	26 Sept.	216 km 275 km	89.1 min 82.3°		Earth resources satellite. Recovered on 9 October 1980
Progress-11 modified *Soyuz* spacecraft without the descent section; mass at launch: 7 tonnes	1980-79-A	USSR (BAI)	28 Sept.	193 km 270 km	88.8 min 51.6°		Cargo-spacecraft. Docked with the *Salyut-6/Soyuz-37* complex on 30 September 1980. Made to re-enter the Earth's atmosphere on 11 December 1980
Cosmos-1213	1980-80-A	USSR (PLE)	3 Oct.	207 km 343 km	89.6 min 72.8°		Photographic reconnaissance satellite. Recovered on 17 October 1980
Raduga-6 (Statsionar-3) 3-axis stabilized satellite; mass: 5 tonnes; solar cells	1980-81-A	USSR (BAI)	6 Oct.	36 000 km geostationary orbit	1444 min (24 h 04 min) 0.4°	5.7-6.2 GHz (reception) 3.4-3.9 GHz (emission)	Carries apparatus for transmitting television programmes and multichannel radiocommunications
Cosmos-1214	1980-82-A	USSR (PLE)	10 Oct.	181 km 368 km	89.7 min 67.2°		Photographic film recovery reconnaissance satellite. Recovered on 23 October 1980
Cosmos-1215	1980-83-A	USSR (PLE)	14 Oct.	499 km 553 km	95.1 min 74.0°		Electronic monitoring satellite

Cosmos-1216	1980-84-A	USSR (PLE)	16 Oct.	209 km 404 km	90.3 min 72.9°		Photographic film recovery reconnaissance satellite. Recovered on 30 October 1980
Cosmos-1217	1980-85-A	USSR (PLE)	24 Oct.	642 km 40 165 km	726 min 62.8°		Early warning satellite
Cosmos-1218	1980-86-A	USSR (PLE)	30 Oct.	178 km 374 km	89.7 min 64.9°		High-resolution photographic reconnaissance satellite. Recovered on 12 December 1980
FLTSATCOM-4 3-axis stabilized hexagonal satellite; width: 2.44 m; overall height: 6.70 m; mass at launch: 1876 kg; mass in orbit: 1005 kg	1980-87-A	United States USN (ETR)	31 Oct.	35 033 km 36 237 km in geostationary orbit at 172° E	1428.4 min 2.5°	240-400 MHz band (communications) 2252.5; 2262.2 MHz 2.4 W (telemetry)	Government communication satellite providing 23 UHF communication channels and one SHF up-link channel. Fourth in a series of five satellites
Cosmos-1219	1980-88-A	USSR (PLE)	31 Oct.	205 km 353 km	89.7 min 72.9°		Medium-resolution photographic reconnaissance satellite. Recovered on 13 November 1980
Cosmos-1220	1980-89-A	USSR (BAI)	4 Nov.	432 km 454 km	93.3 min 65.0°		Ocean surveillance satellite
Cosmos-1221	1980-90-A	USSR (PLE)	12 Nov.	207 km 424 km	90.5 min 72.9°		Medium-resolution photographic reconnaissance satellite
SBS-1 mass: 550 kg	1980-91-A	United States SBS (ETR)	15 Nov.	in geostationary orbit at 106° W		14-12 GHz band	United States domestic communication satellite. First of three all-digital business communications satellites. Transmits point-to-point voice, data, facsimile and telex messages. Ten transponders
48th Molnya-1 hermetically sealed cylinder with conical ends; mass: 1000 kg; 6 solar panels	1980-92-A	USSR (PLE)	16 Nov.	640 km 40 651 km	736 min 62.3°	800 MHz band 40 W (emission) 1000 MHz band (reception) 3400-4100 MHz (retransmission of television)	Television and multichannel radiocommunications

Cosmos-1222	1980-93-A	USSR (PLE)	21 Nov.	624 km 659 km	97.4 min 81.2°		Electronic monitoring satellite
Soyuz-T3	1980-94-A	USSR (BAI)	27 Nov.	253 km 271.5 km	89.6 min 51.6°		For the first time in nine years three cosmonauts were launched aboard a *Soyuz*: L. Kizim, flight commander; O. Makarov, flight engineer, G. Strekalov, research engineer. *Soyuz-T3* docked with *Salyut-6* on 28 November and the crew boarded Salyut-6 on 29 November. Soyuz-T3 was returned to Earth with its crew on 10 December, landing in Kazakhstan
Cosmos-1223	1980-95-A	USSR (PLE)	27 Nov.	614 km 40 165 km	726 min 62.8°		Early warning satellite
Cosmos-1224	1980-96-A	USSR (PLE)	1 Dec.	209 km 403 km	90.3 min 72.9°		Medium-resolution photographic reconnaissance satellite. Recovered on 15 December 1980
Cosmos-1225	1980-97-A	USSR (PLE)	5 Dec.	967 km 1041 km	105.0 min 82.9°		Navigation satellite
Intelsat-V F2 3-axis stabilized satellite; height: 6.60 m; mass at launch: 1950 kg; 2 solar arrays (1.2 kW)	1980-98-A	International INTELSAT (ETR)	6 Dec.	in geostationary orbit at 335.5° E		2202.5 MHz 3.5 W 5764 MHz 1 W (telemetry) 4-6 GHz (communications)	INTELSAT commercial telecommunication satellite: 12 000 telephone channels and two colour television channels
Cosmos-1226	1980-99-A	USSR (PLE)	10 Dec.	982 km 1025 km	105.0 min 83.0°		Navigation satellite
No name	1980-100-A	United States USAF (WTR)	13 Dec.	250 km 39 127 km	63.8°		Satellite data systems spacecraft. Provides UHF communications and relays data and communications between satellite control facility earth stations
Cosmos-1227	1980-101-A	USSR (PLE)	16 Dec.	209 km 325 km	89.5 min 72.9°		Medium-resolution photographic reconnaissance satellite
Cosmos-1228 to **Cosmos-1235** mass: 40 kg each	1980-102-A to 1980-102-H	USSR (PLE)	24 Dec.	1415 km 1491 km	114.6 min 74°		Government communication satellites

Code name Spacecraft description	International number	Country Organization Site of launching	Date	Perigee Apogee	Period Inclination	Frequencies Transmitter power	Observations
Prognoz-8 pressurized central body; 4 solar panels	1980-103-A	USSR (BAI)	25 Dec.	550 km 199 000 km	95 h 23 min 65°		**Automatic satellite to study influence of solar activity on Earth's magnetosphere. Experiments or equipment have been supplied by Czechoslovakia, Poland and Sweden**
Ekran-6 **(Statsionar-T)**	1980-104-A	USSR (BAI)	26 Dec.	35 554 km 35 554 km geostationary orbit	1424 min 0.4	5.7-6.2 GHz (reception) 3.4-3.9 GHz (emission)	**Television relay satellite**
Cosmos-1236	1980-105-A	USSR (PLE)	26 Dec.	180 km 388 km	89.8 min 67.1°		**High-resolution photographic reconnaissance satellite**

Appendix D

Tables and Charts

Satcom I Transponder Frequencies and Polarization Table

Transponder	Frequency	Polarization
1	3720	V
2	3740	H
3	3760	V
4	3780	H
5	3800	V
6	3820	H
7	3840	V
8	3860	H
9	3880	V
10	3900	H
11	3920	V
12	3940	H
13	3960	V
14	3980	H
15	4000	V
16	4020	H
17	4040	V
18	4060	H
19	4080	V
20	4100	H
21	4120	V
22	4140	H
23	4160	V
24	4180	H

Wind Pressure Table

Wind Speed in Miles Per Hour at +60° Fahrenheit	Pressure lbs./sq. ft.
20	1.6
21	1.75
22	1.925
23	2.1
24	2.3
25	2.5
26	2.7
27	2.9
28	3.2
29	3.4
30	3.6
31	3.9
32	4.1
33	4.4
34	4.65
35	4.95
36	5.2
37	5.5
38	5.8
39	6.0
40	6.5
41	6.75
42	7.0
43	7.5
44	7.75
45	8.0
46	8.5
47	8.75
48	9.25
49	9.75
50	10.0
60	14.5
70	20.0

Antenna pointing angles for Front Royal, Virginia

Satellite	Longitude	Azimuth	Elevation	Hour Angle	Declination
EAST U.S. LIMIT	70.00	167.13	44.15	−9.25	−6.12
COMSTAR III	87.00	193.90	44.01	10.00	−6.12
WESTAR III	91.00	199.93	43.00	14.52	−6.11
COMSTAR II	95.00	205.72	41.66	19.03	−6.09
WESTAR I	99.00	211.21	40.01	23.52	−6.07
ANIK A1	104.00	217.62	37.58	29.11	−6.04
ANIK A2	109.00	223.54	34.80	34.66	−6.00
ANIK A3	114.00	228.98	31.73	40.18	−5.96
SATCOM II	119.00	233.99	28.43	45.65	−5.92
WESTAR II	123.50	238.16	25.31	50.53	−5.87
COMSTAR I	128.00	242.07	22.08	55.38	−5.83
SATCOM I	135.00	247.68	16.88	62.83	−5.75
WEST U.S. LIMIT	150.00	258.35	5.39	78.44	−5.58

Earth station latitute (Deg., Min., Sec.) = 38 55 0.00 North

Earth station longitude (Deg., Min., Sec.) = 78 10 0.00 West

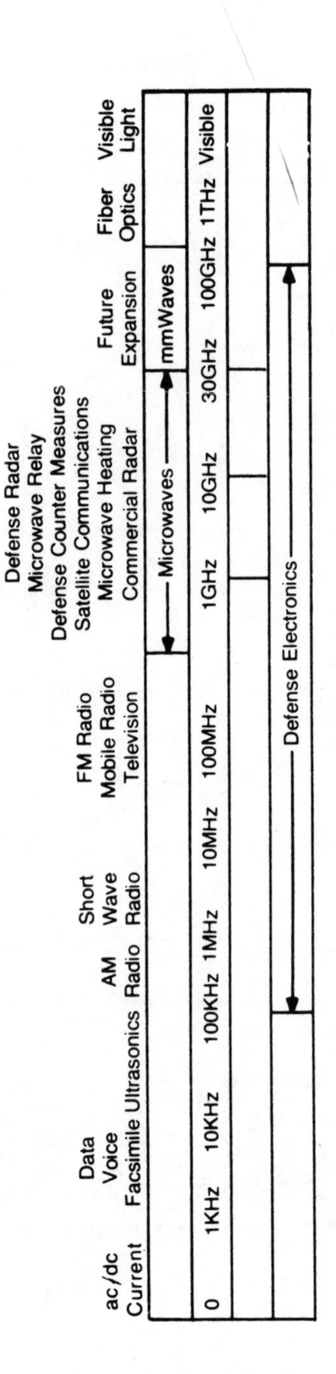
Frequency Spectrum Chart
ac/dc Current
Data Voice Facsimile
Ultrasonics
AM Radio
Short Wave Radio
FM Radio Mobile Radio Television
Defense Radar Microwave Relay Defense Counter Measures Satellite Communications Microwave Heating Commercial Radar
Future Expansion
Fiber Optics
Visible Light
Microwaves
mmWaves
0
1KHz
10KHz
100KHz
1MHz
10MHz
100MHz
1GHz
10GHz
30GHz
100GHz
1THz
Visible
Defense Electronics

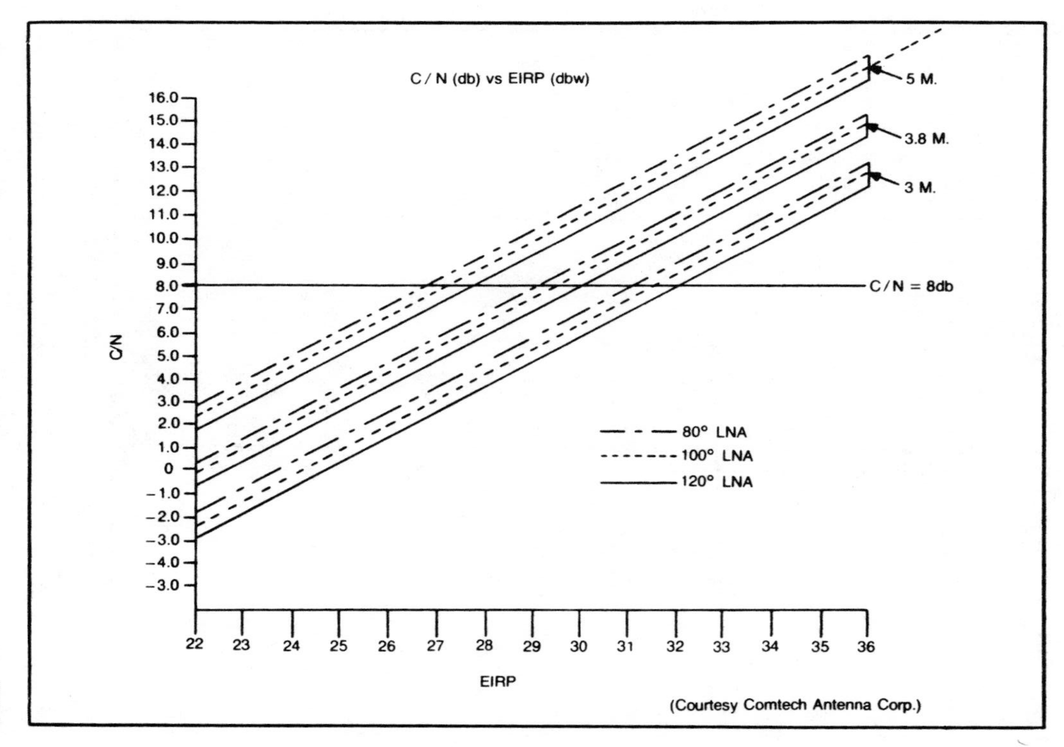

Preliminary Carrier/Noise Vs EIRP Chart

Noise Temperature to Noise Figure Table

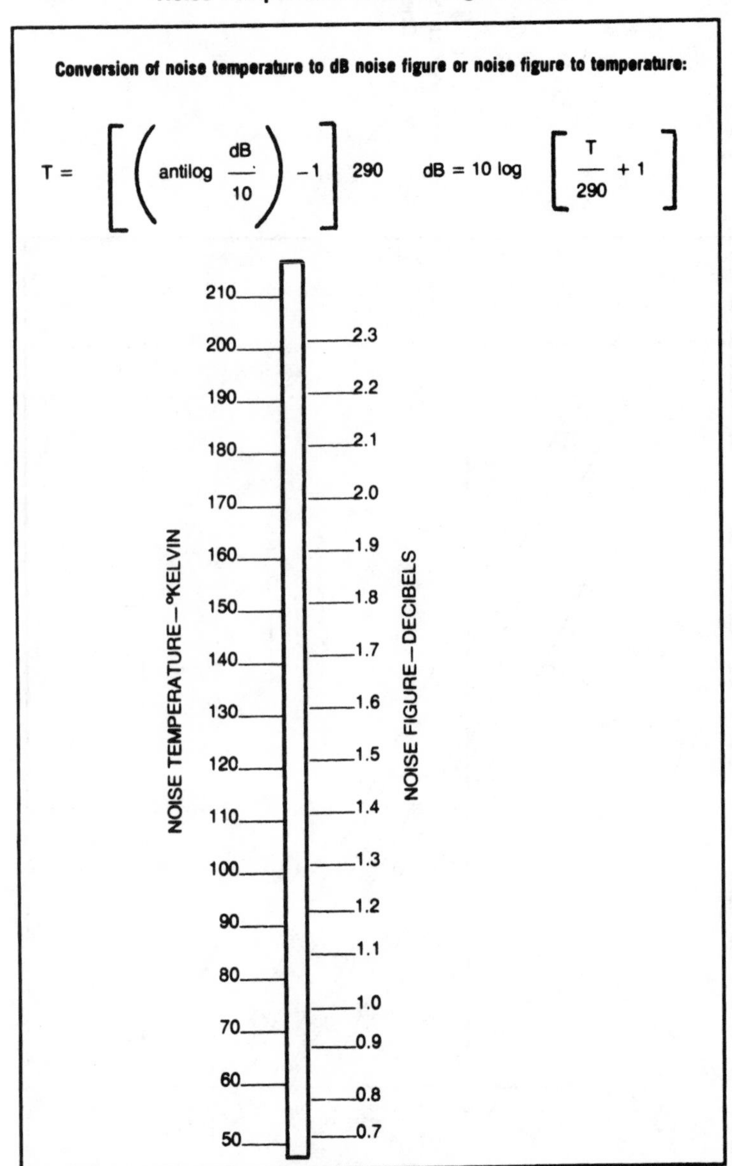

LNA Noise Temperature to Noise Figure Conversion Chart (Approximations)

Noise Temp. in Degrees K.	Noise Figures in dB
69	0.825
70	0.95
75	1.0
80	1.05
85	1.125
90	1.175
100	1.3
110	1.4
120	1.5
130	1.6
140	1.7
150	1.8
160	1.9
170	2.0

Appendix E

Maps

Preliminary EIRP Contour Map

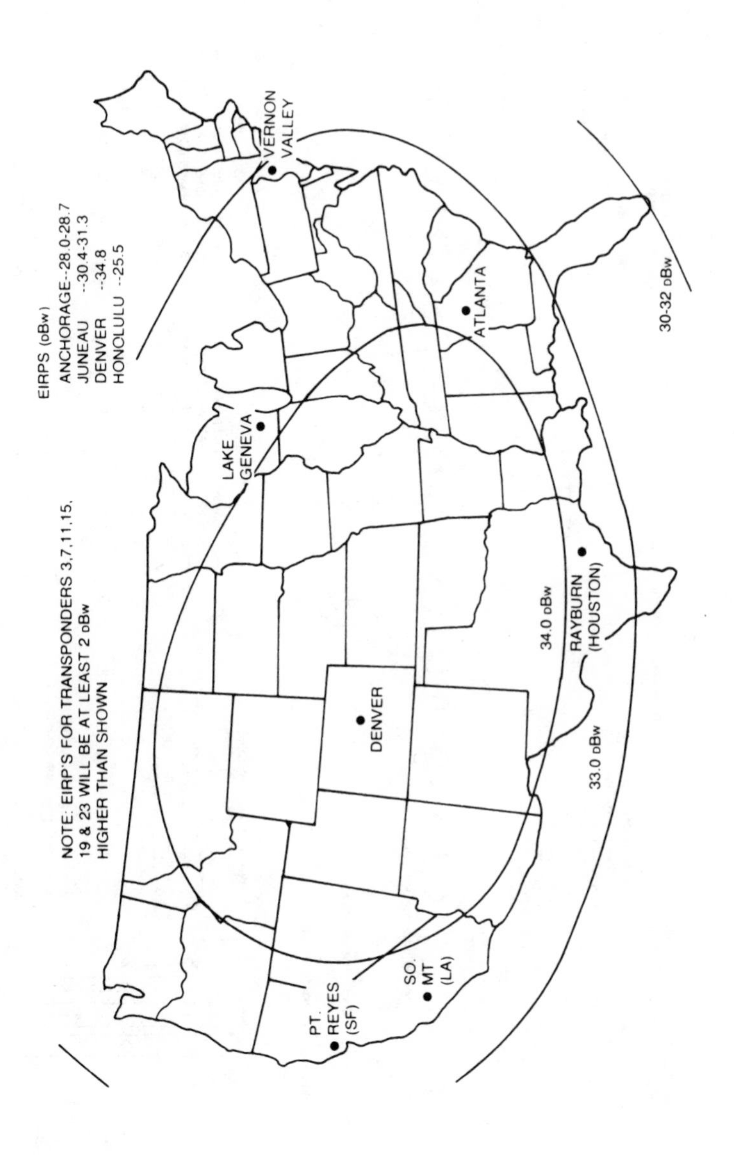

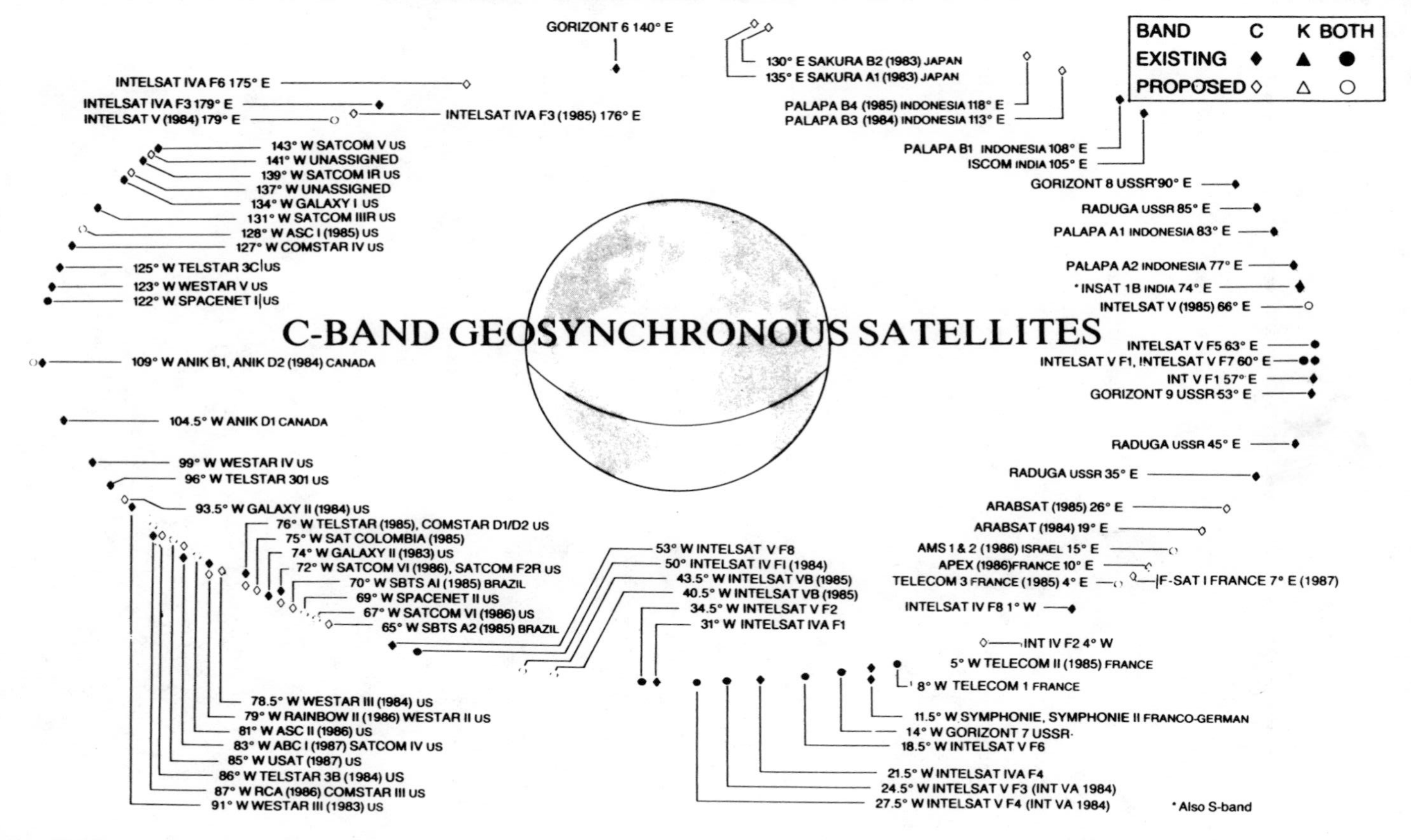
C-BAND GEOSYNCHRONOUS SATELLITES
BAND C K BOTH
EXISTING
PROPOSED
GORIZONT 6 140° E
130° E SAKURA B2 (1983) JAPAN
135° E SAKURA A1 (1983) JAPAN
INTELSAT IVA F6 175° E
INTELSAT IVA F3 179° E
INTELSAT V (1984) 179° E
INTELSAT IVA F3 (1985) 176° E
PALAPA B4 (1985) INDONESIA 118° E
PALAPA B3 (1984) INDONESIA 113° E
PALAPA B1 INDONESIA 108° E
ISCOM INDIA 105° E
GORIZONT 8 USSR 90° E
RADUGA USSR 85° E
PALAPA A1 INDONESIA 83° E
PALAPA A2 INDONESIA 77° E
* INSAT 1B INDIA 74° E
INTELSAT V (1985) 66° E
INTELSAT V F5 63° E
INTELSAT V F1, INTELSAT V F7 60° E
INT V F1 57° E
GORIZONT 9 USSR 53° E
RADUGA USSR 45° E
RADUGA USSR 35° E
ARABSAT (1985) 26° E
ARABSAT (1984) 19° E
AMS 1 & 2 (1986) ISRAEL 15° E
APEX (1986) FRANCE 10° E
TELECOM 3 FRANCE (1985) 4° E
F-SAT I FRANCE 7° E (1987)
INTELSAT IV F8 1° W
INT IV F2 4° W
5° W TELECOM II (1985) FRANCE
8° W TELECOM 1 FRANCE
11.5° W SYMPHONIE, SYMPHONIE II FRANCO-GERMAN
14° W GORIZONT 7 USSR
18.5° W INTELSAT V F6
21.5° W INTELSAT IVA F4
24.5° W INTELSAT V F3 (INT VA 1984)
27.5° W INTELSAT V F4 (INT VA 1984)
* Also S-band
143° W SATCOM V US
141° W UNASSIGNED
139° W SATCOM IR US
137° W UNASSIGNED
134° W GALAXY I US
131° W SATCOM IIIR US
128° W ASC I (1985) US
127° W COMSTAR IV US
125° W TELSTAR 3C US
123° W WESTAR V US
122° W SPACENET I US
109° W ANIK B1, ANIK D2 (1984) CANADA
104.5° W ANIK D1 CANADA
99° W WESTAR IV US
96° W TELSTAR 301 US
93.5° W GALAXY II (1984) US
76° W TELSTAR (1985), COMSTAR D1/D2 US
75° W SAT COLOMBIA (1985)
74° W GALAXY II (1983) US
72° W SATCOM VI (1986), SATCOM F2R US
70° W SBTS AI (1985) BRAZIL
69° W SPACENET II US
67° W SATCOM VI (1986) US
65° W SBTS A2 (1985) BRAZIL
53° W INTELSAT V F8
50° INTELSAT IV FI (1984)
43.5° W INTELSAT VB (1985)
40.5° W INTELSAT VB (1985)
34.5° W INTELSAT V F2
31° W INTELSAT IVA F1
78.5° W WESTAR III (1984) US
79° W RAINBOW II (1986) WESTAR II US
81° W ASC II (1986) US
83° W ABC I (1987) SATCOM IV US
85° W USAT (1987) US
86° W TELSTAR 3B (1984) US
87° W RCA (1986) COMSTAR III US
91° W WESTAR III (1983) US

Appendix F

TVRO Manufacturers and Distributors

A-B Electronics
1782 West 32nd Place
Hialeah, FL 33012
(305) 887-3203
Parabolic Antennas

ADM (Antenna Development and Manufacturing)
Box 1178
Poplar Bluff, MO 63901
Antennas

American Microcom
P.O. Box 1006
Cambridge, Ohio 43275
(614) 439-7771
Earth Stations

American Micro Supply
500 South 9th St.
Cambridge, Ohio 43725
(614) 439-1552
Earth Stations

American Microwave Technology
Box 248
Fairfield, IA 52556
(515) 472-3174
Low Noise Amplifiers

Amplica Inc.
950 Lawrence Drive
Newbury Park, CA 91320
(805) 498-9671
Low Noise Amplifiers/Line Amplifiers

Andrew Corp.
10500 West 153rd St.
Orlando Park, IL 60462
(312) 349-3300
Antennas

Anixter Brothers
4711 Golf Road
Skokie, IL 60076
(312) 298-9420
Complete Earth Stations

Arcfinder
815 W. New Hope
Rogers, AR 72756
(501) 631-2200
Antenna Control Systems

Asia Pacific Satellite Systems
P.O. Box 9614
San Rafael, CA 94416
(415) 461-6245
Residential, Hotel and Commercial Systems

ATV Research
13th and Broadway
Dakota City, NE 68731
(402) 987-3771
Parabolic Antennas

Automation Techniques, Inc.
1846 N. 106th E. Avenue
Tulsa, OK 74116
(918) 836-2584
Satellite Receivers

Avantek Inc.
3175 Bowers Avenue
Santa Clara, CA 95051
(408) 496-6710
Receivers/Downconverters

AVCOM of Virginia Inc.
500 Research Rd.
Richmond, VA 23236
(804) 794-2500
Complete Earth Stations

Best Reception Systems, Inc.
800-321-0281
Complete Earth Stations

Birdview Satellite Communications, Inc.
Box 963
Chaunte, KS 66720
(316) 431-0400
Complete Earth Stations

BR Satellite
35 Lumber Road
Roslyn, NY 11576
(516) 484-6080
Antenna Systems

California Amplifier
3481 Old Conejo Rd.
Newbury Park, CA 91320
(805) 499-8535
Low Noise Amplifiers

Cayson Electronics
Route 3, Box 160
Fulton, MN 38843
(601) 862-2132
Complete Earth Stations

Central Satellite
Box 684
Jasper, IN 47546
Complete Earth Stations

Channel Master
Ellenville, NY 12428
(914) 647-5000
Complete Earth Stations

Channel One Inc.
Willarch Rd.
Lincoln, MA 01773
(617) 259-0333
Complete Earth Stations

Chaparral Communications
Box 832
Los Altos, CA 94022
(415) 941-1555
Feedhorns

Commander Satellite Systems
309 Steeles Avenue E.
Milton, Ontario L9T 1Y2
(416) 876-4707
Antennas

Communications Gigahertz Inc.
4649 des Grandes Prairies
St. Leonard, Quebec H1R 1A5
(514) 323-0030
Complete Systems for Eastern Canada

Communications Plus
3680 Cote Vertu
The Bazaar Center
St. Lawrence, PQ Canada HR41P8
Receivers/Antennas

Comtech Antenna Corp.
895 Central Florida Parkway
Orlando, FL 32769
(305) 892-6111
Receivers/Antennas/Feedhorns

Comtech Data Corp.
350 North Hayden Rd.
Scottsdale, AZ
(602) 968-2433
Complete Line

Conifer Corp.
1400 N. Roosevelt
Burlington, IA 52601
(319) 752-3607
Complete Earth Stations

Continental Satellite Systems
15450 For-Mor Court
Clackamas, OR 97015
(503) 656-2774
Antennas

Cox Enterprises
RR 4, Box 280
Rockwood, TN 37854
(615) 354-1993
Antennas

Cresscom Inc.
10-B Washington Avenue
Fairfield, NJ 07006
(201) 575-4184
Cable and Connectors

DBS Corporation
P.O. Box 26A05
Los Angeles, CA 90026
(805) 652-0255
Complete Earth Stations

Deep Space Communications
Route 1, Box 351A
Chancellor, AL 36316
(205) 393-3211
Complete Earth Stations

Delta-Benco-Cascade Ltd.
124 Belfield Rd.
Rexdale, ON, Canada M9W 1G1
(416) 241-2651
Cable Television and Equipment

Dexcel
2580 Junction Ave.
San Jose, CA 95134
(408) 943-9055
Receivers/LNAs/Downconverters/Stereo Receivers

DH Satellite
P.O. Box 239
Prairie duChien, WI 53821
(608) 326-6705
Antennas

Discom Satellite Systems, Inc.
Box 8699
Independence, MO 64054
(816) 836-2828
Antennas

Draco Laboratories, Inc.
1005 Washington St.
Grafton, WI 53204
(414) 377-0770
Antenna Positioners

Robert L. Drake Co.
540 Richard St.
Miamisburg, OH 45342
(513) 866-2421
Satellite Receivers

DX Communications, Inc.
10 Skyline Dr.
Hawthorne, NY 10532
(914) 347-4040
Satellite Receivers

Earthstar Corp.
16012 Cottage Grove
South Holland, IL 60573
(312) 755-5400
Complete Earth Stations

Earth Terminals
255 Northland Boulevard
Cincinnati, OH 45246
(513) 772-6900
Satellite Receivers

Echosphere Corp.
5315 S. Broadway
Littleton, CO 80120
(303) 797-3231
Complete Earth Stations

Energy Systems Limited
2306 Charles Avenue
Dunbar, WV 25064
(800) 624-9046
Complete Earth Stations

Engineered Communications Inc.
18-A Home News Row
New Brunswick, NJ 08901
(201) 828-5009
Antennas

Gardiner Communications Corp.
1980 South Post Oak Rd.
Suite 240
Houston, TX 77056
(713) 961-7348
Antennas/LNAs/Receivers/Modulators/Power Supplies

Gillaspie and Associates
355 Sinclair Frontage Rd.
Milpitas, CA 95035
(408) 943-8800
Receivers/Downconverters/LNAs

H&G Systems
15109 Chicago Rd.
Dolton, IL 60419
Receivers

Heath Company
Benton Harbor, MI 49022
(616) 982-3210
Kits to Build Complete Earth Station

Helfer's Antenna Service
23 Brookside Pl.
Pleasantville, NY 10570
(914) 769-2588
Complete Earth Stations

Hero Communications
2470 W. 8th Ave.
Hialeah, FL 33010
(305) 887-3203
Complete Earth Stations

High Frontier Distribution
1445 W. 12th Pl.
Tempe, AZ 85281
(602) 966-9824
Complete Earth Stations

Hoosier Electronics
P.O. Box 3300
Terre Haute, IN 47803
(800) 457-3330
Complete Earth Stations

Houssen Tech, Inc.
P.O. Box 2126
Moncton, NB, Canada E1C 8H7
(506) 534-2530
Antennas/Receivers

Houston Satellite Systems, Inc.
9429 Harwin
Houston, TX 77036
(713) 784-8953
Complete Earth Stations

H&R Communications Inc.
Route 3, Box 103 G
Pocahontas, AR 72455
Complete Earth Stations

Hughes Corp.
Microwave Communications Products
Box 2999
Torrance, CA 90509
(213) 534-2146
Complete Earth Stations

Hustler
3275 North B Avenue
Kissimmee, FL 32741
(305) 348-5111
Complete Earth Stations

Hytek Amplifier
3602 E. 20th St.
Joplin, MO 64801
(417) 624-4061
LNAs

Industrial Scientific Inc.
3280 Wynn Rd.
Las Vegas, NV 89102
(702) 362-2405
Antennas

Intermountain Video Systems
1742F Edgemont Ave.
Bristol, TN 37620
(615) 968-2334
Complete Earth Stations

International Crystal Manufacturing (ICM)
10 North Lee
Oklahoma City, OK 73102
(405) 236-3741
Receivers

International Video Communications
4005 Landski Rd.
North Little Rock, AR 72204
(501) 771-2800
Complete Earth Stations

Intersat Corp.
2 Hood Drive
St. Peters, MO 63376
(800) 325-63376
Complete Earth Stations/Receivers

Interstar Satellite Systems
21708 Marilla St.
Chatsworth, CA 91311
(213) 882-6770
Complete Earth Stations

Kaul-Tronics, Inc.
Highway 14 East
Richland Center, WI 53581
(800) 826-5285
Antennas

Kent Research Corporation
1900 Burdett Ave.
Troy, NY 12180
(518) 272-6870
Antenna Positioners

KLM Electronics Inc.
1170025 Laurel Rd.
Morgan Hill, CA 95037
(408) 779-7363
Complete Earth Stations

Kosmos Systems, Inc.
11985 U.S. Highway One, St. 207
Juno, FL 33408
(305) 626-3800
Antennas

Leaming Industries
180 McCormick Ave.
Costa Mesa, CA 92626
(714) 979-4511
FM Stereo Processors

Lindsay America
Consumer Electronics Division
2608 East Hills Dr.
Williamsport, PA 17701
(717) 326-7133
Antennas/Complete Earth Stations

Long's Electronics
2700 Crestwood Blvd.
Birmingham, AL 35210
(800) 633-4984
Complete Earth Stations

Lowrance Electronics
12000 Skelly Dr.
Tulsa, OK 74128
(800) 331-4105
Satellite Receivers

Robert Luly Associates
Box 2311
San Bernardino, CA 92405
(714) 888-7525
Antennas/Complete Earth Stations

Mac Line Inc.
West 3125 Seltice Blvd.
Couer D'Alene, ID 83814
(208) 765-0909
Complete Earth Stations/Antennas

McCullough Satellite Systems Inc.
Box 57
Salem, AR 72576
(501) 895-3176
Spherical Antennas/8-Ball Antenna Kit

Merrimac Industries Inc.
41 Fairfield Pl.
West Caldwell, NJ 07006
(201) 575-1300
Receivers/Down Converters

Microdyne Corp.
Box 7213
Ocala, FL 32672
(904) 687-4633
Complete Earth Station

Micro-Link Systems, Inc.
6012 Dixie Dr.
Dayton, OH 45414
(201) 575-1300
Receivers/Down Converters

M/A COM (Microwave Associates Communication)
63 Third Avenue
Burlington, MA 01803
Complete Earth Stations

Microwave Filter Co.
6743 Kinne St.
E. Syracuse, NY 13057
(800) 448-1666
Filters for Earth Stations

Microwave General
2680 Bayshore Frontage Rd.
Mountain View, CA 90019
Complete Earth Station

Mid-America Video
324 Pershing Blvd.
North Little Rock, AR 72204
(501) 753-3555
Complete Earth Stations

Mini-Casat
Route 3, Box 150
Fulton, MS 38843
(601) 862-2132
Antennas/Receivers

MTI Systems
Salt Lake City, UT 45301
(801) 972-4817
Antenna Positioners

Muntz Electronics Inc.
7700 Densmore
Van Nuys, CA 91408
(213) 782-7511
Complete Earth Stations

National Microtech
Box 417
Grenada, MS 38901
(800) 647-6144
Apollo Earth Stations with Square-parabolic antenna

National Satellite Communications
21st Century Park
Clifton Park, NY 12065
Complete Earth Stations

Newton Electronics
2218 Old Middlefield Way
Mountain View, CA 94043
(415) 967-1473
Receivers

Norsat International Inc.
205-19425 Langley Bypass
Surrey, B.C. Canada V3S 4N9
(604) 533-4371
LNAs

Norsat Systems
Box 1234
Blaine, WA 98230
(604) 585-9428
Two-Part Satellite Receiver

North Supply Co.
10951 Lakeview Ave.
Lenexa, KS 66219
(913) 888-9800
Antennas/Receivers

Orion Cable Systems Inc.
3675 S. Highland
Las Vegas, NV 89103
(703) 362-0005
Transmission Cable

Orlando Antenna Co.
2356 W. Oakridge Rd.
Orlando, FL 32809
(305) 851-8332
Antennas

Paraclipse
Paradigm Mfg Inc.
Redding, CA 96002
Antennas

Paraframe Inc.
Box 423
Monee, IL 60449
Antenna Kits & Design Manuals

PCM INC.
3700 N. Harold
N. Little Rock, AR 72118
(501) 753-5715
Complete Earth Stations

Phantom Engineering
16840 Joleen Way E3
Morgan Hill, CA 95037
Intrusion Alarms for Earth Stations

Pinzone Communications Products, Inc.
10142 Fairmount Rd.
Newbury, OH 44065
(304) 296-4493
Complete Earth Stations

Prodelin Inc. (a division of MA/COM)
Box 131
Hightstown, NJ 08520
(609) 448-2800
Antennas

Quantum Associates Inc.
Box 18
Alpine, WY 83128
(307) 654-2000
Remote Antenna Positioners

Quarles Satellite Systems
1616 Calhoun Rd.
Greenwood, SC 29646
(800) 845-6952
Complete Earth Stations

Raydx Satellite Systems Ltd.
9 Oak Dr.
Silver Springs Shores Ind. Park
P.O. Box 4078
Ocala, FL 32678-4078
(904) 687-2003
Antennas

RSI Inc.
2436 N. Woodruff Ave.
Idaho Falls, ID 83401
(208) 523-5721
Complete Earth Stations

Satcom Canada
1140 Claire Crescent
Montreal, PQ Canada H8S 1A1
Antennas/Receivers

Satelco
5540 Pico Blvd.
Los Angeles, CA 90019
(213) 931-6274
Antennas

Satelinc
Riverside Dr.
Chestertown, NY 12817
(518) 494-4151
Complete Earth Stations

Satellite Communications Inc.
8101 Gallia St.
Wheelersburg, OH 45694
(614) 574-8121
Complete Earth Stations

Satellite Communications International Inc.
1820 Lejeune Rd.
Miami, FL 33126
(305) 871-1554
Complete Earth Stations

Satellite Earth Station Technology
4369 Creditview Rd.
Mississauga, ON, Canada L5M 2B5
(416) 826-8066
Antennas with Wireless Remote

Satellite Reception Systems, Inc.
2370 Morse Rd.
Columbus, OH 43229
(614) 471-6118
Earth Station Components

Satellite Supplies
Box 278
Aldergrove, BC, Canada VoX 1A0
Receivers

Satellite Systems Unlimited
Box 43
Conway, AR 72032
(501) 327-6501
Receivers

Satellite Technology Services
2310-12 Millpark Dr.
Maryland Heights, MO 63043
Receivers

Satellite Television Inc.
Box 817
Pcatello, ID 83204
(208) 232-1926
Complete Earth Stations

Satellite Television Systems
Box 22557, Suite 1140
Denver, CO 80222
(303) 750-0564
Complete Earth Stations

Satellite Television Systems
Box 51837
Lafayette, LA 70505
Complete Earth Stations

Satellite Television Technology Intl Inc.
Box G
Arcadia, OK 73007
(405) 396-2574
Receivers

Satfinder Systems Inc.
6541 E. 40th St.
Tulsa, OK 74145
(718) 664-4466
Complete Earth Stations

SAT Share Inc.
5301 Hollister, Suite 100
Houston, TX 77040
(713) 460-9900
Complete Earth Stations

Sat-Tec Systems
2575 Baird Rd.
Penfield, NY 14526
(716) 586-3950
Receivers

SAT Vision
Box 1490
Miami, OK 74354
(981) 542-1616
Antennas

Scientific Atlanta Inc.
3845 Pleasantville Rd.
Atlanta, GA 30340
(404) 449-2000
Antennas/LNAs

Scientific Communications Inc.
3425 Kingsley Rd.
Garland, TX 75041
(214) 271-3685
Receivers/LNAs

SED Systems Ltd.
2414 Coyle Ave.
Saskatoon, SK, Canada 57K 3P7
(306) 664-1825
Complete Earth Stations

Simcomm Labs
Box 60
Kersey, CO 80644
(303) 352-1020
LNAs

Skyscan Corp.
250 E. 36th St.
Tucson, AZ 85713
(602) 622-2261
Complete Earth Stations

Southern Fiberglass Supply
1105 Tuckahoe Dr.
Nashville, TN 37207
(615) 868-4094
Antennas and Mounts

Space-Age Industries Inc.
35 Progress Ave.
Nashua, NH 03062
(603) 882-6032
Pedestal Positioning Systems

Standard Communications
Box 92151
Los Angeles, CA 90009
(213) 532-5300
Receivers

Startech
1403 Mechanical Blvd.
Garner, NC 27529
(919) 779-0273
Complete Earth Stations

Star Trak Systems
404 Arrawanna St.
Colorado Springs, CO 80909
(303) 475-7050
Receivers/Antennas

Star Vision Systems
611 N. Wymore Rd.
Winterpark, FL 32789
(305) 628-5458
Complete Earth Stations

SWAN TVRO Satellite Systems
614 Cimarron
Stockton, CA 95210
(209) 478-2756
Complete Earth Stations/Antenna Kits

Telecom Industries Corp.
27 Beonventura Dr.
San Jose, CA 95134
(408) 262-3100
Receivers

Television Satellite Communication Systems Inc.
9800 S. Dixie Highway
Miami, FL 33156
305-662-2020
Antennas

Tel Vi Communications
1307 W. Lark Industrial Boulevard
St. Louis, MO 63026
(314) 343-9977
Antenna Drive Systems

Third Wave Communications Corp.
2600 Gladstone
Ann Arbor, MI 48104
(313) 996-1483
Complete Earth Stations

TL Systems
1101 E. Chestnut Ave., Suite C
Santa Ana, CA 92701
(714) 547-1981
Receivers/Modulators

Transtar Communications Corp.
1408 Columbia Ave.
Franklin, TN 37064
(615) 244-6112
Complete Earth Stations

Transvision
2100 Redwood Highway
Greenbrai, CA 94904
(415) 924-6963
Complete Earth Stations

Uniden Satellite Technology
200 Park Avenue
New York, NY 10166
Complete Earth Stations

VSC Corporation
P.O. Box 327
Columbus, NB 68601
(402) 563-3625
Antennas

Wagner Industries
Box 559
Alva, OK 73717
(405) 327-1877
Eagle Spherical Antennas

Western Satellite
Box 22959, Wellshire Station
Denver, CO 80222
Complete Earth Stations

Wilson Microwave Systems, Inc.
1 Sunset Way
Henderson, NV 89015
(702) 458-0144
Complete Earth Stations

Glossary

Glossary

absolute scale—See *Kelvin scale.*

ac—An abbreviation for alternating current, a current whose polarity changes at a specific frequency, usually measured in hertz. Standard house current is 110 Vac. The current polarity reverses itself at a rate of sixty times per second, or 60 hertz.

active component—Any component or circuit capable of amplifying a signal. An active device generally requires its own operating current and its output amplifies its input. The amplification factor may be a negative one, whereby the output is less than the input, or a positive one, whereby the output is greater than the input. An active device with an amplification factor of 1 will produce an output that is equivalent in amplitude to the input.

active satellite—An artificial satellite that receives a signal from Earth and retransmits it, usually back to Earth. The active satellite contains its own electronic circuitry and power supply that receives and processes an Earth-based transmission before retransmitting it. Usually, the receiving frequency and the retransmitting frequency are significantly different.

af—An abbreviation for audio frequency. This describes that portion of the spectrum that ranges from 20 Hz to 20 kHz. Audio frequency is generally considered to be those frequencies that lie within the human range of hearing, although the lower and upper ends of the established audio frequency band will generally fall slightly outside the range humans can detect.

afc—See *automatic frequency control.*
aft—See *automatic fine tuning.*
agc—see *automatic gain control.*
alc—see *automatic level control.*
AM—see *amplitude modulation.*

amplifier—Any electronic device or circuit that is generally designed to accept a low amplitude signal at its input and produce a higher amplitude equivalent of the same signal at the output.

amplitude modulation—A method of impressing intelligent information on a radio frequency carrier. The carrier amplitude is changed in accordance with the changes in the intelligence impressed. The modulating signal that contains the intelligence is superimposed upon a radio frequency carrier. The information is then transmitted at the carrier frequency. At the receiving end, the carrier is removed and only the intelligence remains.

ANIK—The name given to Canadian television, and more recently, to Canadian television satellites. ANIK is an Eskimo word meaning "brother".

aperture—The square area of a parabolic antenna, although this term is more commonly used to describe the diameter of a parabolic antenna.

apogee—In geosynchronous satellite terminology, the point in an orbit that is highest from the Earth.

attenuation—A measurement of a decrease in signal strength often purposely created by passing the signal through an electronic network to purposely decrease signal strength. Attenuation also occurs in coaxial cable and is a product of the frequency passed and the cable length.

attenuator—A usually passive device designed to reduce the amplitude of a signal that travels through it. Attenuators are often adjustable, which allows them to vary the amount of attenuation affecting the passed signal.

automatic fine tuning—A general description of an electronic circuit that automatically fine tunes another circuit portion. Automatic fine tuning and automatic frequency control can be the same thing, although the former often addresses minor, but critical, frequency fluctuations.

automatic frequency control—A circuit that maintains a constant selected frequency or band of frequencies automatically. Abbreviated afc, such a circuit locks onto a chosen frequency and maintains that frequency, often by anticipating drift and making suitable corrections.

automatic gain control—An electronic circuit that maintains a constant, preset amplitude within a larger circuit that may be receiving varying input signal levels. The input signals may arrive at the automatic gain control circuit at many different levels, but the output of these signals will be held at one constant level.

automatic level control—Synonymous with automatic gain control.

AZ-EL—An abbreviation for azimuth-elevation.

AZ-EL mount—A supporting structure for a parabolic antenna that allows the dish to be adjusted by two separate controls, one of which affects azimuth, the other elevation. This allows the parabolic dish to be infinitely adjusted in order to point to all available geosynchronous satellites.

azimuth—The angular measurement left or right of true north or south in a horizontal plane. In satellite antenna terminology, azimuth is the horizontal rotation of the antenna expressed as a compass heading in degrees.

bandpass filter—An electronic circuit that allows a selected band of frequencies to pass unattenuated. All frequencies that lie outside of the passband are severely attenuated. Bandpass filters are often used to eliminate interference from strong signals that lie near the intended reception frequency.

bandwidth—A range of frequencies from the lowest to the highest in a selected group. For instance, the bandwidth of the C-band satellite transmissions is 3.7 to 4.2 GHz. However, bandwidth is more often used to describe the width of the specific transmission. For instance, a signal transmitted at exactly 3.8 GHz has a center frequency of 3.8 GHz, but its bandwidth will probably lie in a range somewhat below 3.8 to slightly above 3.8 GHz.

baseband—A pure audio or video signal minus a carrier. This is the group of video and audio signals in addition to the synchronizing and blanking pulses and color subcarrier that make up a standard television transmission. This is the modulating signal or intelligence that is superimposed upon the radio frequency carrier.

beamwidth—The pattern angle of an antenna. Generally speaking, a beamwidth is the area in degrees that a transmitted signal presents to a receiving antenna. The beamwidth of a transmitting antenna spreads outward as it leaves the transmitting source. The amount of spread is dependent on the antenna type, which relates to its size and gain and also to the frequency. The transmitted signal spreads as does the beam from an electric spotlight.

bird—A slang name that applies to any satellite.

blanking pulse—A portion of a television transmission that is a rectangularly shaped pulse used to cut off the scanning action of television receivers.

block down converter—An active electronic device or circuit that accepts a signal input at a high frequency and retransmits it as the same signal at a lower frequency. In TVRO Earth stations, a block down converter attaches directly to or near the low noise amplifier. It accepts the 3.7 to 4.2 GHz amplified signal from the satellite at its input and retransmits the same signal at a lower frequency of 70 MHz. This lower frequency is also known as the intermediate frequency, which is the one detected by the satellite receiver. It is called a block down converter because it is a part of the receiving system located in a separate unit at some distance from the receiver proper. On a block diagram, the down converter would be shown in one position and the receiver in another. Block down converters enable the signal picked up by the antenna to be quickly converted to a lower frequency at the antenna. This lower frequency can be passed through coaxial cable with much less attenuation than the initial satellite signal.

BNC connector—A weatherproof, twist-lock connector used with coaxial cable. It is a low-loss connector whose presence produces relatively low attenuation of microwave frequencies.

boresight—In satellite television terminology, an area on the Earth's surface that serves as the aiming point for the satellite antenna transmissions. The boresight area receives the strongest signal from the satellite. As distance increases from the boresight, signal strength drops.

cable television—Abbreviated CATV, the distribution of television signals via cable from a central receiving point. Cable television operations maintain their own receiving stations, usually at high elevations for Earth-based television reception. The signals that are received at the cable installation's antennas are boosted by line amplifiers and then routed to each home that is on the service by low-loss coaxial cable.

carrier—A radio frequency wave consisting of one constant frequency. It is called this because the radio wave is often used to "carry" intelligence in the from of video and/or audio signals. These signals are superimposed on the carrier through modulation.

carrier-to-noise ratio—A comparison of the signal strength of a radio frequency carrier to unwanted signals or noise. In satellite television installations, a carrier-to-noise ratio is often measured at the output of the LNA. The carrier is transmitted by the satellite while noise is a product of extraneous frequencies generated externally to the Earth station and within its circuits as well.

Cassegrain antenna—An antenna design often used for operation at microwave frequencies that consists of a parabolic reflector to initially receive a signal. This signal is reflected to a hyperbolic subreflector located at the focal point. The subreflector reflects the collected signal back to the middle of the parabolic main antenna. This is the point where the feedhorn is located. Thus, the signal can be routed to the receiving equipment through a channel cut in the middle of the main reflector. Some satellite television antennas are of this design, but most use only a single parabolic reflector with a feedhorn located at the focal point rather than at the physical center of the parabolic dish.

CATV—A acronym for cable television. CATV stands for community antenna television.

C-band—A portion of the frequency spectrum used to transmit and receive satellite signals. The C-band uses frequencies in the range of 3.7 to 4.2 GHz for downlink (transmission from satellite to Earth) and frequencies of 5.9 to 6.4 GHz for uplink (transmission from Earth to satellite).

characteristic impedance—The impedance exhibited by a coaxial cable of theoretically infinite length. Generally, this is the impedance exhibited by the coaxial cable, regardless of its physical length.

circular polarization—A description of a transmitted signal that takes on a helical form. Most satellites transmit in horizontally and vertically polarized modes. This means that the feedhorn must be situated in a vertical configuration to receive signals that are vertically polarized and in a horizontal configuration to receive those that are horizontally polarized. Circular polarization has components of both vertical and horizontal polarization. A circularly polarized signal is polarized in a corkscrew pattern.

coaxial cable—A type of transmission line that usually consists of a stranded or solid center conductor surrounded by a circular foam dielectric that, in turn, is surrounded by a braided conductor. This describes an unbalanced coaxial line. A balanced coaxial line consists of one extra center conductor surrounded by a braid. Characteristic impedance of typical unbalanced coaxial lines is usually 50 or 75 ohms, although many values are available but are less commonly seen.

cold lead—The outer conductor of an unbalanced coaxial cable. It serves as a signal carrier and also as a shield for the inner conductor. Typically, the outer conductor is grounded and serves as the return signal path in addition to its shielding properties.

common carrier—Any service that transmits signals but does not originate programming. A common carrier is more or less a radio transmission service that is for hire by companies that do originate programming and wish to sell such a service to cable companies or to the general public. In a way, when you place an ad on radio or television, these services become your own personal common carrier.

composite video signal—A signal that contains video information, horizontal, vertical, and blanking pulses. Often, a composite video signal will also carry audio information as well. In effect, a composite video signal is what remains when the carrier is removed from a standard or satellite television broadcast. It is the intelligence that is superimposed on a radio frequency carrier.

cps—An abbreviation for cycles per second, which during the past decade has been generally replaced by the term hertz. Sixty hertz is equivalent to 60 cycles per second.

cross polarization—A method of broadcasting signals that are both horizontally and vertically polarized. Often, this is done using antennas that have both horizontal and vertical elements. On the receiving end, cross polarization is a situation that occurs when the feedhorn is located 45° between vertical and horizontal polarization. This causes the simultaneous reception of portions of two different signals.

crt—An abbreviation for cathode ray tube, which is a television picture tube or other type of vacuum tube that is capable of putting a visual trace on a viewing surface.

dB—See *decibel.*

DBS—An abbreviation for Direct Broadcast Satellite. This describes a service whose transmissions will fall within the K-band (around 12 GHz) and intended to be broadcast directly to its customers as opposed to a central cable television facility.

dBw—The ratio between a pair of values using 1 watt as the reference.

dc—An abbreviation for direct current as opposed to alternating current or ac. Direct current polarities are fixed and do not alternate.

decibel—Abbreviated dB, a unit for describing the ratio of two signal levels. The decibel is used often as a measurement of antenna gain and levels in general, as in noise levels, signal amplifications levels, carrier strength levels, etc.

decoder—An electronic device that reprocesses a coded or scrambled video and/or audio signal so that it may be properly received at the television set.

demodulator—An electronic circuit that separates the video and audio information (the intelligence) from the radio frequency carrier. In TVRO Earth stations, the carrier is usually the 70 MHz i-f signal, which is the result after the original satellite transmission has been amplified and passed through the down converter.

Direct Broadcast Satellite—See *DBS.*

dish—The reflector portion of a parabolic antenna used to collect and focus microwave energy received directly from an overhead satellite.

down converter—An electronic circuit that receives a signal at a high frequency and outputs it at a lower frequency. In TVRO Earth stations, the down converter accepts its input directly from the output of the low noise amplifier. At the input, the frequency is in the 3.7-4.2 GHz range. The down converter outputs a carrier or intermediate frequency of 70 MHz. This lower frequency carrier contains the same video and audio information as did the higher frequency.

downlink—Satellite to Earth transmissions. In TVRO Earth station operation, the downlink frequency is between 3.7 and 4.2 GHz.

downlink frequency—See *downlink.*

dual feed—An antenna feed assembly that is capable of utilizing both vertically and horizontally polarized signals. Dual feed arrangements sometimes consist of two feedhorns, two LNAs, and two cable outputs that may later be combined into one receiver feed.

Earth station—Any terrestrial station that is used to either receive or transmit signals from and to geosynchronous satellites.

EIRP—An abbreviation for effective isotropic radiated power.

This is a measurement of the signal strength that a satellite transmits toward its target area on the Earth. In other words, it's the energy level of a tightly beamed radio frequency signal. EIRP is measured in dBw.

elevation—The vertical angle of a TVRO antenna measured in degrees above the horizon.

encoder—An electronic device used at a transmitting end that in some way rearranges the video and/or audio portion of the signal before it is passed on to the transmitter proper. At the receiving end, a decoder is required to reprocess the signal into a proper format for reception. See *decoder.*

equatorial orbit—The path a satellite assumes in relation to the Earth, whereby the satellite is placed in an equatorial plane.

FCC—An abbreviation for Federal Communications Commission. This is the Federal regulatory agency that sets standards for and governs all United States broadcasting.

feedhorn—A section of waveguide into which signals from a parabolic dish reflector are aimed. In most parabolic antenna designs, the feedhorn is mounted at the reflector's focal point. The output from the feedhorn directly feeds the low noise amplifier.

field of view—That portion of the sky that can be seen by the parabolic reflector. Field of view usually refers to the unobstructed area of the sky from which signals may be gathered by the reflector.

FM—See *frequency modulation.*

footprint—Generally, the area on the Earth that a satellite transmission strikes. This especially refers to the contour lines of EIRP, which indicate signal strength of downlink signals. The footprint describes the signal strength pattern of a satellite transponder's transmission.

frequency—A measurement of the number of completed waves per second measured in cycles per second, or more recently, in hertz. This describes the number of oscillations of an electromagnetic signal within any one-second period. A frequency of 70 MHz produces 70 million cycles in a one-second period.

frequency modulation—A method of superimposing intelligence on a carrier whereby the frequency of the carrier changes in response to the amplitude of the intelligence.

frequency response—The ability of a circuit to amplify, pass, or in general, respond to a signal of specific frequency. A frequency response of 20 to 30 MHz would indicate that signals within this range are freely acted upon by the circuit, while frequencies that lie below and above this range are generally attenuated or ignored.

gain—A measurement of the amplification of a signal. More appropriately, this is a ratio of an output compared to the input. If the output and input are identical, the gain is 1. Gain is often measured in decibels.

geostationary orbit—See *geosynchronous orbit.*

geosynchronous orbit—An orbit in which the speed and direction of the satellite match the speed and direction of the turning Earth. Geosynchronous orbits fix the satellite into a fairly stable position in relation to the Earth. Therefore, the satellite always seems to be in the same position when viewed from the Earth.

gigahertz—Abbreviated GHz, the equivalent of 1,000 MHz, or 1 billion Hz. Most TVRO Earth stations receive signals in the range of 3.7 to 4.2 GHz.

global beam—A very broad or wide radiation pattern from a geosynchronous satellite. As opposed to a concentrated beam, a global beam spreads out tremendously to cover very large sections of the Earth. However, due to the signal dispersion the received signals on Earth are much weaker.

global beam antenna—An antenna specifically designed to efficiently receive a global beam transmission. Such antennas typically contain very large reflectors on the order of 50 to 100 feet in diameter. They are designed to gather far more signal than is necessary in most TVRO Earth station systems, as the latter signals are more concentrated and thus stronger.

g-t—The ratio of the gain of a parabolic dish compared to the noise temperature. g-t is expressed in decibels per 1 degree Kelvin.

hardline—A very efficient type of coaxial cable that induces low losses at microwave frequencies. Instead of being composed of a braid, the outer conductor consists of a jacket of solid metal. Hardline is extremely difficult to work with when compared with standard coaxial cable, but its losses are several times lower.

heliax—A type of low-loss coaxial cable used at high frequencies. See *hardline.*

heterodyning—The process of mixing an incoming signal with a signal produced by a local oscillator in order to arrive at an intermediate frequency. The mixture of two signals produces two other signals, one which is the sum of the two, another which is the difference between the two. For instance, if a 100 MHz signal was mixed with a 125 MHz signal, two other signals will be produced at 225 MHz (100 + 125) and at 25 MHz (125 − 100). The frequency that is selected to be passed to other circuits is known as the i-f or intermediate frequency.

high Z—An abbreviation for high impedance.

horn—See *feedhorn.*

hot lead—Generally, the center conductor of an unbalanced coaxial cable. The hot lead is sometimes called the positive lead and is the ungrounded portion of a coaxial cable, or for that matter, any other circuit.

Hz—An abbreviation for hertz, which is identical to the older cycles per second (cps).

IC—An abbreviation for integrated circuit, a miniaturized crystalline chip containing an entire circuit or mini-circuit in a single unit. Today, IC technology has greatly reduced the number of discrete components that are used in any electronic circuit. One IC may take the place of hundreds or even thousands of discrete components such as transistors, diodes, resistors, etc.

i-f—An abbreviation for intermediate frequency. An i-f is usually the result of a down conversion or superheterodyning process. Usually, it contains the same information as the original input signal, but carries this information at a different frequency. In most TVRO Earth stations, the i-f is 70 MHz and the information contained on this 70 MHz carrier is identical to the information that was transmitted by the satellite.

isolator—Any device that allows a signal or current to flow in one direction unattenuated while attenuating a similar signal of opposite polarity.

K-band—In the radio frequency spectrum, a continuous band of frequencies that ranges from 11.7 to 12.2 GHz. The K-band has been reserved for Direct Broadcast Satellite services. See also *C-band.*

Kelvin scale—A scientific temperature scale in which absolute zero equates to −459° Fahrenheit. At absolute zero, all molecular motion theoretically ceases. Low noise amplifiers are rated in performance by using the Kelvin scale. The lower the Kelvin temperature rating, the higher the overall gain. Technically, the Kelvin rating of the low noise amplifier indicates the amount of noise present on the output signal.

kHz—An abbreviation for kilohertz (formerly kilocycle), a frequency measurement equivalent to 1000 Hz or cycles per second.

latitude—The distance north or south of the equator that is also known as lines of latitude and is measured in degrees, minutes, and seconds. The equator itself is the reference point and is at latitude 0°.

LCD—See *liquid crystal display.*

LED—See *light-emitting diode.*

level—Generally, a measurement of signal strength that is usually expressed in volts. More specifically, a description of the signal strength of the audio and video signals as input to amplifiers in a TVRO Earth station.

light-emitting diode—A solid-state diode that glows when it passes current in one direction. Abbreviated LED, these devices are often incorporated in digital displays whereby several sections of LEDs can be used to display all alphanumeric characters. Many satellite receivers and antenna positioners use light-emitting diode displays as indicators.

line amplifier—Generally, a small amplifier placed in series with a wire conductor to boost the signal this conductor carries, thus overcoming losses induced by the line. More specifically, an rf amplifier that is placed in coaxial line to boost the signal of the down converted signal picked up by the TVRO antenna. Line amplifiers used in TVRO Earth stations generally accept a 70 MHz input and also output at 70 MHz, but at a higher level. Line amplifiers are used to overcome severe losses at the receiver caused by long coaxial cable runs.

liquid crystal display—A type of electronic digital display that uses liquid crystal technology. Liquid crystal displays typically do not require as much operating current as LED displays. In an LCD display, a single slab of crystal is sensitized by small amounts of current. These sensitized areas no longer reflect light and a darkened area appears. This area is used to form letters and numbers. Liquid crystal displays are often used in battery-powered equipment where current drain is a prime concern. Most equipment powered from the ac line, however, will continue to use LED displays. LCDs require the presence of ambient light to be seen, whereas LED displays generate their own fluorescent light.

LNA—See *low noise amplifier.*

LNC—See *low noise converter.*

longitude—The distance east or west of the prime meridian measured in degrees, minutes, and seconds. Drawn graphically, these are the lines extending from the north pole to the south pole. The prime meridian is at 0° longitude.

low noise amplifier—An extremely sensitive and noise-free amplifier designed to receive a very weak signal and amplify it for use by other equipment. Specifically, the component in a TVRO Earth station that attaches to the output of the feedhorn by means of a small section of waveguide. The extremely weak signals gathered by the parabolic dish and passed on to the feedhorn are accepted at the input to the low noise amplifier and boosted in strength. This is done while inducing very little external noise to the circuit. Abbreviated LNA, the performance levels of these amplifiers are rated in degrees Kelvin. The lower the degree rating, the better the performance.

low noise converter—A single electronic component used in TVRO Earth stations that attaches directly to the output of the feedhorn and combines the low noise amplifier and the down converter in one housing. LNCs are to be preferred over separate LNA/down converter arrangements because the connection between the LNA and the down converter will always produce a degradation in performance. While the losses are very small, an LNC is certainly to be preferred when installations are made in areas of extremely weak satellite signals.

low Z—An abbreviation for low impedance.

magnetic deviation—The error in degrees between a magnetic compass reading that indicates north and true north. Adjustments must be made for these deviations when aligning TVRO antennas.

MATV—An abbreviation for master antenna television. Generally, a receiving system used to receive television signals at a signal antenna or group of antennas and then distribute the signals to separate television receivers. MATV is differentiated from CATV (cable television) in that the former usually addresses a smaller number of receivers. Typical MATV operations serve hotels, motels, office buildings, and large apartment complexes.

MDS—An abbreviation for multipoint distribution system. MDS is a form of broadcast pay TV that uses a single channel for sending its video programs to subscribers. MDS is not Direct

Broadcast Satellite television; it can be better described as a cable system that replaces the cable hookups with a microwave transmitter. For satellite reception, a standard TVRO Earth station receives the satellite signal. The output from the TVRO receiver is fed to an MDS transmitter, which then microwaves the rebroadcast of the original satellite transmission to homes equipped with MDS receivers.

MHz—An abbreviation for megahertz, or one million Hz.

microwave—A band within the frequency range that starts at 400 MHz and extends to 30 GHz. The microwave region of the frequency spectrum is the one used for TVRO uplink and downlink transmissions.

modulation—A process whereby audio or video intelligence is superimposed on a radio frequency carrier for transmission to distant points.

modulator—The electronic circuits or components that accept video and audio information and superimpose this intelligence on a radio frequency carrier. See *AM, FM,* and *modulation.*

monochrome—Another name for black and white. A monochrome TV receiver is simply a black and white TV receiver.

N-connector—An extremely low-loss coaxial cable connector that is used at microwave frequencies to avoid connector losses in coaxial cable runs.

noise—Any undesired electrical signal, regardless of where it is generated. Noise may be thought of as that portion of a received signal that does not carry desired intelligence and often interferes with the reception of audio and video signals. Noise can be generated in space, in the Earth's atmosphere, or by external and internal electrical components and circuits.

noise factor—The noise induced in a system by a piece of internal equipment. It is the signal to noise ratio of an input signal compared with the same ratio at the output. The result is a noise factor of that particular piece of equipment or circuit.

noise figure—The total amount of noise induced by a system. Similar to noise factor, the noise figure of a TVRO Earth station would be the difference between the noise factor at the antenna to the noise factor at the receiver.

noise temperature—The noise of a system measured in degrees Kelvin. More specifically, the noise introduced by a low noise amplifier or the noise rating of that amplifier.

NTSC—An abbreviation for National Television Standards Committee. This Committee sets standards for all North American television transmissions.

offset angle—In polar antenna mounts, the measurement in degrees that the antenna must be tilted away from its polar axis, presumably to lock onto a geostationary satellite. This angle increases as the antenna site is moved north of the equator.

orbital day—The time of one complete orbit of a satellite. In geosynchronous satellites, this is equivalent to one standard Earth day.

orbital slot—The specific position in space a satellite assumes in relation to the Earth.

PAL—An abbreviation for phase alternate line. This is a television broadcast format not commonly known in the United States and Canada. It is a European color format whereby the color burst is inverted on alternate lines. This helps to minimize color transmission errors. The PAL format is not compatible with standard NTSC format.

parabola—A geometric curve produced when a cutting plane is sent through a right cone parallel to a slant edge. See *parabolic antenna.*

parabolic antenna—An antenna used mostly at microwave frequencies which consists of a parabolic dish (a reflector whose surface has a parabolic shape) that gathers signals from a large physical area and focuses them on the feedhorn, which is located at the small focal point. Parabolic antennas are those most often used for TVRO Earth station applications.

perigee—The point nearest the Earth in an elliptical orbit. See *apogee.*

petal—One portion of a parabolic reflector designed to be assembled by coupling several sections. Petalized dish construction is quite common today. Usually, the parabolic dish is cut into four different sections, which facilitates shipping and installation.

polarization—The orientation of a broadcast signal. Most satellite transmissions are vertically or horizontally polarized, while some are circularly polarized.

polar mount—An antenna mount with only one adjustment that moves the antenna so that it may target different satellites.

prime focus feed—A type of microwave feed that places the feedhorn at the focal point. This is the standard feed arrangement used with most TVRO antennas whereby the reflector concentrates all received signals at the focal point and thus at the feedhorn.

reception window—A rectangular area at the front of a dish antenna whose outline defines that area through which satellite signals may be received. The reception window indicates the area of deviation that can be tolerated while still adequately maintaining reception of the signals from a specific satellite.

rf—An abbreviation for radio frequency. Rf describes that portion of the electromagnetic spectrum that lies between 10 kHz and 100 GHz.

satellite—Any object that rotates around another. More specifically, any body, artificial or manmade, which orbits the Earth. The Moon, as well as the artificial orbiters launched from the Earth, are said to be satellites of Earth. The Earth is a satellite of the Sun.

satellite bandwidth—The frequency width established for transmissions from satellites. Also, the total frequency width from the low frequency end of a broadcast spectrum to its high frequency end. The total bandwidth of the C-band frequency is 500 MHz, as this band ranges from 3.7 to 4.2 GHz. The bandwidth for transponders operating in the C-band is 40 MHz.

satellite receiver—A complex integration of circuits to form a single component that is used to select and tune satellite transponder channels. Typically, a satellite receiver accepts an input frequency of 70 MHz, which is the frequency of the satellite information once it has exited the down converter. In some instances, the satellite receiver may be thought of as the LNA, down converter, and receiver proper. Generally, the satellite receiver outputs composite video based on the original signals passed to it from the antenna. The composite video is then passed on to a modulator to produce an acceptable signal for the television receiver. In some equipment, the modulator is a part of the receiver. In others, it is a separate unit.

scrambler—See *encoder.*

snow—Electrical interference that manifests itself on a television picture tube as numerous black and white blips.

spherical antenna—A microwave antenna that has seen usage in the TVRO Earth station field. It uses a reflector that is spherically shaped (a portion of a sphere) to focus incoming satellite signals to a focal point. Usually, spherical antennas will exhibit several different major focal points. Signal concentration at these various focal points is efficient enough to allow the reflector to be aimed at one satellite but to receive signals from several different satellites simply by moving the feedhorn. Like the parabolic dish antenna, the spherical antenna has only one prime focal point, but several other focal points are usable. This type of arrangement makes a good tradeoff in installations where it is not practical to rotate the entire reflector. Generally, spherical antennas must be of large dimensions to equal the performance of smaller parabolic dishes.

stereo—Generally, two channels. More specifically, stereo refers to two-channel sound that is impressed upon the carrier by means of subcarriers. The left channel contains audio signals that are different from those of the right channel, although the two may be combined (and usually are) at the receiver.

stereo processor—A complex electronic circuit used for separating the two audio channels in a stereo broadcast. Recently, stereo processors have become popular for TVRO Earth stations as more and more satellites begin to broadcast in stereo.

subcarrier—An additional carrier separated from the main carrier that may be modulated to contain discrete information. Subcarriers are used for stereophonic broadcasting and also for close-caption viewing.

thermal noise—Electrical noise that is generated by the heat agitation of molecules in the Earth's atmosphere. Also, the noise generated by the heating effects of electrical components in circuits.

translation frequency—In satellite broadcasting, the frequency difference between the uplink and the downlink. In the C-band, the difference is 2.22 GHz, since the signal is transmitted at 5.9 to 6.4 GHz (uplink) and the same signal is received at 3.7 to 4.2 GHz (downlink). The translation frequency is the difference between the higher and the lower frequency in any one transmission band.

transmission line—The conductor or conductors used to carry signals. More specifically, the cable or waveguide used to transfer radio frequency signals from the antenna to the LNA, from the LNA to the down converter, and from the down converter to the receiver. Any physical device used to carry signals.

transponder—A component in a satellite that is a combination receiver, amplifier,and transmitter. It may be thought of as a transceiver. The satellite receives the signal from the uplink via the receiver portion of its transponder. The amplifier portion then boosts the strength of the received signal, which is then fed to the transmitter portion for rebroadcast (downlink).

TVRO—An abbreviation for television receive-only. This is the term used to reference most personally owned receiving stations used for the reception of satellite signals.

uplink power—The amount of power usually expressed in watts, kilowatts, or megawatts, used to transmit a signal from the Earth to an orbiting satellite.

uplink signal—The modulation information that is superimposed upon a radio frequency carrier beamed at a satellite from an Earth station.

uplink station—Any ground station that is used specifically for transmitting signals from the Earth to an orbiting satellite.

video—That portion of a signal that contains information that has been gathered from a camera or other sensing device designed to convert the appearance of physical objects into electrical signals. Also, that portion of a received signal that is designed to be fed to a cathode ray tube for visual display.

video inversion—A type of signal scrambling in which the video signals are reversed in polarity. At the descrambling end, the received video signals are inverted once again, which results in a faithful reproduction of the original unscrambled broadcast.

waveguide—A completely enclosed metallic duct or tube that acts as a transmission line, usually for microwave frequencies. The dimensions of the waveguide are critical to a fraction of a centimeter in order to avoid transmission line loss.

wavelength—The distance while traveling at the speed of light that a radio wave will travel during a single cycle.

window—See *reception window*

zero tuning meter—A logarithmic meter containing a zero at the center of its scale that is used to indicate receiver tuning. The idea is to adjust the receiver until the needle points to zero. This indicates maximum signal strength or some other desired control fixation.

Index

Index

A

active communications satellite, 19, 347
AM station, 13
amateur radio, 9, 26
ANIK 3, 286
ANIK B, 286
ANIK, 217, 348
antenna alignment, 80
antenna actuator installation, 126
antenna assembly, 113
antenna base, 114
antenna controller, 126, 153, 162
antenna elevation angle, 242
antenna mounting site, 113
antenna obstructions, 216
antenna pointing angles, 311
antenna positioner, 102, 110, 162
antenna system, 29
antenna trackers, 102
antenna, Cassegrain, 351
antenna, dish, 101
antenna, fiberglass, 102
antenna, Ground Star, 105
antenna, paraclipse, 110
antenna, perforated aluminum, 107
antenna, helix, 95
antenna, multi-element array, 90
antenna, parabolic, 87
antenna, quad array, 91
antenna, spherical, 81, 87
antenna, yagi, 21, 94
antennas, 71
antennas, Channel Master, 104
antennas, horn, 75
aperture, 348
apogee, 348
astronomy, 31
ATT/GTE COMSTAR 1, 286
ATT/GTE COMSTAR 2
Avcom, 175
azimuth angles, 218
azimuth, 349

B

BDC 60 Block Down Converter, 178
block down converter, 350
boresight, 350
boresighting, 81
broadcast band, AM, 16
broadcast band, FM, 16
broadcast station, AM, 14
broadcast station, FM, 14
broadcast stations, foreign, 16
broadcast, line of sight, 14
broadcasting services, 9

C

C band geosynchronous satellites, 320
C band, 251, 351
cable television, 350
cable TV company, 66, 225
cable, coaxial, 33, 44, 72, 351

cable, helix, 38, 355
carrier noise vs. EIRP chart, 313
carrier to noise ratio, 351
Cassegrain antenna, 351
CATV, 351
CB stations, 9
Channel Master, 104, 159
circular polarization, 351
coaxial cable, 33, 44, 72, 351
color television pictures, 25
COM 20T Receiver, 179
COM 23T Receiver, 179
COM 2A Receiver, 175
COM 2B Receiver, 175
COM 3 Receiver, 176
COM 3R Receiver, 176
COM 65T Receiver, 176
communications satellite, 19
connectors, 44, 45
contour adjustment, 118
cross polarization, 352

D

DA 5000P Distribution Amplifier, 171
DBS, 249, 251, 264, 352
DBS, upgrading to, 261
DC 60 Down Converter, 173
demodulator, 353
descrambling, 250, 266, 267
Direct Broadcast Satellite, 249
dish antenna, 101
dish site, 238
dish, 30
dish, fiberglass, 102
dish, parabolic, 78
dish, wire-mesh, 102
down converter, 35, 38, 173, 181, 353
downlink, 353

E

earth station assembly, 46
earth station, 29, 353
earth station, personal, 7
earth-based television signal, 37
EIRP contour map, 319
EIRP, 353
electromagnetic energy, 72
elevation angles, 218
elevation, 354
entertainment programming, 45, 88, 281
equatorial orbit, 354
equipment selection, 272
Explorer I, 25

F

feedhorn positioning, 78, 79, 80, 85
feedhorn, 30, 354
fiberglass, 102, 104
focal point, 31, 72
footprint, 60, 354
frequency spectrum chart, 312

G

Gagarin, Major Yuri, 24
GCI 2001C Antenna Controller, 153
GCI 2001R Satellite Receiver, 151
GCI 3742 LNA, 185
GCI 8200 Antenna Control System, 126
GCI 8300 Satellite Receiver, 143
GCI 9600 Receiver, 159
geostationary orbit, 71, 355
geostationary, 205
geosynchronous orbit, 355
geosynchronous satellites, 205, 320
Gillaspie Communications, Inc., 126, 142, 143, 185
Ground Star antenna, 105

H

H-1 Receiver, 181
HBO, 263, 265
Heathkit, 207
helix antenna, 95
helix cable, 38, 355
Home Box Office, 206, 263
Houston Tracker Systems, 102

I

ICM Video, 166
information, 5
interference, 279

K

K band, 251, 356
kits, 46, 63, 99, 141
KLM Electronics, 142

L

LNA noise figures, 184
LNA noise temperature to noise figure conversion chart, 315
LNA, 32, 38, 60, 181, 185, 357
LNC, 358
low noise amplifier, 32, 358
low noise converter, 358

M

M/A-COM Cable Home Group, 181
magnetic deviation, 218, 219, 358

manufacturer support, 278
media, 5
mesh dish, 107
microwave lenses, 31
microwave, 32, 359
Model 6131 Receiver, 162
Model 6136 Receiver, 162
Model 6255 Antenna Controller, 162
modulator, 35, 59
moon bounce, 18
motor drives, 110
motor power leads, 132
MTI 2100, 103
MTI 4100, 102
MTI 2800, 103

N

noise temperature to noise figure table, 314
noise temperature, 359
NTSC, 360

O

Ohm's Law, 73
Olympics, 25

P

PAL, 360
parabolic antenna, 360
paraboloid reflectors, 77
paraclipse antenna, 110
Paradigm Manufacturing, Inc., 111
passive communications satellite, 19
passive reflector, 15, 18
pay television, 262
PenTec Enterprises, Inc., 102
perforated aluminum, 107
perigee, 360
polarization, 351, 352, 360
power losses, 73
preamplifier, 32

Q

QSL cards, 89

R

radiation losses, 74
radio repeater, 19
radio waves, 13
radio, 25
RCA SATCOM 2, 286
RCA SATCOM I, 283
receivers, 34, 60, 102, 103, 143, 151, 159, 162, 166, 175, 176, 179, 181
receiving stations, 12
reflector, parabolic, 78
repeater, earth-based, 20
repeater, radio, 19
robots, 253
rockets, 25

S

SA 50 Signal Purifier, 173
SATCOM I transponder frequencies, 309
SATCOM I, 213, 217
satellite bulletin board, 260
Satellite Computer Service, 207
satellite frequencies, 22,23
satellite industries, new, 252
satellite look angles, 207
satellite orbital positions, 287
satellite program sources, 281
satellite receiver installation, 192
satellite receiver, 34, 60, 102, 103, 143, 151, 159, 162, 166, 175, 176, 179, 181
satellite TV, future of, 249
satellite window, 223, 238, 363
satellite, 18, 19, 21, 25, 361
satellites in orbit, 293
satellites, entertainment, 24, 45, 88, 281
satellites, geosynchronous, 71, 205, 320, 355
satellites, stationary, 24, 71, 205
scrambling, 250, 261
sheet-molding composition, 104
Shephard, Alan B., 24
Showtime, 206
signal losses, 42,43
site determination, 271
site selection, 237
site survey package, 207
skin effect, 74
skip, 17, 207
skywave communications, 16
SMC, 104
Soviet Union, 24
space communications, 18, 26
Space Shuttle, 26
spherical antenna, 361
SPM 3 Stereo Processor, 180
Sputnik II, 25
Sputnik, 24
spy satellites, 88
SR 4600P Receiver, 166
stereo processor, 362
Superstation, 206
surplus components, 63
system assembly, 273, 274, 275

system checkout, 276
system problems, 277

T

T-1 Receiver, 181
TA 30 Tunable Audio, 173
Telaco, Inc, 225
telemetry, 25
television receive-only earth stations, 20, 29, 30, 363
television receive-only, 29
television, 1-4, 263
television, audience-participation, 5
television, satellite, 7, 45, 249, 281
Telstar, 26
The Movie Channel, 206
Tracker III Plus, 102
tracking, 25
transmitting stations, 9
transponder, 20, 362
TVRO earth station, 20, 29, 36, 363
TVRO manufacturers and distributors, 321
TVRO, 20, 29, 36, 363

U

Uniden, 181
United States, 24, 88
uplink, 363
UST 1000 Receiver, 181
UST 500 Down Converter, 181

V

video signal, 352

W

waveguide, 72
weather, 244
Wilson Microwave Systems, 113
wind pressure table, 310
wind resistance, 101, 102
wind, 244
window, 223, 363
wire mesh dish, 102
WU WESTAR 1, 285
WU WESTAR 2, 285
WU WESTAR 3, 285

Y

yagi antenna, 2, 94

Edited by Brint Rutherford

Other Bestsellers From TAB

☐ **GETTING THE MOST OUT OF YOUR VIDEO GEAR—Quinn**

Document the celebrations of family and friends—to treasure and enjoy again and again . . . Make learning fun for your children by using video . . . Receive your initial investment back many times over by using your video to earn extra income. You can do all these, and countless other imaginative projects using this up-to-the-minute guide—the ULTIMATE sourcebook for every video owner who wants to take full advantage of his equipment. With it, you can be sure you're getting every dollar's worth out of your VCR, TV, video camera, and home computer! 256 pp., 88 illus., 7″ × 10″.
Paper $12.95 Hard $19.95
Book No. 2641

☐ **TELEVISION—FROM ANALOG TO DIGITAL—Prentiss**

Discover the leading edge of television technology in this unique sourcebook that brings together, for the first time, the information serious TV experimenters and professional TV engineers and technicians have been asking for. Written by Stan Prentiss—one of the leading electronics writers in the U.S. and author of more than 40 highly regarded books for hobbyists and professionals—this is a thorough and comprehensive look at the most recent developments affecting TV broadcasting and receiving. 352 pp., 258 illus. 7″ × 10″.
Hard $25.00 Book No. 1972

☐ **TROUBLESHOOTING AND REPAIRING SATELLITE SYSTEMS—Maddox**

This first-of-its kind troubleshooting and repair manual can mean big bucks for the professional service technician (or electronics hobbyist looking for a way to start his own servicing business) . . . *plus big savings for the TVRO owner who wants to learn the ins and outs of maintaining and servicing his own receiver!* You'll find service information plus complete schematics for most popular receivers as well as full explanations of every aspect of installation, testing, servicing, and alignment, plus much more. This is definitely a hands-on reference that no TVRO or TV service technician can afford to miss! 400 pp., 479 illus. 7″ × 10″.
Paper $18.95 Hard $26.95
Book No. 1977

☐ **AM STEREO AND TV STEREO—NEW SOUND DIMENSIONS—Prentiss**

Includes the most up-to-date AM & FM stereo information available. An in-depth look at the new sound in AM radio and TV broadcasting that gives much needed advice on equipment availability and operation and provides insight into FCC regulations and equipment specifications. Treating AM stereo radio and TV multi-channel sound separately, this guide details the various systems and equipment developments with plain language descriptions of the final hardware available in both receivers and transmitters. This is truly a double-barreled bargain for all stereo enthusiasts. 192 pp., 127 illus. 7″ × 10″.
Paper $12.95 Hard $17.95
Book No. 1932